Handbook of Mathematical and Digital Engineering Foundations for Artificial Intelligence

Artificial intelligence (AI) and digital engineering have become prevalent in business, industry, government, and academia. However, the workforce still has a lot to learn on how to leverage them. This handbook presents the preparatory and operational foundations for the efficacy, applicability, risk, and how to take advantage of these tools and techniques.

Handbook of Mathematical and Digital Engineering Foundations for Artificial Intelligence: A Systems Methodology provides a guide for using digital engineering platforms for advancing AI applications. The book discusses an interface of education and research in the pursuit of AI developments and highlights the facilitation of advanced education through AI and digital engineering systems. It presents an integration of soft and hard skills in developing and using AI and offers a rigorous systems approach to understanding and using AI.

This handbook will be the go-to resource for practitioners and students on applying systems methodology to the body of knowledge of understanding, embracing, and using digital engineering tools and techniques.

The recent developments and emergence of Chatbots (AI tools) all have mathematical foundations for their efficacy. Such AI tools include ChatGPT, GPT-4, Bard, Tidio Support Bot, Kuki AI Companion, Meena, BlenderBot, Rose AI Chatbot, Replika: AI Friend, Eviebot, and Tay. This handbook highlights the importance of mathematical and digital foundations for AI developments. The handbook will enhance the understanding and appreciation of readers about the prevailing wave of artificial intelligence products, and, thereby, fitting the current market needs.

Systems Innovation Book Series
Series Editor: Adedeji Badiru

Systems Innovation refers to all aspects of developing and deploying new technology, methodology, techniques, and best practices in advancing industrial production and economic development. This entails such topics as product design and development, entrepreneurship, global trade, environmental consciousness, operations and logistics, introduction and management of technology, collaborative system design, and product commercialization. Industrial innovation suggests breaking away from the traditional approaches to industrial production. It encourages the marriage of systems science, management principles, and technology implementation. Particular focus will be the impact of modern technology on industrial development and industrialization approaches, particularly for developing economics. The series will also cover how emerging technologies and entrepreneurship are essential for economic development and society advancement.

Conveyors
Application, Selection, and Integration
Patrick M McGuire

Innovation Fundamentals
Quantitative and Qualitative Techniques
Adedeji B. Badiru and Gary Lamont

Global Supply Chain
Using Systems Engineering Strategies to Respond to Disruptions
Adedeji B. Badiru

Systems Engineering Using the DEJI Systems Model®
Evaluation, Justification, Integration with Case Studies and Applications
Adedeji B. Badiru

Handbook of Scholarly Publications from the Air Force Institute of Technology (AFIT), Volume 1, 2000-2020
Edited by Adedeji B. Badiru, Frank Ciarallo, and Eric Mbonimpa

Project Management for Scholarly Researchers
Systems, Innovation, and Technologies
Adedeji B. Badiru

Industrial Engineering in Systems Design
Guidelines, Practical Examples, Tools, and Techniques
Brian Peacock and Adedeji B. Badiru

Leadership Matters
An Industrial Engineering Framework for Developing and Sustaining Industry
Adedeji B. Badiru and Melinda Tourangeau

Handbook of Mathematical and Digital Engineering Foundations for Artificial Intelligence

A Systems Methodology

Adedeji B. Badiru
Olumuyiwa S. Asaolu

CRC Press
Taylor & Francis Group
Boca Raton London New York

CRC Press is an imprint of the
Taylor & Francis Group, an **informa** business

First edition published 2023
by CRC Press
6000 Broken Sound Parkway NW, Suite 300, Boca Raton, FL 33487-2742

and by CRC Press
4 Park Square, Milton Park, Abingdon, Oxon, OX14 4RN

CRC Press is an imprint of Taylor & Francis Group, LLC

ISBN: 978-1-032-16181-5 (hbk)
ISBN: 978-1-032-16182-2 (pbk)
ISBN: 978-1-003-24739-5 (ebk)

DOI: 10.1201/9781003247395

Typeset in Times
by SPi Technologies India Pvt Ltd (Straive)

Dedication

Dedicated to the memory of late Professor Oyewusi Ibidapo-Obe, whose unquenchable thirst for mathematical-based innovation and artificial intelligence motivated our interest in writing this book. He left a mathematical legacy that we must continue to carry forward.

Contents

Preface

The rapid growth of new technological tools has created new opportunities for operational efficiency, effectiveness, and productivity improvement. Artificial intelligence (AI) has emerged with new technological tools to provide new opportunities for how we live and work. From an operational perspective, the digital era consists primarily of digital-based science, technology, engineering, and mathematics (STEM) that have enabled new AI capabilities and expanded application opportunities. Learning to use the new developments requires rigorous and adaptive understanding of the mathematical foundations for AI. The present wave of interest in AI is facilitated by the rapid growth in computational tools, represented by powerful software and hardware resources. This book presents a collection of mathematical roots, digital techniques, and statistical tools that are essential for developing AI-based applications. This is done from a systems perspective, whereby cooperating and diverse tools and techniques come together to facilitate building powerful new AI products.

Acknowledgments

We thank everyone around us, at home, work, and play, for the support and inspiration to write this monumental contemporary book on the widely spreading topic of artificial intelligence. We particularly thank Mr. Abubakar Ademola Raji and Mrs. Iswat Badiru for their painstaking help in typing and formatting many of the ugly-complicated equations in this book.

Authors

Dr. Adedeji B. Badiru is a professor of Systems Engineering at the Air Force Institute of Technology (AFIT). He is a registered professional engineer and a fellow of the Institute of Industrial Engineers as well as a fellow of the Nigerian Academy of Engineering. He has a BS degree in Industrial Engineering, MS in Mathematics, and MS in Industrial Engineering from Tennessee University, and PhD in Industrial Engineering from the University of Central Florida. He is the author of several books and technical journal articles and has received several awards and recognitions for his accomplishments. His special skills, experience, and interests center on research mentoring of faculty and graduate students.

Dr. Olumuyiwa S. Asaolu is a professor of Systems Engineering in the Faculty of Engineering at the University of Lagos, Nigeria. He obtained a BSc in Civil Engineering from the University of Lagos and MSc and PhD in Engineering Analysis from the University of Lagos. He has successfully supervised and mentored several doctoral candidates. His research interests and publications cover Artificial Intelligence, Systems Modeling and Analysis. He has conducted and published original work in robotics path planning, solving simultaneous nonlinear equations, word-processing and translation of African languages, etc. He is a recipient of several scholarly awards, including Nigerian Best Researcher Software Developer (2001), the World Academy of Sciences & Inter Academy Panel Young Scientist for 2009. His research team has obtained a patent for laughter as a biometric. Several outlets acclaimed the news of the published research, including US Homeland Security Newswire, US Department of Defense on Twitter, Russian TechNews, and UK Sciencespot. Dr. Asaolu is founder/chairman of Lainos international Limited, a co-founder and non-executive director of Performance Evaluation and Selection Limited (PEAS Ltd), as well as a minority investor in Alajo Technologies Limited. He is a registered engineer and member/fellow of several professional bodies.

1 Artificial Intelligence within Industrial and Systems Engineering Framework

1.1 INTRODUCTION

Everything around us is a system. Embracing a systems-thinking approach is the first order of possibilities for Artificial Intelligence (AI). Nothing brings this fact to the fore more than the robust discipline of Industrial and Systems Engineering (ISE). The traditional discipline of industrial engineering uses a foundation of systems thinking to facilitate process improvement, efficiency, effectiveness, and productivity in both technical and non-technical areas. The forward-looking approach of industrial engineering makes the embrace of AI natural for industrial engineers. Over the decades, industrial engineering has been at the forefront of the application of new and emerging techniques and technologies. This is evidenced in the early development and utilization of mathematical operations research in the military and the operational leveraging of electronic spreadsheets and computational tools in business and industry in the early days of the emergence of microcomputers (Badiru and Bauer, 1985; Badiru, 1985, 1988; Badiru and Whitehouse, 1989; Badiru and Whitehouse 1986; Badiru et al., 1987).

AI is all about fast computation, approximation, estimation, and prediction (See Badiru et al., 1987). In mimicking the human brain, an AI tool must quickly calculate outputs on the order of millions of combinations and permutations. Such a feat is only possible through mathematical foundations, both simple and complex. Every part of the vast expanse of the entire field of mathematics has something to offer to how AI works. One reference is estimation tricks to arrive at a quick answer, similar to using mental math strategies (Balch et al., 1993). Johann Carl Friedrich Gauss (1777–1855), of the Gaussian fame, is reputed to have demonstrated amazingly fast math calculation techniques. Gauss was a German mathematician and physicist who made significant contributions to many fields in mathematics and science. Gauss had an exceptional influence in many fields of mathematics and science, and he is ranked among history's most influential mathematicians. He is reported to have impressed his teacher at the age 10 with his ability to quickly mentally calculate answers to calculation questions. A story goes that when the teacher asked students to find the sum of the numbers from 1 through 100, young Gauss amazed his teacher and the class by writing the answer immediately. Gauss used a mental math strategy to find the sum. One mental math strategy is compensation. In compensation, you change a

DOI: 10.1201/9781003247395-1

1

problem so that it is easy to solve mentally. Then, you make an adjustment to compensate for the change you made. There are many ways to compensate in finding an answer. In many AI techniques, estimation and compensation strategies are used. AI is not just one modeling or computational approach. The diverse applications of AI call for diverse mathematical formulations. This chapter and the chapters that follow present a potpourri of mathematical representations that are directly applicable to AI modeling, programming, and applications. The expectation is that the collection will motivate additional advancements in the area of mathematical foundations for AI in multidisciplinary fields of application. See Badiru (2022), Badiru and Ibidapo-Obe (2008), Badiru, et al. (2012), Doran, et al. (2022), Ibidapo-Obe, et al. (1995), and Ibidapo-Obe et al. (2001).

1.2   QUANTUM POTENTIAL FOR AI

Artificial intelligence is becoming more pervasive in our modern society. The interest in AI has heightened in recent years, particularly because of the availability of more powerful hardware and software tools. The emerging field of Quantum Computing bodes well for more robust AI developments in the future. A quantum computer is a computer which makes use of the quantum states of subatomic particles to store information. The better we can store and retrieve data, the more achievable our AI goals can be.

Quantum computing is the processing of information that's represented by special quantum states. The process taps into quantum phenomena like "superposition" and "entanglement," thereby creating a fundamentally different way of handling information. When practical quantum computing finally arrives, we will see dramatic improvement in the operations of computers, laptops, smartphones, and others. Quantum computers can tackle complex problems involving humongous numbers of variables and potential outputs. Simulations and optimization modeling in industrial engineering applications will become faster and more effective. Although quantum computing is not yet where advocates are touting it to be, there are rapid advancements happening almost daily. Quantum now is where AI used to be some 20 years ago. We now see what AI can do in the present research and development landscape. We expect quantum to approach a similar level of growth and practical applications in just a few years more.

1.3   OLD AND NEW AI ACHIEVEMENTS

What was hypothesized as being with the realm of possibilities several decades ago are now being realized. Although the capabilities have not yet peaked, we can already see applications in the home, at work, in leisure pursuits, in sports, in technical undertakings, and so on. Consider the days of physical peep holes on wooden front doors compared to today's digital front-door ring cameras. Our society has come a long way in a short period of time. There is a strong wave of interest in AI in every corner of business, industry, government, academia, and military in our society. Sadiku et al. (2022) present several practical AI applications in smart cities, smart electrical grids, healthcare, engineering, education, business, industries, manufacturing, agriculture, vehicle autonomy, social media, and the military. The premise of this book is to focus on selected mathematical framework and foundation for developing sustainable AI tools, techniques, and processes. This is to emphasize that a technology that is based on a mathematical foundation will have a basis for sustainment, very much in the same way that mathematical formulations can stand the test of time.

For years, detractors have derided the prospects for AI as a serious and sustainable tool. But AI can no longer be ignored as more and more realistic AI-based products are developed. AI is not yet where it could be, but it is getting there, albeit slowly.

AI is actually not new. It actually originated as a research curiosity in the 1950s. As far back as 1992, Badiru (1992) highlighted the potential applications of AI in manufacturing, with an explicit emphasis on Expert Systems, the legacy software component of AI. It has taken this long to get to where we are today with AI. Adedeji Badiru developed, directed, and ran a rudimentary AI laboratory at the University of Oklahoma, School of Industrial Engineering, as far back as 1986. It was the first such ambitious university-based technical pursuit in the State of Oklahoma at that time. New AI-based courses were developed, and thesis and dissertation products emanated from the AI efforts of those days. The prediction then was that AI would someday emerge from the laboratory to become a viable asset for the society. We are now beginning to realize that prediction. Some of the scholarly AI products that the authors have been involved in over the past decades are summarized below:

- 1987: Arif, Alaa E., and A. B. Badiru, "An Integrated Expert System with a Fuzzy Linguistic Model for Facilities Layout," Proceedings of the Sixth Oklahoma Symposium on Artificial Intelligence, Tulsa, Oklahoma, November 11–12, 1992, pp. 185–194.
- 1987: Badiru, A. B., "Applications of Artificial Intelligence in Economic Development," Oklahoma State Legislature, February 1987.
- 1988: Badiru, A. B., Janice M. Mathis, and Bob T. Holloway, "Knowledge Base Design for Law Enforcement," Proc of 10th Annual Conf on Computers and Industrial Engineering, Dallas, Texas, March 1988, Computers & Industrial Engineering, Vol. 15, Nos. 1–4, 1988, pp. 78–84.
- 1988: Datar, Neetin N., and A. B. Badiru, "A Prototype Knowledge Based Expert System for Robot Consultancy – ROBCON," in Proceedings of the 1988 Oklahoma Symposium on Artificial Intelligence, Norman, Oklahoma, November 1988, pp. 51–68.
- 1988: Joshi, Ajay P.; Neetin N. Datar; and A. B. Badiru, "Knowledge Acquisition and Transfer," in Proceedings of the 1988 Oklahoma Symposium on Artificial Intelligence, Norman, Oklahoma, November 1988, pp. 355–378.
- 1990: Badiru, A. B., "Artificial Intelligence Applications in Manufacturing," in The Automated Factory Handbook: Technology and Management, Cleland, David I. and Bopaya Bidanda, Editors, TAB Professional and Reference Books, New York, 1990, pp. 496–526.
- 1991: Badiru, A. B., "OKIE-ROOKIE: An Expert System for Industry Relocation Assessment in Oklahoma," Proceedings of the Fifth Oklahoma Symposium on Artificial Intelligence, Norman, Oklahoma, November 1991.

- 1991: Somasundaram, S. and A. B. Badiru, "An Expert System for External Cylindrical Grinding: Planning, Diagnosing, and Training," Proceedings of the Fifth Oklahoma Symposium on Artificial Intelligence, Norman, Oklahoma, November 1991, pp. 20–29.
- 1992: Badiru, A. B. (1992), Expert Systems Applications in Engineering and Manufacturing, Prentice-Hall, New Jersey.
- 1992: Badiru, A. B. (1992), Expert Systems Applications in Engineering and Manufacturing, Prentice-Hall, New Jersey.
- 1992: Badiru, A. B.; Le Gruenwald; and Theodore Trafalis, "A New Search Technique for Artificial Intelligence Systems," Proceedings of the Sixth Oklahoma Symposium on Artificial Intelligence, Tulsa, Oklahoma, November 11–12, 1992, pp. 91–96.
- 1992: Joshi, Shirish; S. Raman; A. B. Badiru; B. L. Foote, "Assembly Interface Determination for Symmetric Rotational Parts," Proceedings of the Sixth Oklahoma Symposium on Artificial Intelligence, Tulsa, Oklahoma, November 11–12, 1992, pp. 20–29.
- 1993: Badiru, A. B., "Development of a New Search Algorithm for Artificial Intelligence," University of Oklahoma PI Research Investment Program, January 1, 1993 to December, 1993.
- 1993: Badiru, A. B., "The Role of Artificial Intelligence and Expert Systems in New Technologies," in Management of New Technologies for Global Competitiveness, Madu, Christian N., Editor, Quorum Books, Greenwood Publishing Company, Westport, Connecticut, 1993, pp. 301–317.
- 1993: Chandler, D., G. Abdelnour, S. Rogers, J. Huang, A. Badiru, J. Cheung, C. Bacon, "Fuzzy Rule-Based System for Early Fault Detection Prediction," Proceedings of the 7th Oklahoma Symposium on Artificial Intelligence, Stillwater, Oklahoma, November 18–19, 1993, pp. 62–66.
- 1993: McCauley-Bell, Pamela, "A Fuzzy Linguistic Artificial Intelligence Model for Assessing Risks of Cumulative Trauma Disorders of the Forearm and Hand," Ph.D. Dissertation, University of Oklahoma, 1993.
- 1993: Oliver, Gaugarin E., and A. B. Badiru, "An Expert System Model for Supplier Development Program in a Manufacturing Firm," Proceedings of the 7th Oklahoma Symposium on Artificial Intelligence, Stillwater, Oklahoma, November 18–19, 1993, pp. 135–141.
- 1993: Rogers, Steve, "Fuzzy Inferencing Approach to Modeling and Simulation," Ph.D. Dissertation, University of Oklahoma, 1993. Senior Artificial Intelligence Engineering, Seagate Technologies, Oklahoma City.
- 1993: Sieger, David Bruce, "A Methodology for A Real-Time Artificial Intelligence Surveillance System," M.S., School of Industrial Engineering, University of Oklahoma, 1993.
- 1993: Wei, Hong, "Improving Cantor Set Search for Applications in Database and Artificial Intelligence," M.S. (Computer Science), University of Oklahoma, 1993 (co-chair with Dr. Le Gruenwald).
- 1993: Wei, Hong; L. Gruenwald; and A. B. Badiru, "Improving Cantor Set Search for Applications in Database and Artificial Intelligence,"

Proceedings of the 7th Oklahoma Symposium on Artificial Intelligence, Stillwater, Oklahoma, November 18–19, 1993, pp. 250–259.

- 1995: Ibidapo-Obe, O.; A. B. Alonge; and A. B. Badiru, "On Active Controls for a Biped Mechanism," Applied Mathematics and Computation, Vol. 69, 1995, pp. 159–183.
- 1996: Milatovic, Milan, "Development of Mode Estimating Technique for Cantor Search of Sorted Data in Artificial Intelligence Systems and Manufacturing Design," M.S. Thesis, School of Industrial Engineering, University of Oklahoma, 1996.
- 1997: Milatovic, M. and A. B. Badiru, "Mode Estimating Procedure for Cantor Searching In Artificial Intelligence Systems," Proceedings of First International Conference on Engineering Design and Automation, Bangkok, Thailand, March 1997.
- 2001: Ibidapo-Obe, O. Sunday Asaolu, and Adedeji B. Badiru (2001), "Generalized solutions of the pursuit problem in three-dimensional Euclidean space," Applied Mathematics and Computation, Vol. 119, No. 3, pp. 35–45.
- 2001: Milatovic, M. and A. B. Badiru, "Control Sequence Generation in Multistage Fuzzy Control Systems for Design Process," AIEDAM (Artificial Intelligence for Engineering Design, Analysis, and Manufacturing), Vol. 15, 2001, pp. 81–87.
- 2002: Badiru, A. B. and John Cheung (2002), Fuzzy Engineering Expert Systems with Neural Network Applications, John Wiley & Sons, New York.
- 2002: Badiru, A. B. and John Cheung (2002), Fuzzy Engineering Expert Systems with Neural Network Applications, John Wiley & Sons, New York.
- 2004: Badiru, A. B. (2004), "Hybrid NeuroFuzzy Expert System for Crime Investigation," US Department of Justice/Law Enforcement Innovation Center, June 1, 2003–May 31, 2004.
- 2008: Badiru, A. B. and O. Ibidapo-Obe (2008), "On Set-Theoretic Fundamentals of Systems Integration," African Journal of Computing & ICT (Journal of the IEEE Nigeria Computer Chapter), Vol. 1, No. 2, December 2008, pp. 3–25.
- 2008: Olunloyo, V. O. S.; Abayomi Ajofoyinbo; and A. Badiru, "An Alternative Approach for Computing the Union and Intersection of Fuzzy Sets: A Basis for Design of Robust Fuzzy Controller," in Proceedings of 7th WSEAS International Conference on Artificial Intelligence, Knowledge Engineering and Data Bases (AIKED '08), Cambridge, UK, February 23–25, 2008. Note: This paper won best student paper at the conference.
- 2012: Badiru, A. B.; Oye Ibidapo-Obe; and B. J. Ayeni (2012), Industrial Control Systems: Mathematical and Statistical Models and Techniques, Taylor & Francis Group / CRC Press, Boca Raton, FL.
- 2022: Badiru, A. B. (2022), Artificial Intelligence and Digital Systems Engineering, Taylor & Francis Group / CRC Press, Boca Raton, FL.
- 2022: Badiru, A. B. (2022), Artificial Intelligence and Digital Systems Engineering, Taylor & Francis Group / CRC Press, Boca Raton, FL.

- 2022: Doran, Eric, Sharon Bommer, and Adedeji Badiru, "Integration of Human Factors, Cognitive Ergonomics, and Artificial Intelligence in the Human-Machine Interface for Additive Manufacturing," International Journal of Quality Engineering and Technology, International Journal of Mechatronics and Manufacturing Systems, Vol. 15, No. 4, 2022, pp. 310–330.

Some of these intellectual products culminated in seminal AI-themed textbooks (Badiru and Cheung, 2002; Badiru, 1992). So, why is AI now bursting into the limelight in light of its long tenure in the dark recesses of many university laboratories? This renewed emergence is due to the emergence of new powerful computational tools and techniques. Below of its process orientation, the discipline of industrial engineering actually had many early influences on the development of AI then and now. Those who know AI inside and out know how much mathematical programming affects optimization of AI products. In fact, there is a noticeable application of industrial engineering and operations research techniques of non-linear programming to machine learning and artificial neural networks. Further, the people orientation of industrial engineering paves the way for the consideration of social science and humanities in the practical uses of AI products by people. Nowadays, we are more sensitized to how people use AI products, from a humanistic standpoint rather than from mechanical standpoint. Similarly, the emergence of more powerful data manipulation techniques created additional inroads into predictive analytics, data envelopment, data quality control, decision science, process efficiency, and path optimization. All of these have foundational linkages to industrial engineering.

An interesting example of recent AI application is its adoption by FIFA for decision-making in the 2022 Football World Cup in Qatar. AI is deployed to ascertain when the ball has crossed the goal line, and also for detection of "offside" via usage of sensors and intelligent image/video processing. The Virtual Assistant Referee (VAR) help obviate controversies such as England's 1966 cross-bar related goal in the final as well as Maradona's (in)famous "hand of God" goal in 1986.

1.4   INDUSTRIAL ENGINEERING LINKAGE

So, what does industrial engineering have to do with the surge and resurgence of AI? The answer can be seen in the very definition of industrial engineering, as presented below:

> Industrial engineering is concerned with the design, installation, and improvement of integrated systems of people, materials, information, equipment, and energy by drawing upon specialized knowledge and skills in the mathematical, physical, and social sciences, together with the principles and methods of engineering analysis and design to specify, predict, and evaluate the results to be obtained from such systems.

The above definition demonstrates the versatility and diversity of industrial engineering to enmesh with AI. For the purpose of this book, we are focusing on the "mathematical" angle of industrial engineering. The fact that the most fruitful applications of AI can be found in "industry" further solidifies the linkage of the discipline of industrial engineering to the field of AI.

Artificial intelligence is not just one single thing. It is a conglomerate of various elements, involving software, hardware, data platform, policy, procedures, specifications, rules, and people intuition. How we leverage such a multifaceted system to do seemingly intelligent things, typical of how humans think and work, is a matter of systems implementation. This is why the premise of this book centers on a systems methodology. In spite of the recent boost in the visibility and hype of AI, it has actually been around and toyed with for decades. What has brought AI more to the forefront nowadays is the availability and prevalence of high-powered computing tools that have enabled the data-intensive processing required by AI systems. The resurgence of AI has been driven by the following developments:

- Emergence of new computational techniques and more powerful computers
- Machine learning techniques
- Autonomous systems
- New/Innovative applications
- Specialized techniques: Intelligent Computational Search Technique Using Cantor Set Sectioning
- Human-in-the-loop requirements
- Systems integration aspects

As long ago as the mid-1980s, the author has led many research and development projects that embedded AI software and hardware into conventional human decision processes. AI has revolutionized and will continue to revolutionize many things we see and use around us. So, we need to pay attention to the emerging developments.

The comprehensive definition of industrial engineering epitomizes what AI is expected to accomplish.

1.5 HISTORICAL BACKGROUND OF AI

The background of AI has been characterized by controversial opinions and diverse approaches. The controversies have ranged from the basic definition of intelligence to questions about the moral and ethical aspects of pursuing AI. However, despite the unsettled controversies, the technology continues to generate practical results. With increasing efforts in AI research, many of the prevailing arguments are being resolved with proven technical approaches.

"Artificial intelligence" is a controversial name for a technology that promises much potential for improving human productivity. The phrase seems to challenge human pride in being the sole creation capable of possessing real intelligence. All kinds of anecdotal jokes about AI have been offered by casual observers. A speaker once recounted his wife's response when he told her that he was venturing into the new technology of AI. "Thank God, you are finally realizing how dumb I have been saying you were all these years," was alleged to have been the wife's words of encouragement. One whimsical definition of AI refers to it as the "Artificial Insemination of knowledge into a machine." Despite the deriding remarks, serious embracers of AI may yet have the last laugh. It is being shown again and again that AI may hold the key to improving operational effectiveness in many areas of application. Some observers have suggested changing the term "Artificial Intelligence" to a less controversial one such as "Intelligent Applications (IA)." This refers more to the way that computers and software are used innovatively to solve complex decision problems.

Natural Intelligence involves the capability of humans to acquire knowledge, reason with the knowledge, and use it to solve problems effectively. It also refers to the ability to develop new knowledge based on existing knowledge. By contrast, *Artificial Intelligence* is defined as the ability of a machine to use simulated knowledge in solving problems.

1.6   ORIGIN OF ARTIFICIAL INTELLIGENCE

The definition of intelligence had been sought by many ancient philosophers and mathematicians, including Aristotle, Plato, Copernicus, and Galileo. These great philosophers attempted to explain the process of thought and understanding. The real key that started the quest for the simulation of intelligence did not occur, however, until the English philosopher Thomas Hobbes put forth an interesting concept in the 1650s. Hobbes believed that thinking consists of symbolic operations and that everything in life can be represented mathematically. These beliefs directly led to the notion that a machine capable of carrying out mathematical operations on symbols could imitate human thinking. This is the basic driving force behind the AI effort. For that reason, Hobbes is sometimes referred to as the grandfather of AI.

While the term "Artificial Intelligence" was coined by John McCarthy relatively recently, the idea had been considered centuries before. As early as 1637, Rene Descartes was conceptually exploring the ability of a machine to have intelligence when he said:

> For we can well imagine a machine so made that it utters words and even, in a few cases, words pertaining specifically to some actions that affect it physically. However, no such machine could ever arrange its words in various different ways so as to respond to the sense of whatever is said in its presence—as even the dullest people can do.

Descartes believed that the mind and the physical world are on parallel planes that cannot be equated. They are of different substances following entirely different rules and can, thus, not be successfully compared. The physical world (i.e., machines) cannot imitate the mind because there is no common reference point.

Hobbes proposed the idea that thinking could be reduced to mathematical operations. On the other hand, Descartes had the insight into functions that machines might someday be able to perform. But he had reservations about the concept that thinking could be simply a mathematical process.

The 1800s was an era that saw some advancement in the conceptualization of the computer. Charles Babbage, a British mathematician, laid the foundation for the construction of the computer, a machine defined as being capable of performing mathematical computations. In 1833, Babbage introduced an Analytical Engine. This computational machine incorporated two unprecedented ideas that were to become crucial elements in the modern computer. First, it had operations that were fully programmable, and second, the engine could contain conditional branches. Without these two abilities, the power of today's computers would be inconceivable. Babbage was never able to realize his dream of building the analytic engine due to a lack of financial support. However, his dream was revived through the efforts of later researchers. Babbage's basic concepts could be observed in the way that most computers operate today.

Another British mathematician, George Boole, worked on issues that were to become equally important. Boole formulated the "laws of thought" that set up rules of logic for representing thought. The rules contained only two-valued variables. By this, any variable in a logical operation could be in one of only two states: yes or no, true or false, all or nothing, 0 or 1, on or off, and so on. This was the birth of digital logic, a key component of the AI effort.

In the early 1900s, Alfred North Whitehead and Bertrand Russell extended Boole's logic to include mathematical operations. This not only led to the formulation of digital computers but also made possible one of the first ties between computers and thought process.

However, there was still a lack of an acceptable way to construct such a computer. In 1938, Claude Shannon published "A Symbolic Analysis of Relay and Switching Circuits." This work demonstrated that Boolean logic consisting of only two-variable states (e.g., on-off switching of circuits) can be used to perform logic operations. Based on this premise, the ENIAC (Electronic Numerical Integrator and Computer) was built in 1946 at the University of Pennsylvania. The ENIAC was a large-scale fully operational electronic computer that signaled thebeginning of the first generation of computers. It could perform calculations 1000 times faster than its electromechanical predecessors. It weighed 30 tons, stood two stories high, and occupied 1500 square feet of floor space. Unlike today's computers that operate in binary codes (0s and Is), the ENIAC operated in decimal (0, 1, 2,..., 9) and it required 10 vacuum tubes to represent one decimal digit. With over 18,000 vacuum tubes, the ENIAC needed a great amount of electrical power, so much that it was said that it dimmed the lights in Philadelphia whenever it operated.

1.7 HUMAN INTELLIGENCE VERSUS MACHINE INTELLIGENCE

Two of the leading mathematicians and computer enthusiasts during the 1900 to 1950 timeframe were Alan Turing and John Von Neumann. In 1945, Von Neumann insisted that computers should not be built as glorified adding machines, with all their operations specified in advance. Rather, he suggested that computers should be built as general-purpose logic machines capable of executing a wide variety of programs. Such machines, Von Neumann proclaimed, would be highly flexible and capable of being readily shifted from one task to another. They could react intelligently to the results of their calculations, could choose among alternatives, and could even play checkers or chess. This represented something unheard of at that time: a machine with built-in intelligence, able to operate on internal instructions.

Prior to Von Neumann's concept, even the most complex mechanical devices had always been controlled from the outside, for example, by setting dials and knobs. Von Neumann did not invent the computer, but what he introduced was equally significant: computing by use of computer programs, the way it is done today. His work paved the way for what would later be called AI in computers.

Alan Turing also made major contributions to the conceptualization of a machine that can be universally used for all problems based only on variable instructions fed into it. Turing's universal machine concept, along with Von Neumann's concept of a storage area containing multiple instructions that can be accessed in any sequence, solidified the ideas needed to develop the programmable computer. Thus, a machine was developed that could perform logical operations and could do them in varying orders by changing the set of instructions that were executed.

Due to the fact that operational machines were now being realized, questions about the "intelligence" of the machines began to surface. Turing's other contribution to the world of AI came in the area of defining what constitutes intelligence. In 1950, he designed the Turing test for determining the intelligence of a system. The test utilized the conversational interaction between three players to try and verify computer intelligence.

The test is conducted by having a person (the interrogator) in a room that contains only a computer terminal. In an adjoining room, hidden from view, a man (Person A) and a woman (Person B) are located with another computer terminal. The interrogator communicates with the couple in the other room by typing questions on the keyboard. The questions appear on the couple's computer screen and they respond by typing on their own keyboard. The interrogator can direct questions to either Person A or Person B, but without knowing which is the man and which is the woman.

The purpose of the test is to distinguish between the man and the woman merely by analyzing their responses. In the test, only one of the people is obligated to give truthful responses. The other person deliberately attempts to fool and confuse the interrogator by giving responses that may lead to an incorrect guess. The second stage of the test is to substitute a computer for one of the two persons in the other room. Now, the human is obligated to give truthful responses to the interrogator, while the computer tries to fool the interrogator into thinking that it is human. Turing's contention is that if the interrogator's success rate in the human/computer version of the game is not better than his success rate in the man/woman version, then the computer can be said to be "thinking." That is, the computer possesses "intelligence." Turing's test has served as a classical example for AI proponents for many years.

By 1952, computer hardware had advanced far enough that actual experiments in writing programs to imitate thought processes could be conducted. The team of Herbert Simon, Allen Newell, and Cliff Shaw organized to conduct such an experiment. They set out to establish what kinds of problems a computer could solve with the right programming. Proving theorems in symbolic logic such as those set forth by Whitehead and Russell in the early 1900s fit the concept of what they felt an intelligent computer should be able to handle.

It quickly became apparent that there was a need for a new, higher level computer language than was currently available. First, they needed a language that was more user-friendly and could take program instructions that are easily understood by a human programmer and automatically convert them into machine language that could be understood by the computer. Second, they needed a programming language that changed the way in which computer memory was allocated. All previous languages would pre-assign memory at the start of a program. The team found that the type of programs they were writing would require large amounts of memory and would function unpredictably.

To solve the problem, they developed a list processing language. This type of language would label each area of memory and then maintain a list of all available memory. As memory became available, it would update the list, and when more memory was needed, it would allocate the amount necessary. This type of programming also allowed the programmer to be able to structure his or her data so that any information that was to be used for a particular problem could be easily accessed.

The end result of their effort was a program called Logic Theorist. This program had rules consisting of axioms already proved. When it was given a new logical expression, it would search through all of the possible operations in an effort to discover a proof of the new expression. Instead of using a brute force search method, they pioneered the use of heuristics in the search method.

The Logic Theorist that they developed in 1955 was capable of solving 38 of 52 theorems that Whitehead and Russell had devised. It was not only capable of the proofs but also did them very quickly. What took a Logic Theorist a matter of minutes to prove would have taken years to do if it had been done by simple brute force on a computer. By comparing the steps which it went through to arrive at a proof to those that human subjects went through, it was also found that it had a remarkable imitation of the human thought process.

1.8   NATURAL LANGUAGE DICHOTOMIES

Despite the various successful experiments, many observers still believe that AI does not have much potential for practical applications. There is a popular joke in the AI community that points out the deficiency of AI in natural language applications. It is said that a computer was asked to translate the following English statement into Russian and back to English: *The spirit is willing but the flesh is weak.* The reverse translation from Russian to English yielded: *The vodka is good but the meat is rotten.*

From my own author perspective, AI systems are not capable of thinking in the human sense. They are great in mimicking based on the massive amounts of data structures and linkages available. For example, consider the following natural language interpretations of the following ordinary statements:

"No salt is sodium free."
A human being can quickly infer the correct interpretation and meaning based on the prevailing context of the conversation. However, an "intelligent" machine may see the same statement in different ways, as enumerated below:
"No (salt) is sodium free," which negates the property of the object, salt. This means that there is no type of salt that is sodium free. In other words, all salts contain sodium.

Alternately, the statement can be seen as follows:

"(No-salt) is sodium free," which is a popular advertisement slogan for the commercial kitchen ingredient named (No-salt). In this case, the interpretation is that this product, named No-salt does not contain sodium.

Here is another one:

"No news is good news."
This is a common saying that humans can easily understand regardless of the context. In AI reasoning, it could be subject to the following interpretations:
"(No news) is good news," which agrees with the normal understanding that the state of having no new implies the absence of bad news, which is good (i.e., desirable). In this case, (No-news), as a compound word, is the object.

Or, an AI system could see it as:

"No (news) is good news," which is a contradiction of the normal interpretation. In this case, the AI system could interpret it as a case where all pieces of news are bad (i.e., not good). This implies that the object is the (news).
Here is another one from the political arena:
"The British parliament wants no deal off the table."

The AI interpretations could see the objects as follow:

(No deal), as a condition of negotiation, is off the table.

Alternately, it could be seen, in the negation sense, as all deals are acceptable to be on the table.

Pattern recognition is another interesting example that distinguishes human intelligence from machine intelligence. For example, when I park my vehicle in a large shopping center parking lot, with many similarly colored and shaped vehicles, I can always identify my vehicle from a far distance by simply seeing a tiny segment of the body of the vehicle. This could be by seeing half of the headlight jotting out from among several vehicles. It could be by seeing a portion of the tail light. It could even be by seeing the luggage carriage on top of the vehicle barely visible above other vehicles around mine. For an AI system to use pattern recognition to correctly identify my vehicle, it would have to use a tremendous amount data collection, data manipulation, interpolation, extrapolation, and other complex mathematical algorithms to consider the options to provide a probable match. Considering example such as the above, it can be seen that human intelligence and natural perception still trump machine's simulated intelligence. In spite of this deficiency, machine intelligence, under the banner of AI, can be useful to supplement human intelligence to arrive at a more efficient and effective decision processes. Consequently, AI is a useful and desirable ally in human operations. This belief was what drove the early efforts in defining and advancing the science, technology, engineering, and mathematics foundations for AI.

1.9 THE FIRST CONFERENCE ON ARTIFICIAL INTELLIGENCE

The summer of 1956 signified the first attempt to establish the field of machine intelligence into an organized effort. The Dartmouth Summer Conference, organized by John McCarthy, Marvin Minsky, Nathaniel Rochester, and Claude Shannon, brought together people whose work and interest formally founded the field of AI. The conference, held at Dartmouth College in New Hampshire, was funded by a grant from the Rockefeller foundation. It was at that conference that John McCarthy coined the term "artificial intelligence." It was the same John McCarthy who developed the LISP programming language that has become a standard tool for AI development. In attendance at the meeting, in addition to the organizers, were Herbert Simon, Allen Newell, Arthur Samuel, Trenchard More, Oliver Selfridge, and Ray Solomon off.

The Logic Theorist (LT), developed by Newell, Shaw, and Simon, was discussed at the conference. The system, considered the first AI program, used heuristic search to solve mathematical problems in *Principia Mathematica*, written by Whitehead and Russell (Newell and Simon 1972). Newell and Simon were far ahead of others in actually implementing AI ideas with their Logic Theorist. The Dartmouth meeting served mostly as an avenue for the exchange of information and, more importantly, as a turning point in the main emphasis of work in the AI endeavor. Instead of concentrating so much on the hardware to imitate intelligence, the meeting set the course for examining the structure of the data being processed by computers, the use of computers to process symbols, the need for new languages, and the role of computers for testing theories.

1.10 EVOLUTION OF SMART PROGRAMS

The next major step in software technology came from Newell, Shaw, and Simon in 1959. The program they introduced was called General Problem Solver (GPS). GPS was intended to be a program that could solve many types of problems. It was capable of solving theorems, playing chess, or doing various complex puzzles. GPS was a significant step forward in AI. It incorporates several new ideas to facilitate problem solving. The nucleus of the system was the use of means-end analysis. Means-end analysis involves comparing a present state with a goal state. The difference between the two states is determined and a search is done to find a method to reduce this difference. This process is continued until there is no difference between the current state and the goal state.

In order to further improve the search, GPS contained two other features. The first is that, if while trying to reduce the deviation from the goal state, it finds that it has actually complicated the search process, it was capable of backtracking to an earlier state and exploring alternate solution paths. The second is that it was capable of defining sub-goal states that, if satisfied, would permit the solution process to continue. In formulating GPS, Newell and Simon had done extensive work studying human subjects and the way they solved problems. They felt that GPS did a good job of imitating the human subjects. They commented on the effort by saying (Newell and Simon 1961):

The fragmentary evidence we have obtained to date encourages us to think that the General Problem Solver provides a rather good first approximation to an information processing theory of certain kinds of thinking and problem-solving behavior. The processes of 'thinking' can no longer be regarded as completely mysterious.

GPS was not without critics. One of the criticisms was that the only way the program obtained any information was to get it from human input. The way and order in which the problems were presented was controlled by humans, thus, the program was doing only what it was told to do. Newell and Simon argued that the fact that the program was not just repeating steps and sequences but was actually applying rules to solve problems it had not previously encountered, is indicative of intelligent behavior.

There were other criticisms also. Humans are able to devise new shortcuts and improvise. GPS would always go down the same path to solve the same problem, making the same mistakes as before. It could not learn. Another problem was that GPS was good when given a certain area or a specific search space to solve. The problem with this limitation was that in the solution of problems, it was difficult to determine what search space to use. Sometimes solving the problem is trivial compared to finding the search space. The problems posed to GPS were all of a specific nature. They were all puzzles or logical challenges: problems that could easily be expressed in a symbolic form and operated on in a pseudo-mathematical approach. There are many problems that humans face that are not so easily expressed in a symbolic form.

Also during the year 1959, John McCarthy came out with a tool that was to greatly improve the ability of researchers to develop AI programs. He developed a new computer programming language called LISP (list processing). It was to become one of the most widely used languages in the field.

LISP is distinctive in two areas: memory organization and control structure. The memory organization is done in a tree fashion with interconnections between memory groups. Thus, it permits a programmer to keep track of complex structural relationships. The other distinction is the way the control of the program is done. Instead of working from the prerequisites to a goal, it starts with the goal and works backwards to determine what prerequisites are required to achieve the goal.

In 1960, Frank Rosenblatt did some work in the area of pattern recognition. He introduced a device called PERCEPTRON that was supposed to be capable of recognizing letters and other patterns. It consisted of a grid of 400 photo cells connected with wires to a response unit that would produce a signal only if the light coming off the subject to be recognized crossed a certain threshold.

During the latter part of the 1960s, there were two efforts in another area of simulating human reasoning. Kenneth Colby at Stanford University and Joseph Weizenbaum at MIT wrote separate programs that were capable of interacting in a two-way conversation. Weizenbaum's program was called ELIZA. The programs were able to sustain very realistic conversations by using very clever techniques. For example, ELIZA used a pattern matching method that would scan for keywords like "I," "you," "like," and so on. If one of these words was found, it would execute rules associated with it. If there was no match found, the program would respond with a request for more information or with some noncommittal response.

It was also during the 1960s that Marvin Minsky and his students at MIT made significant contributions toward the progress of AI. One student, T. G. Evans, wrote a program that could perform visual analogies. The program was shown two figures that had some relationship with each other and was then asked to find another set of figures from a set that matched the same relationship. The input to the computer was not done by a visual sensor (like the one worked on by Rosenblatt), but instead the figures were described to the system.

In 1968, another student of Minsky's, Daniel Bobrow, introduced a linguistic problem solver called STUDENT. It was designed to solve problems that were presented to it in a word problem format. The key to the program was the assumption that every sentence was an equation. It would take certain words and turn them into mathematical operations. For example, it would convert "is" into "=" and "per" into "/ ".

Even though STUDENTS responded very much the same way that a real student would, there was a major difference in depth of understanding. While the program was capable of calculating the time two trains would collide given the starting points and speeds of both, it had no real understanding or even cared what a "train" or "time" was. Expressions like "per chance" and "this is it" could mean totally different things than what the program would assume. A human student would be able to discern the intended meaning from the context in which the terms were used.

In an attempt to answer the criticisms about understanding, another student at MIT, Terry Winograd, developed a significant program named SHRDLU. In setting up his program, he utilized what was referred to as a micro-world or blocks-world. This limited the scope of the world that the program had to try to understand. The program communicated in what appeared to be natural language.

The operation of SHRDLU consisted of a set of blocks of varying shapes (cubes, pyramids, etc.), sizes, and colors. These blocks were all set on an imaginary table. Upon request, SHRDLU would rearrange the blocks to any requested configuration. The program was capable of knowing when a request was unclear or impossible. For instance, if it was requested to put a block on top of the pyramid, it would request that the user specify more clearly what block and what pyramid. It could also recognize that the block would not sit on top of the pyramid.

Two other approaches that the program took that were new to programs were the ability to make assumptions and the ability to learn. If asked to pick up a larger block, it would assume a larger block than the one it was currently working on. If asked to build a figure that it did not know, it would ask for an explanation of what it was and, thereafter, it would recognize the object. One major sophistication that SHRDLU added to the science of AI programming was its use of a series of expert modules or specialists. There was one segment of the program that specialized in segmenting sentences into meaningful word groups, a sentence specialist to determine the relationship between nouns and verbs, and a scenario specialist that understood how individual scenes related to one another. This sophistication added much enhancement to the method in which instructions were analyzed.

As sophisticated as SHRDLU was at that time, it did not escape criticism. Other scholars were quick to point out its deficiencies. One of the shortcomings was that

SHRDLU only responded to requests; it could not initiate conversations. It also had no sense of conversational flow. It would jump from performing one type of task to a totally different one if so requested. While SHRDLU had an understanding of the tasks it was to perform and the physical world in which it operated, it still could not understand very abstract concepts. Some of the latest developments in the evolution of AI systems is the emergence of ChatBots, which are AI-based chat tools. The most popular tool in this genre is ChatGPT (Generative Pre-trained Transformer), which has been hailed as a very astute ChatBot. It uses deep learning to produce human-like text. Given an initial text as a prompt, it will produce text that continues the prompt. More developments in this regard will further transform human-AI interfaces.

1.11   BRANCHES OF ARTIFICIAL INTELLIGENCE

The various attempts to formally define the use of machines to simulate human intelligence led to the development of several branches of AI. Current sub-specialties of AI include:

(1) *Natural language processing:* This deals with various areas of research such as database inquiry systems, story understanders, automatic text indexing, grammar and style analysis of text, automatic text generation, machine translation, speech analysis, and speech synthesis.
(2) *Computer vision:* This deals with research efforts involving scene analysis, image understanding, and motion derivation.
(3) *Robotics:* This involves the control of effectors on robots to manipulate or grasp objects, locomotion of independent machines, and use of sensory input to guide actions.
(4) *Problem solving and planning:* This involves applications such as refinement of high-level goals into lower-level ones, determination of actions needed to achieve goals, revision of plans based on intermediate results, and focused search of important goals.
(5) *Learning:* This area of AI deals with research into various forms of learning, including rote learning, learning through advice, learning by example, learning by task performance, and learning by following concepts.
(6) *Expert systems:* This deals with the processing of knowledge as opposed to the processing of data. It involves the development of computer software to solve complex decision problems.

1.12   NEURAL NETWORKS

Neural networks, sometimes called connectionist systems, represent networks of simple processing elements or nodes capable of processing information in response to external inputs. Neural networks were originally presented as being models of the human nervous system. Just after World War II, scientists found out that the physiology of the brain was similar to the electronic processing mode used by computers. In both cases, large amounts of data are manipulated. In the case of computers, the elementary unit of processing is the bit, which is in either an "on" or "off" state. In the case of the brain, *neurons* perform the basic data processing. Neurons are tiny cells that follow a binary principle of being either in a state of firing (on) or not firing (off). When a neuron is on, it fires a signal to other neurons across a network of synapses.

In the late 1940s, Donald Hebb, a researcher, hypothesized that biological memory results when two neurons are active simultaneously. The synaptic connection of synchronous neurons is reinforced and given preference over connections made by neurons that are not active simultaneously. The level of preference is measured as a weighted value. Pattern recognition, a major strength of human intelligence, is based on the weighted strengths of the reinforced connections between various pairs of simultaneously active neurons.

The idea presented by Hebb was to develop a computer model based on the way in which neurons form connections in the human brain. But the idea was considered to be preposterous at that time since the human brain contains 100 billion neurons and each neuron is connected to 10,000 others by a synapse. Even by today's computing capability, it is still difficult to duplicate the activities of neurons. In 1969, Marvin Minsky and Seymour Pappert wrote the book entitled *Perceptrons*, in which they criticized existing neural network research as being worthless. It has been claimed that the pessimistic views presented by the book discouraged further funding for neural network research for several years. Funding was, instead, diverted to further research of expert systems, which Minsky and Pappert favored. It is only recently that neural networks are beginning to make a strong comeback.

Because neural networks are modeled after the operations of the brain, they hold considerable promise as building blocks for achieving the ultimate aim of AI. The present generation of neural networks use artificial neurons. Each neuron is connected to at least one other neuron in a synapse-like fashion. The networks are based on some form of learning model. Neural networks learn by evaluating changes in input. Learning can be either supervised or unsupervised. In supervised learning, each response is guided by given parameters. The computer is instructed to compare any inputs to ideal responses, and any discrepancy between the new inputs and ideal responses is recorded. The system then uses this databank to guess how much the newly gathered data is similar to or different from the ideal responses. That is, how closely the pattern matches. Supervised learning networks are now commercially used for control systems and for handwriting and speech recognition.

In unsupervised learning, input is evaluated independently and stored as patterns. The system evaluates a range of patterns and identifies similarities and dissimilarities among them. However, the system cannot derive any meaning from the information without human assignment of values to the patterns. Comparisons are relative to other results, rather than to an ideal result. Unsupervised learning networks are used to discover patterns where a particular outcome is not known in advance, such as in physics research and the analysis of financial data. Several commercial neural network products are now available. An example is NeuroShell from Ward Systems Group. The software is expensive, but it is relatively easy to use. It interfaces well with other software such as Lotus 1-2-3 and dBASE, as well as with C, Pascal, FORTRAN, and BASIC programming languages.

Despite the proven potential of neural networks, they drastically oversimplify the operations of the brain. The existing systems can undertake only elementary pattern-recognition tasks, and are weak at deductive reasoning, math calculations, and other computations that are easily handled by conventional computer processing. The difficulty in achieving the promise of neural networks lies in our limited understanding of how the human brain functions. Undoubtedly, to accurately model the brain, we must know more about it. But a complete knowledge of the brain is still many years away.

Google's ANN-based "deepmind" is arguably one of the strongest AI engines available at the moment: it has been used to tackle numerous problems in Mathematics, Chess, etc.

1.13 EMERGENCE OF EXPERT SYSTEMS

In the late 1960s to early 1970s, a special branch of AI began to emerge. The branch, known as expert systems, has grown dramatically in the past few years and it represents the most successful demonstration of the capabilities of AI. Expert systems are the first truly commercial application of work done in the AI field and as such have received considerable publicity. Due to the potential benefits, there is currently a major concentration in the research and development of expert systems compared to other efforts in AI.

Not driven by the desire to develop general problem-solving techniques that had characterized AI before, expert systems address problems that are focused. When Edward Feigenbaum developed the first successful expert system, DENDRAL, he had a specific type of problem that he wanted to be able to solve. The problem involved determining which organic compound was being analyzed in a mass spectrograph. The program was intended to simulate the work that an expert chemist would do in analyzing the data. This led to the term expert system.

The period of time from 1970 to 1980 saw the introduction of numerous expert systems to handle several functions from diagnosing diseases to analyzing geological exploration information. Of course, expert systems have not escaped the critics. Given the nature of the system, critics argue that it does not fit the true structure of AI. Because of the use of only specific knowledge and the ability to solve only specific problems, some critics are apprehensive about referring to an expert system as being intelligent. Proponents argue that if the system produces the desired results, it is of little concern whether it is intelligent or not.

A controversy of interest surfaced in 1972 with a book published by Hubert Dreyfus called *What Computers Can't Do: A Critique of Artificial Reason*. Views similar to those contained in the book were presented in 1976 by Joseph Weizenbaum. The issues that both authors raised touched on some of the basic questions that prevailed way back in the days of Descartes. One of Weizenbaum's reservations concerned what should ethically and morally be handed over to machines. He maintained that the path that AI was pursuing was headed in a dangerous direction. There are some aspects of human experience, such as love and morality, that could not adequately be imitated by machines.

While the debates were going on over how much AI could do, the work on getting AI to do more continued. In 1972, Roger Shrank introduced the notion of script; the set of familiar events that can be expected from an often encountered setting. This enables a program to quickly assimilate facts. In 1975, Marvin Minsky presented the idea of frames. Even though both concepts did not drastically advance the theory of AI, they did help expedite research in the field.

In 1979, Minsky suggested a method that could lead to a better simulation of intelligence. He presented the "society of minds" view, in which the execution of knowledge is performed by several programs working in conjunction simultaneously. This concept helped to encourage interesting developments such as present-day parallel processing.

As time proceeded through the 1980s, AI gained significant exposure and interest. AI, once a phrase restricted to the domain of esoteric research, has now become a practical tool for solving real problems. While AI is enjoying its most prosperous period, it is still plagued with disagreements and criticisms. The emergence of commercial expert systems on the market has created both enthusiasm and skepticism. There is no doubt that more research and successful applications developments will help prove the potential of expert systems. It should be recalled that new technologies sometimes fail to convince all initial observers. IBM, which later became a giant in the personal computer business, hesitated for several years before getting into the market because the company never thought that those little *boxes* called personal computers would ever have any significant impact on the society. How wrong they were!

The effort in AI is a worthwhile endeavor as long as it increases the understanding that we have of intelligence and as long as it enables us to do things that we previously could not do. Due to the discoveries made in AI research, computers are now capable of doing things that were once beyond imagination.

Embedded Expert Systems: More expert systems are beginning to show up, not as stand-alone systems, but as software applications in large software systems. This trend is bound to continue as systems integration takes hold in many software applications. Many conventional commercial packages such as statistical analysis systems, data management systems, information management systems, project management systems, and data analysis systems now contain embedded heuristics that constitute expert systems components of the packages. Even some computer operating systems now contain embedded expert systems designed to provide real-time systems monitoring and troubleshooting. With the success of embedded expert systems, the long-awaited payoffs from the technology are now beginning to be realized.

Because the technology behind expert systems has changed little over the past decade, the issue is not whether the technology is useful, but how to implement it. This is why the integrated approach of this book is very useful. The book focuses not only on the technology of expert systems, but also on how to implement and manage the technology. Combining neural network technology with expert systems, for example, will become more prevalent. In combination, the neural network might be implemented as a tool for scanning and selecting data, while the expert system would evaluate the data and present recommendations.

While AI technology is good and amenable to organizational objectives, we must temper it with human intelligence. The best hybrid is when machine intelligence is integrated with human intelligence. The human can handle the intuition part, while the machine, as AI, handles the data-intensive and number-crunching parts.

Over the past several years, expert systems have proven their potential for solving important problems in engineering and manufacturing environments. Expert systems are helping major companies to diagnose processes in real time, schedule operations, troubleshoot equipment, maintain machinery, and design service and production facilities. With the implementation of expert systems in industrial environments, companies are finding that real-world problems are best solved by an integrated strategy involving the management of personnel, software, and hardware systems.

Solutions to most engineering and manufacturing problems involve not only heuristics, but also mathematical calculations, large data manipulations, statistical analysis, real-time information management, system optimization, and man machine interfaces. These issues and other related topics are addressed in detail in this book. In addition to the basic concepts of expert systems, guidelines are presented on various items ranging from problem selection, data analysis, knowledge acquisition, and system development to verification, validation, integration, implementation, and maintenance.

Can artificial intelligence systems and products live up to the hype?

In general, the expectations for all products, systems, and processes of AI include effectiveness, efficiency, ease of use, elegance, safety, security, sustainability, and satisfaction. A systems view can, indeed, bring us closer to realizing these expectations. In terms of a practical and readily seen example of AI, look no further than your mobile phone. The present generation of smart phones is readily a common example of an AI system. So, AI is already all around us on a daily basis. Figure 1.1 illustrates a comprehensive view of IE and DE for AI. It can be seen that every element within the definition of industrial engineering fits within the wide scope of AI. Liebowitz (2021) presents a wide coverage of the digital platforms using data analytics to enable AI applications in diverse fields. Some of the topics include autonomy, machine intelligence, managerial decision-making, deep learning, natural language

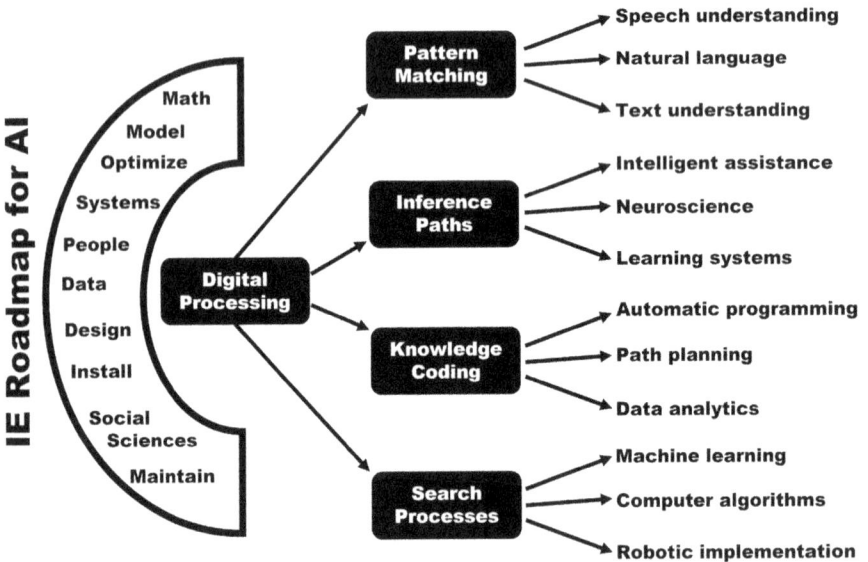

FIGURE 1.1 Industrial engineering and digital engineering roadmap for artificial intelligence.

processing, smart materials, predicting, and cognitive neuroimaging. Lohr (2022) describes how one entrepreneur, David Ferrucii, is fusing common sense into AI applications. It is recalled that common sense applications and implications are among the bastions of industrial engineering.

REFERENCES

Badiru, A. B. (1985). "Process Capability Analysis on a Microcomputer," in *Softcover Software*, Industrial Engineering & Management Press, Norcross, Georgia, 1985, pp. 7–14.

Badiru, A. B. (1988). "Expert Systems and Industrial Engineers: A Practical Guide for a Successful Partnership". *Computers & Industrial Engineering* 14(1): 1–13.

Badiru, A. B. (1992). *Expert Systems Applications in Engineering and Manufacturing*, Prentice-Hall, Hoboken, NJ.

Badiru, A. B. (2022). *Artificial Intelligence and Digital Systems Engineering*, Taylor & Francis Group/CRC Press, Boca Raton, FL.

Badiru, A. B. and Christian S. Bauer (1985). "Stochastic Rate of Return Analysis on a Microcomputer," in *Softcover Software*, IIE Press, Norcross, GA, pp. 127–135.

Badiru, A. B. and John Cheung (2002). *Fuzzy Engineering Expert Systems with Neural Network Applications*, John Wiley & Sons, New York.

Badiru, A. B. and O. Ibidapo-Obe (2008). "On Set-Theoretic Fundamentals of Systems Integration". *African Journal of Computing* & ICT (Journal of the IEEE Nigeria Computer Chapter) 1(2): 3–25.

Badiru, A. B., Oye Ibidapo-Obe and B. J. Ayeni (2012). *Industrial Control Systems: Mathematical and Statistical Models and Techniques*, Taylor & Francis Group/CRC Press, Boca Raton, FL.

Badiru, A. B., J. Mathis and B. Holloway, 1987. "AREST Expert System To Aid Crime Investigations". *AI Interactions, Texas Instruments Data Systems* 3(3): 5–6.

Badiru, A. B. and Gary E. Whitehouse (1986). "The Impact of the Computer on Resource Allocation Algorithms," Presented at the *ORSA/TIMS Fall Conference*, Miami, FL, October 1986.

Badiru, A. B. and Gary E. Whitehouse (1989). *Computer Tools, Models, and Techniques for Project Management*, TAB Professional & Reference Books, Inc., Blue Ridge Summit, PA.

Balch, Kay et al (1993). *Mathematics: Applications and Connections*, MacMillan, New York.

Doran, Eric, Sharon Bommer and Adedeji Badiru (2022). "Integration of Human Factors, Cognitive Ergonomics, and Artificial Intelligence in the Human-Machine Interface for Additive Manufacturing". *International Journal of Quality Engineering and Technology, International Journal of Mechatronics and Manufacturing Systems* 15(4): 310–330.

Ibidapo-Obe, O., A. B. Alonge and A. B. Badiru (1995). "On Active Controls for a Biped Mechanism". *Applied Mathematics and Computation* 69: 159–183.

Ibidapo-Obe, O., Sunday Asaolu and Adedeji B. Badiru (2001). "Generalized Solutions of the Pursuit Problem in Three-Dimensional Euclidean Space". *Applied Mathematics and Computation* 119(3): pp. 35–45.

Liebowitz, Jay (ed) (2021). *Data Analytics and AI*, CRC Press/Taylor & Francis Group, Boca Raton, FL.

Lohr, Steve (2022). "One Man's Dream of Fusing AI, Common Sense," *The New York Times*, Technology Section, November 13, 2022.

Newell, Allen and Herbert A. Simon (1961). "Computer Simulation of Human Thinking". The RAND Corporation, April 20, 1961, p. 2276.

Sadiku, Matthew N. O., S. M. Musa and S. R. Nelatury (2022). *Applications of Artificial Intelligence*, Gotham Books, Sheridan, WY.

2 Mathematics of Cantor Set for AI Searches

2.1 INTRODUCTION

Everything about artificial intelligence (AI) is about conducting a search among a myriad of alternatives. The human brain is able to do this naturally and quickly. For us to design AI machines that can mimic human brain requires that we develop representational mathematics to encode the available alternatives, considering various permutations and combinations of options. Over the years, our ability to come up with the required representational coding has been increasing rapidly. More remains to be explored and done in this regard. There is a rich collection of mathematical formulations and foundations that AI researchers can draw from. It is on the basis of this mathematical quest that Badiru (1993) introduced the Cantor Set as a viable basis for database search in AI applications. Cantor Set is often referred to as the set of "deleted middle thirds." In successive iterations, the middle third of whatever is left in the set is deleted. Good advances were made with the technique in the early years, but the methodology was stymied by the computational limitations of hardware tools of that era. Now, that more powerful computational techniques (hardware and software) are available, the technique is worthy of reviving as an AI research topic. The emergence of Quantum Computing and Neuro-Fuzzy techniques bodes well for a new exploration of Cantor Set as an AI search tool.

The research promise of Cantor Set involves the development of an intelligent search algorithm to optimize search patterns in AI databases. The search technique is based on the fractal property of the Cantor Dust. Any new research of Cantor Set Search will be an advancement of the original Cantor Set Search technique presented in several publications (Badiru, 1992, 1993; Badiru and Cheung, 2002; Badiru et al., 1992; Milatovic and Badiru, 1998; Milatovic, 1996; Nakada, 1993; Wei, 1993). The previous Cantor tri-sectioning algorithm was efficient only for domains where the distribution of the data searched was approximately bell-shaped, but performed poorly when searching through skewed data. It is believed that new computational techniques may alleviate the previous impediment.

It is expected that a new enhancement to the cantor search algorithm will estimate the position of the peak of the density trace curve and always start the very first comparison with the search key approximately at the modal concentration of the data distribution. A new formula was derived by Milatovic and Badiru (1998), which, in terms of only five specific percentile values, calculates the position of the modal concentration in unimodal curves with an accuracy of more than 95%. In addition, the five percentile values (in particular, 5th, 16th, 50th, 84th, and 95th percentile) can be easily accessed for any given distribution, regardless of the number of records in the database.

DOI: 10.1201/9781003247395-2

2.2   INTELLIGENT SEARCHES

In as much as we search for results, whether qualitative or quantitative, it will serve us well to pursue intelligent searches. The intellectual merit of using Cantor Set search is that for a sorted array of data points, Cantor Set search is strictly superior to the binary search for a range between 10% and 20% of data points closest to the modal concentration, regardless of the type of distribution and its skewness. Cantor Set search is further competitive for the range of data points between 30% and 45% around the modal concentration. As the distance of the search key further increases, the performance of the Cantor search severely deteriorates, and the binary search becomes more effective. This is an interesting property worthy of further research investigation that the readers of this book are encouraged to consider.

There is an increasing broad appeal in the development of an intelligent $1/n$ sectioning of the search space in terms of the peakedness (kurtosis) of the data distribution. The intelligent aspect of any new methodology will come from its being neural network adaptive to the specific distribution being searched. Potential broad applications of Cantor Set Search in modern AI applications include the following:

- Detection and categorization of elements in large data sets.
- Taxonomical organization of training sets in large-scale neural network applications.
- Exploratory research in information generation for industrial applications.
- Analysis and effective utilization of massive data sets in data mining applications.
- Pruning of massive data sets collected with modern data gathering tools.
- Segmentation of large information bases for transmission.
- Determination of path search pattern in autonomous vehicles.

2.3   BACKDROP FOR AI SEARCHES

Stone and Wagner (1980) define three basic elements of any optimal search problem: (1) The searcher should possess at least a rough estimate of the *probability distribution* of the search key and space. An example of this would be a construction of a circular bell-shaped distribution of a submarine position, based on its previously reported positions. Since the direction of the target is unknown, its probability distribution would have a standard deviation of σ in any outward direction, but would remain constant along its [0,360] degree circular path. (2) The notion of the *detection function*, which relates the amount of effort placed in a search space to the probability of detecting the target, given that it is located in that search space. (3) *Constraint on effort*, which implies that the searcher is limited in either the amount of effort or resources when carrying out a search. There is an ongoing need to find new ways to improve search tools and techniques. The Cantor Set search is one such improvement approach. While the original cantor set technique (Badiru, 1992, 1993) was widely accepted for its sound mathematical basis, it has, for many years, been considered to be too radical, too unconventional, risky or speculative for practical applications. However, recent developments in information technology research have made it possible to embrace new, radical, and risky exploratory mathematical methodologies. Contents of this chapter can lead to new search paradigms to create novel uses of emerging AI techniques.

Many search techniques have been proposed over the years (Ahlswede and Wegener, 1987; Gonnet, 1984; Goodman and Hedetniemi, 1977; Knuth, 1980; Mehlhorn, 1984; Sedgewick, 1992; and references therein). Some of the basic search strategies include binary searching, radix search, balanced trees search, hashing techniques, string searching procedures, range searching, etc. Lately, search techniques have become a major part of data manipulation techniques such as file compression, pattern recognition, and cryptology.

Baase (1988) shows that any algorithm to find a key x in a list of n entries by comparison must exercise at least $(\lg n) + 1$ comparisons for any input. Consequently, any search algorithm based on comparisons of the order of *log n* should be considered as optimal. And many of the search strategies used today are indeed of the log n order. Examples include the binary search, balanced tree search, radix search, Fibonaccian search (as given by Knuth, 1980), etc. In many instances, however, the search time could be further reduced by considering the frequency of appearance of elements in the search set, their position within the set, and the frequency of their access. For example, one enhancement of the binary search suggested by Sedgewick (1992) is to guess more closely where the key being sought occurs within the current interval of interest. This could greatly reduce the time and effort of the search, as opposed to the strategy of splitting the interval in exactly two at each step. Such a method, called *Interpolation (binary) search*, requires only a minor modification to the original search procedure, and the new "middle" of the interval is computed by adding half the size of the interval to one of the endpoints.

2.4 MATHEMATICS OF CANTOR SET

The original Cantor Set Search technique is based on the geometrical properties of the Cantor Set, which is often referred to as the "set of the excluded middle thirds." Badiru (1992) first proposed using the Cantor Set interval deletions to represent the pruning of a search space. A unique property of the cantor set is that it contains an infinite number of elements, but its representative points occupy no space, in the geometric sense, on the real number line. The concept of excluded middle thirds may have relevance in certain problem domains requiring specialized search strategies. While binary search divides the search space into halves, the Cantor Set search procedure divides the search space into thirds. The relevant search domains for applying Cantor Set search are those where associative property inheritance relationships exist among the elements of the knowledge base such that the elements can be stored in an ordered fashion using a key property. Examples of such domains can be found in computer-aided design, group technology, mineral prospecting, and chemical analysis. Mathematically, the cantor set is denoted as:

$$C = \left\{ \chi \in \Omega \,\middle|\, \chi \in \bigcup_{k=0}^{\infty} \delta_k \right\}$$

where $\Omega = [0,1]$. We consider the closed interval $\Omega = [0,1]$ and the open intervals generated by successive removal of the middle thirds of intervals left after previous removals. The interval deletions are shown geometrically in Figure 2.1.

Note that:

$$\bigcup_{k=0}^{\infty} \delta_k = [0,1]$$

The interval δ_k is the union of the open intervals deleted from Ω after the kth search iteration. The deleted intervals are represented mathematically below:

$$\delta_0 = \phi \, (\text{null set})$$

$$\delta_1 = \left[\frac{1}{3}, \frac{2}{3}\right]$$

$$\delta_2 = \left[\frac{1}{9}, \frac{2}{9}\right] \cup \left[\frac{1}{3}, \frac{2}{3}\right] \cup \left[\frac{7}{8}, \frac{8}{9}\right]$$

If Ω is considered as the universal set, then we may also express the Cantor set as the complement of the original set C. That is, alternately,

$$C = \left(\bigcup_{k=0}^{\infty} \delta_k\right)^c,$$

which, by DeMorgan's Law, implies

$$C = \bigcap_{k=0}^{\infty} (\delta_k)^c$$

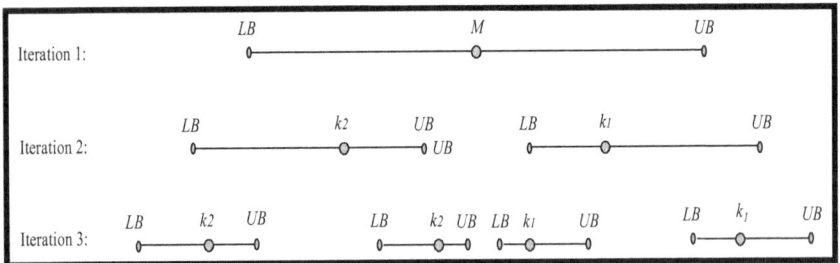

FIGURE 2.1 Composite binary—Cantor search technique.

If we use the following simplifying notation for the complement, $(\delta_k)^c = \lambda_k$, we would obtain the following alternate representation:

$$C = \bigcap_{k=0}^{\infty} \lambda_k,$$

where

$$\lambda_0 = \left[0,1\right]$$

$$\lambda_1 = \left[0, \frac{1}{3}\right] \cup \left[\frac{2}{3}, 1\right]$$

$$\lambda_2 = \left[0, \frac{1}{9}\right] \cup \left[\frac{2}{9}, \frac{1}{3}\right] \cup \left[\frac{2}{3}, \frac{7}{9}\right] \cup \left[\frac{8}{9}, 1\right], \text{ and so on.}$$

It should be noted that λ_k is the remaining search space available for the kth search iteration. Also note that λ_k consists of 2^k closed and non-overlapping intervals each of real length $(1/3^k)$. In the Cantor search procedure, the desired item is found only when it is located in the middle third of some interval. If the value of interest is in the interval (4.333, 5.666), then only one interval search would be needed to find it. If it is in the interval (3.444, 3.888) $\cup$ (6.110, 6.554), then at most three interval searches would be needed. If the value is in the interval (3.148, 3.296) $\cup$ (4.036, 4.184) $\cup$ (5.814, 5.962) $\cup$ (6.702, 6.85), then at most seven interval searches would be needed. In general, the maximum number of interval searches, N, needed to locate an item in a Cantor set search strategy is one of the following: $N_k = 1, 3, 5, 7, 15, 31, \ldots$, where k is the iteration number. That is,

$$N_0 = 0$$

$$N_k = N_{k-1} + 2^{k-1}$$

$$= \sum_{j=0}^{k-1} 2^j$$

$$= \left(2^k\right) - 1$$

It should be noted that if the distribution of the items to be searched is skewed to the right or left (e.g., Chi-squared or lognormal distributions), then the basic Cantor set search strategy would not be appropriate. Hence, there is a need to conduct further research to enhance and expand the applicability of the basic technique.

The actual Cantor Set search scheme is based on attempts to filter out the middle third part of the data set (assuming a bell-shaped distribution), and search through it first, before examining the tails of the distribution. Both Nakada (1993) and Wei (1993) have conducted extensive research in defining and achieving an improved utilization of the Cantor Set search. Their results have shown that applications of Cantor search under certain restrictions are highly competitive to binary search.

As indicated by the two studies above, the Cantor search performed well when both the data and the search key were approximately normally distributed with relatively small standard deviation.

2.4.1 SET SECTIONING TECHNIQUE OF CANTOR SET

In many instances, we may have a modal distribution of data whose spread is not necessarily normal or bell-shaped. Examples of such data may occur in modeling of product failures, time required to complete a task, etc. The application of the existing Cantor Set search technique in such cases would be inefficient. Therefore, the objective of the proposed research is to develop an enhanced Cantor search procedure, where the search interval at each iteration would be more cognizant of the actual dispersion of the data of interest. In other words, the technique is based on defining the position and the size of the initial Cantor search interval as a function of the location of the mode and measures of the skewness and peakedness of the data set. An effective heuristic procedure for fast and accurate location of the modal concentration within a database has been formulated (Milatovic and Badiru, 1998) and serves as the starting point of the proposed enhancement. The term "modal concentration" is defined as a single data element or value around which the majority of the other data elements, observations, or measurements are concentrated. The proposed heuristic estimates the position of this value in terms of only five percentile values regardless of the size of the database, while retaining minimal overhead cost.

There are many application domains for the original Cantor search. A simple example would be the design of fishing hooks, given the size of fish in a certain geographical environment. Very big (or very small) hooks would be functional for only a small portion of fish available, whereas we would want to design one or more hook sizes that could catch the widest range of fish size available and set the production volume (for each hook size) accordingly. Therefore, we would want to search for the most common fish size, based on which, the adequate hook(s) sizes would be designed. The above example is analogous to potential applications in security and warfare searches.

The original Cantor Set Search technique is based on the geometrical properties of the Cantor Set, which is often referred to as the "set of the excluded middle thirds." Badiru (1992) first proposed using the Cantor Set interval deletions to represent the pruning of a search space. A unique property of the cantor set is that it contains an infinite number of elements, but its representative points occupy no space, in the geometric sense, on the real number line. The concept of excluded middle thirds has relevance in certain problem domains requiring specialized search strategies. While binary search divides the search space into halves, the Cantor Set search procedure divides the search space into thirds. The relevant search domains for applying Cantor Set search are those where associative property inheritance relationships exist among the elements of the knowledge base such that the elements can be stored in an ordered fashion using a key property. Examples of such domains can be found in computer-aided design, group technology, mineral prospecting, and chemical analysis. Mathematically, the cantor set is denoted as:

$$C = \left\{ \chi \in \Omega \mid \chi \in \bigcup_{k=0}^{\infty} \delta_k \right\}$$

where $\Omega = [0,1]$. We consider the closed interval $\Omega = [0,1]$ and the open intervals generated by successive removal of the middle thirds of intervals left after previous removals. Note that:

$$\bigcup_{k=0}^{\infty} \delta_k = [0,1]$$

The interval δ_k is the union of the open intervals deleted from Ω after the kth search iteration. The deleted intervals are represented mathematically as:

$$\delta_0 = \phi \text{ (null set)}; \delta_1 = \left[\frac{1}{3}, \frac{2}{3}\right]; \delta_2 = \left[\frac{1}{9}, \frac{2}{9}\right] \cup \left[\frac{1}{3}, \frac{2}{3}\right] \cup \left[\frac{7}{8}, \frac{8}{9}\right], \text{and so on.}$$

If Ω is considered as the universal set, then we may express the Cantor set as a complement:

$$C = \left(\bigcup_{k=0}^{\infty} \delta_k\right)^c \text{ or } C = \bigcap_{k=0}^{\infty} (\delta_k)^c$$

If we denote $(\delta_k)^c = \lambda_k$, we would obtain the following simplified representation:

$$C = \bigcap_{k=0}^{\infty} \lambda_k,$$

where

$$\lambda_0 = [0,1];$$
$$\lambda_1 = \left[0, \frac{1}{3}\right] \cup \left[\frac{2}{3}, 1\right]$$
$$\lambda_2 = \left[0, \frac{1}{9}\right] \cup \left[\frac{2}{9}, \frac{1}{3}\right] \cup \left[\frac{2}{3}, \frac{7}{9}\right] \cup \left[\frac{8}{9}, 1\right], \text{and so on.}$$

In this case, λ_k represents the search space available for the kth search iteration. Also, λ_k consists of 2^k closed and non-overlapping intervals each of real length $(1/3^k)$. In the Cantor search procedure, the desired item is found only when it is located in the middle third of some interval. If the value of interest is in the interval (4.333, 5.666), then only one interval search would be needed to find it. If it is in the interval (3.444,

3.888) ∪ (6.110, 6.554), then at most three interval searches would be needed. If the value is in the interval (3.148, 3.296) ∪ (4.036, 4.184) ∪ (5.814, 5.962) ∪ (6.702, 6.85), then at most seven interval searches would be needed. In general, the maximum number of interval searches, N, needed to locate an item in a Cantor set search strategy is one of the following: $N_k = 1, 3, 5, 7, 15, 31, \ldots$, where k is the iteration number.

$$N_0 = 0$$

$$N_k = N_{k-1} + 2^{k-1} = \sum_{j=0}^{k-1} 2^j = \left(2^k\right) - 1$$

It should be noted that if the distribution of the items to be searched is skewed to the right or left (e.g., Chi-squared or lognormal distributions), then the basic Cantor set search strategy would not be appropriate. Hence, the need to conduct further research to enhance and expand the applicability of the basic technique. One of the major limitations of the original Cantor Set search is that the actual search through-out each iteration is conducted within a fixed "excluded middle third" interval. Nakada (1993) and Wei (1993) have shown that the original Cantor Set search per-forms worse than the binary search unless the two techniques are combined. One approach, proposed and investigated by Nakada (1993) (named as CSS 8 in his the-sis) and Wei (1993) (named as CSS 4 in her thesis), is given below. The search key is compared first with the middle key in the array. If not found, the rest of the array is split into thirds, and each is explored iteratively until the key is either found or the search is terminated unsuccessfully. The procedure is illustrated earlier in Figure 2.1.

Clearly, this approach enhances the search around the mean when the data is nor-mally distributed. Badiru (1993) and Wei (1993) discuss the conditions under which this technique is superior to the common binary search. In a preliminary study of the proposed research, a major improvement to the above technique was achieved by replacing M (as the middle element of the search space) by the estimated position of the modal concentration. An effective approximation of the location of modal con-centration in any unimodal curve is presented below.

2.4.2   SEARCH OF ASYMMETRICALLY DISTRIBUTED DATA

Assume a data search through an asymmetric distribution. The information about the skewness of the data distribution would be of great assistance in determining the starting point of the Cantor search, since it is clear that the position of the modal ten-dency, or more precisely, the peak of the curve is a function of the skewness. In other words, a strongly positive skewness would indicate that the search should start some-where within the left (Cantor Set) interval, rather than the middle one. However, the calculations involved in determining the skewness, that is, the third moment about the mean, may generally add a significant overhead cost to the procedure. In addition, as Inman (1952) pointed out, the moment measures always depend on the entire dis-tributions, which pose auxiliary limitations in cases when the search space is a subset of a bigger database of elements. An alternative approach to estimating the skewness of a distribution, as well as all of its first four moments (about the mean), is through a combination of certain critical values of distribution percentiles. This idea has been explored by many statisticians, Gardiner and Gardiner (1978), McCammon (1962), Otto (1939), and Trask (1932). Inman (1952) proposed two formulas that estimated the skewness by measuring the deviation of the median from the mean and dividing the result by the standard deviation:

$$\text{skewness} = \frac{P84 + P16 - 2 \times P50}{P84 - P16} = \frac{P95 + P5 - 2 \times P50}{P95 - P5},$$

where *PXX* represents *XX*th percentile.

Folk and Ward (1957) extended Inman's work and proposed a more robust skewness formula that would account for the asymmetry of the distribution due to both its central part and its endpoints. The formula, called the *Inclusive Graphic Skewness (IGS)*, is the average of the two equations proposed by Inman, and is denoted as:

$$IGS = \frac{P16 + P84 - 2 \times P50}{2(P84 - P16)} + \frac{P5 + P95 - 2 \times P50}{2(P95 - P5)}$$

This measure yields zero for symmetrical distributions and classification scales for skewed data below:

$-1.0 < IGS < -0.3 \Rightarrow$ data is very negatively skewed
$-0.3 < IGS < -0.1 \Rightarrow$ data is negatively skewed
$-0.1 < IGS < 0.1 \Rightarrow$ data is nearly symmetrical
$< IGS < 0.3 \Rightarrow$ data is positively skewed
$< IGS < 1.0 \Rightarrow$ data is very positively skewed

2.4.3 DERIVATION OF THE MODE ESTIMATING FORMULA IN TERMS OF INCLUSIVE GRAPHIC SKEWNESS

The intelligent aspect of the proposed methodology comes from its being neural network adaptive to the data distribution to be searched. Although there are no precise techniques of detecting the mode of a distribution, a careful examination of Folk and Ward's equation reveals that there is, indeed, a very close relationship between *IGS* and the particular percentile value at the location of modal concentration (Milatovic, 1996). In other words, the median value of any distribution is always at the 50th percentile, whereas the mode-percentile value will vary with the skewness. When *IGS* equals zero, the distribution is perfectly symmetrical, and the modal value should coincide with the median, or reside at the 50th percentile. When *IGS* is at its negative extreme of −1, the modal value is likely to be at the 100% percentile. Likewise, when *IGS* is +1, the high readings cluster toward the left of the distributions, so that the modal value would be very near or at the 0th percentile. Therefore, from these boundary conditions, we could derive the following mathematical relationship between IGS and the modal concentration percentile value, *mod% = a(IGS) + b*.

The boundary conditions yield the following values for the parameters a and b:

$0.0 = a(+1) + b \rightarrow$ (absolutely positive skewness shifts the mode towards 0th percentile)

$1.0 = a(-1) + b \rightarrow$ (absolutely negative skewness drives the mode towards the 100th percentile)

Solving for a and b, we obtain $mod\% = -0.5(IGS) + 0.5$. Expanding the above formula yields:

$$\mod \% = \frac{1}{2} - \frac{P16 + P84 - 2 \times P50}{4(P84 - P16)} + \frac{P5 + P95 - 2 \times P50}{4(P95 - P5)}$$

2.4.4 GRAPHICAL VERIFICATION OF THE MOD%–IGS RELATIONSHIP

Figure 2.2 displays a plot of both calculated and observed modal percentile values of a lognormal distribution. Twenty sets of 1000 generated random numbers were taken. The observed modal percentile value was taken as the average between the highest and the lowest cumulative relative frequency of the modal class. The calculated values were obtained by the IGS–$mod\%$ relation as derived previously. The number of classes in each random number generation (rng) was varied to envelope the calculated value. A simple paired comparison test between the data in $mod\%$ and $class\ avg.$ columns yielded:

$$\bar{d} = -0.00011, \qquad S_d = 0.0381, \qquad t_0 = \frac{\bar{d}}{S_d\sqrt{20}} = -0.012911$$

Since the table look-up value for $t_{\alpha/2,\ n-1}$ at 0.01 significance level is 2.861, we may conclude with 99% confidence that $mod\%$ resembles the average of the cumulative relative frequency for the modal class. To further ensure that the calculated $mod\%$ value is also effective for smaller sample sizes, as well as in cases when a distribution

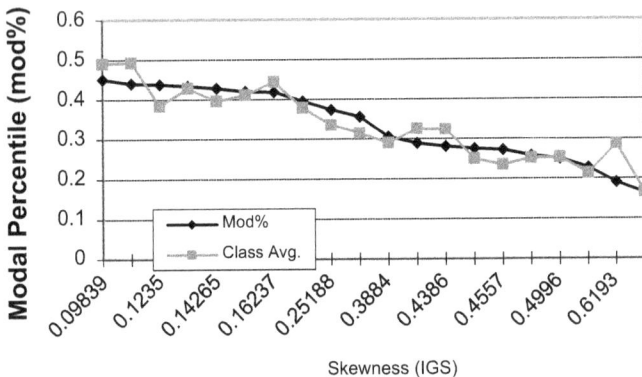

FIGURE 2.2   Plot of the avg. modal class cumulative frequency versus $mod\%$.

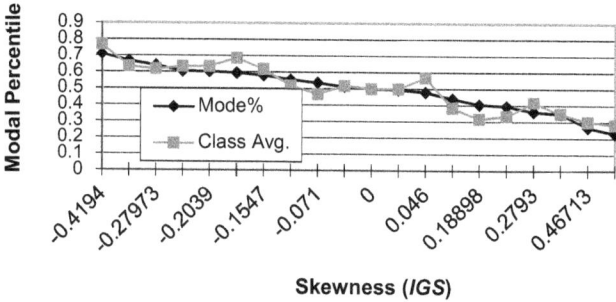

FIGURE 2.3 Plot of *mod%* and the average cumulative frequency of the modal class vs. *IGS*.

is negatively skewed, 20 sets of 30 random numbers generated from a Weibull distribution were taken, such that its scale parameter δ was held fixed at the value of 1, while its shape parameter β was varied between 1 and 12.

The plot of the modal class average cumulative relative frequency (i.e., the midpoint of the modal class) versus the calculated *mod%* is shown in Figure 2.3.

The paired comparisons test yielded $\bar{d} = -.00611$, $S_d = 0.052$, and $t_0 = -0.52521$. Therefore, for any shape and size of a unimodal distribution, we are able to accurately estimate the address (by multiplying its percentile, *mod%*, by the number of elements in the database), as well as the modal value itself (by actually accessing the data element at the specified address). It is noted that the calculations of *mod%* require access to only five particular percentile values of a distribution, regardless of the number of elements it contains.

2.5 RESULTS OF PRELIMINARY RESEARCH

In this section, we compare the expected efficiency of a modification of the Cantor search to the performance of the binary search, and present the interpreted results and conclusions to lay the foundation for future research. The effectiveness of the Cantor search within normally distributed data has already been researched and documented by Wei (1993) and Nakada (1993). The experimentation procedure of the preliminary work investigated the behavior of the modified Cantor search, when the data values searched are drawn from non-normal or moderately skewed distributions. For this particular research, the comparisons between the Cantor search and the binary search were conducted on two different sizes of databases: 150 and 1500 data elements. The random numbers generated for each database size were drawn from three different (generally skewed) distributions: Weibull, gamma, and lognormal. Only distributions with *IGS* (Inclusive Graphic Skewness) between zero and one were considered, since the results could easily be mirrored for the negatively skewed data.

Statistical software was used to generate the data files for each of the three distributions, and their input parameters. For each file, the different input parameters of the distribution produced different skewness *IGS*. The same software was used for sorting the data, since that is a prerequisite for both binary and Cantor search. Two types of files (large and small) were generated: one with 1500 and the other with

150 data points. These files were used as the input to a computer code. For each input file, the code calculated the five specific percentile values (5th, 16th, 50th, 84th, and 95th), skewness (*IGS*), modal percentile (*mod%*), modal address in the database, the data value at the modal address, given a database of any pre-specified size. For each element in the database, the program recorded its relative distance away from the modal concentration and the number of comparisons for both modified Cantor search and binary search needed to locate the datum.

2.6   CANTOR AND BINARY SEARCH COMPARISON OF 1500 DATA POINT FILES

For each of the three distributions, the computer code output files were imported into a spreadsheet and processed, so that, for a datum at a given distance away from the modal concentration and a given *IGS* (Inclusive Graphic Skewness), the difference in number of comparisons needed to successfully locate the datum between the binary search and modified Cantor search has been calculated and entered. The results showed no distinctive pattern or structure to indicate any relationship between the difference in number of comparisons (between the binary search and Cantor search) and the skewness. On the other hand, there was an obvious relationship between the differences in the number of comparisons, as the distance away from the modal concentration increases.

Figure 2.4 shows the plot of the average values across each row (i.e., the distance away from the modal concentration) of the difference in comparison between the binary and Cantor search for numbers generated from the Weibull distribution. It is noted that the average difference at the modal concentration is approximately 8.85. At the next row (one place away from the modal concentration), the average difference drops down to 1.74. The average difference does imply the superiority of the Cantor search until approximately 70 places away from the modal concentration. After that, the majority of the points on the graph are below zero indicating the superiority of the binary search over Cantor search. It is noted that 70 places away from the modal concentration is only in one direction, meaning that we should expect similar behavior on the other side of the modal concentration. In other words, we

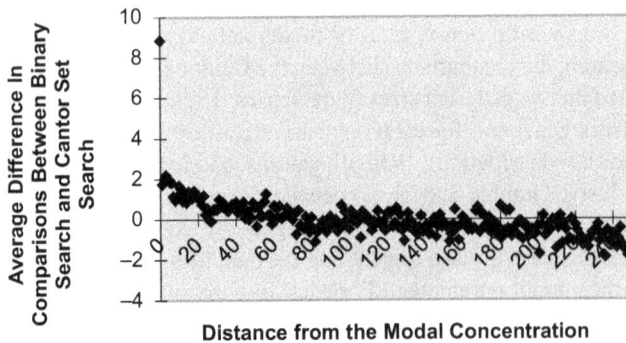

Distance from the Modal Concentration

FIGURE 2.4   Average difference in comparisons versus distance from the modal concentration.

could expect Cantor search to be more effective for about 140 data elements concentrated around the modal concentration which, in essence represents about 10% of the 1500 element database.

The same investigation on data drawn from gamma and lognormal distributions produced identical results. This similarity is not by accident. The reason is that the actual bisection (during the binary search) and trisection (during the Cantor search) is performed on the array indices, and not on the values inside the particular array. In other words, we may have 10 deviates from a Weibull distribution, smallest being, say, 6 and the largest being, for example, 22, which are sorted and stored in a 10-element array. The binary search (as defined to start in the middle) would not start at $(22-6)/2 = 8$, but rather at the fifth (which is 10/2) element of the array. And this principle holds regardless of the distribution stored in the array. This observation is important because, as seen from the above plots, we might be able to predict the behavior of the modified Cantor search for any sorted database.

2.7 CANTOR AND BINARY SEARCH COMPARISON OF 150 DATA POINT FILES

In this case, similar to the 1500 data point search comparisons, there were no apparent patterns or structures indicating any relationship between the difference in number of comparisons and the skewness *IGS*. However, as expected, we did again encounter the relationship between the difference in number of comparisons and the distance away from the modal concentration. Figure 2.5 shows the plot of the average distance away from the modal concentration of the difference in comparisons between the binary and Cantor search for 150 data point search from the Weibull distribution. The average difference in number of comparisons at the modal concentration itself is 5.84.

Figure 2.6 presents the plot of the cumulative sum of the values shown in Figure 2.5. The peak of the curve is at the distance of 15. Therefore, for about 30% or 20% of the data elements around the modal concentration, the Cantor search is superior to the binary search. The curve crosses zero at the distance of 34. Almost identical plots were obtained when searching through data drawn from gamma and lognormal

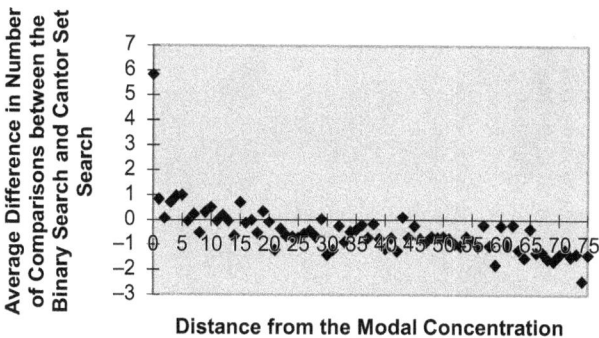

FIGURE 2.5 Average difference in comparisons vs. distance from the modal concentration.

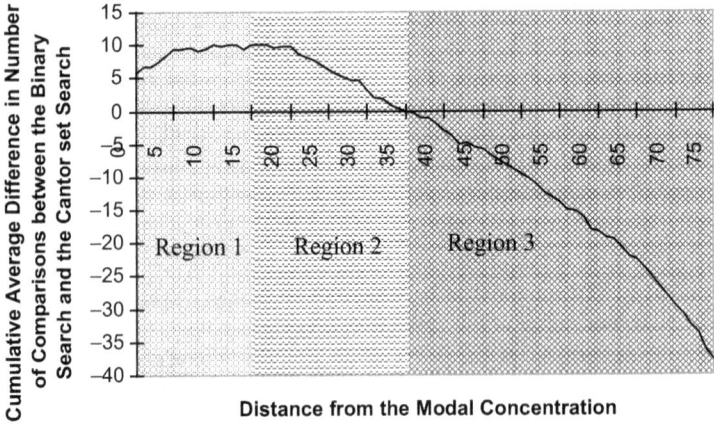

FIGURE 2.6   Cum. avg. difference in comparisons vs. distance from modal concentration.

distributions. For most of the data analyzed, the difference in number of comparisons at the modal concentration is approximately 5.5, the peak of the cumulative difference curve is at the distance of about 15, while the curve crosses zero at the distance of 35.

It is useful to explore the behavior of the above three parameters (the difference in number of comparisons at the modal concentration, the peak of the cumulative difference curve, and the distance from the modal concentration at which the cumulative difference in comparisons curve crosses zero) for the various numbers of data points. The next section presents the plots and the linear regression models observations (Figure 2.6).

2.8   COMPARISON OF THE BINARY SEARCH AND THE CANTOR SEARCH FOR THE VARIOUS DATABASE SIZES

Figure 2.7 shows the plot of the difference in the number of comparisons between the binary search and Cantor search at the modal concentration as a function of the

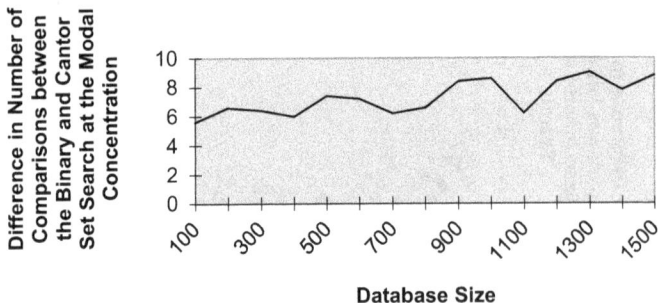

FIGURE 2.7   Avg. difference in comparisons.

database size. Although alternating, the upward trend is evident. The linear regression returned a slope of 0.002 and the y-intercept of 5.74857. However, the coefficient of determination is 0.561. This parameter ranges between zero and one, so that the value of one resembles perfect correlation between the estimated and the actual value, whereas the value of zero suggests that the regression is not an appropriate tool in predicting the y-value. In our case, the value of 0.561 indicates that over 56% of the data can be explained by the regression function.

Figure 2.8 shows the plot of the maximum distance of the search key from the modal concentration for which Cantor search is strictly superior to the binary search. The upward trend with an increase in database size is evident again. The estimated least squares slope value is 0.039, while the y-intercept value was calculated to be 9.142857. The coefficient of determination value is 0.874, indicating more confidence in the predicted y-values. This plot is very important in revealing the percentage of the data points to the size of the whole database for which the Cantor search is superior to the Binary search. For example, for the size of 100, the maximum distance of Cantor search superiority is 12, or 24 on both sides of the modal concentration. In terms of percentage, that would yield 24/100, or 24%. However, for a database size of 1500, the maximum distance is 71, or 141 on both sides of the modal concentration. This yields only 141/1550, or 9.1%. Therefore, based on the following plots, we may conclude that Cantor search compares better to the binary search for smaller database sizes. Figure 2.9 traces the plot of the search key distance from the modal

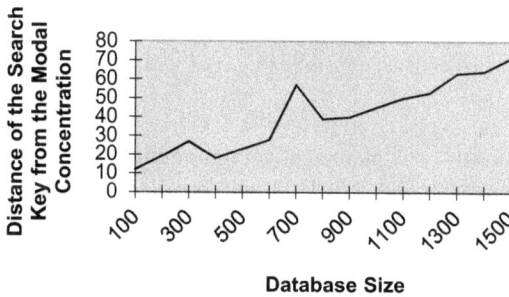

FIGURE 2.8 Maximum distance of search key from the modal concentration.

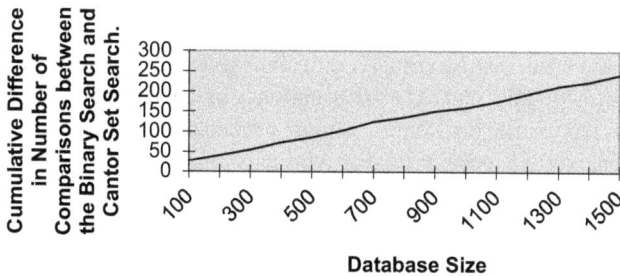

FIGURE 2.9 Plot of the distance of the search key from the modal concentration.

concentration for which the cumulative average difference in comparisons curve crosses zero. It is interesting to note that the plot resembles almost a perfect straight line. The estimated slope value from the linear regression is 0.153, and the y-intercept value is 11.4667. The coefficient of determination, as expected, is very high with a value of 0.998.

A part of an envisioned future research is to investigate the relaxation of the 1/3 sectioning approach of Cantor search. The 1/3 sectioning is a good compromise between the binary search and the sequential search when trying to restrict the search among the data points close to the estimated modal concentration. However, a more intelligent procedure could be investigated to use $1/n$ sectioning, where $n = 1, 2, 3,$ ..., depending on the kurtosis (peakedness) and the size of the data distribution. This can then be implemented in an adaptive neural network training implementation.

2.9   INTELLIGENT 1/N SECTIONING OF THE SEARCH SPACE

The choice of 1/3 search space sectioning, as used in the original Cantor Set search, is a rigid limitation, especially in many cases when the distribution is not bell shaped. One way of adjusting this is to consider the peakedness of the data distribution. Generally, the peakedness of a distribution is evaluated by calculating the kurtosis or the third moment about the mean. Such computations are quite tedious. A more suitable measure of distribution peakedness is proposed by Folk and Ward (1957) and referred to as the *Graphic Kurtosis, K, shown below*:

$$K = \frac{P95 - P5}{2.44(P75 - P25)}$$

Normal curves are expected to have $K = 1$. Curves having K under 0.67 are considered to be platykurtic. The mathematical minimum for the measure is 0.47. There is no mathematical maximum, but Folk and Ward (1957) concluded that $K = 8.0$ could be considered to be the highest practical value. To simplify statistical analysis, K can be normalized such that:

$$K_n = \frac{K}{K + 1}$$

This way, bell-shaped curves should have $K_n = 0.50$, while the whole range is approximately between 0.30 and 0.90. Thus, a distribution having $K_n = 0.3$ is extremely platykurtic, that is, virtually uniform, while a distribution having $K_n = 0.9$ is very "spiky." In this regard, a distribution can be leptokurtic, platykurtic, or mesokurtic. A leptokurtic distribution is more peaked about the mean than a platykurtic distribution, while a mesokurtic distribution is more spread out around the mean.

Binary search is generally the most appropriate searching technique for uniformly distributed data, where all the observations are evenly spread throughout the search space. However, for a very "spiky" distribution, the peak of the density curve is very distinct and evident. In such instances, we may need to section the search space and proceed with the search so that comparisons remain close to the modal concentration.

Therefore, as suggested by Milatovic (1996), we could develop our new sectioning heuristic as:

$$1/n = (K_n)a + b$$

where
    $n \equiv$ sectioning interval ($n = 2$ for binary search)
    $K_n \equiv$ Normalized Graphic Kurtosis as proposed by Folk and Ward (1957)
    $a \equiv$ slope
    $b \equiv$ Y-intercept.

For uniformly distributed data, as mentioned above, n is 2, and $K_n \approx 0.30$, whereas for "spiky" or very leptokurtic distributions, $n \rightarrow \infty$, and $K_n \approx 0.90$. Formulating two equations with two unknowns, we have:

$$\text{Eq1}: 1/2 = (0.30)a + b \text{ and Eq2}: 1/\infty = 0 = (0.90)a + b, \text{which yield}$$
$$a = -5/6 \text{ and } b = 3/4. \text{ In general}, 1/n = (-5/6)K_n + (3/4).$$

At this point, it is desirable to compute the n for the case of normal distribution. As indicated previously, bell-shaped curves have K_n of approximately 0.5. Substituting this value into the equation above, we obtain:

$$1/n = (-5/6)(0.5) + 3/4 = 1/3$$

In other words, the heuristic suggests $n = 3$ for normal distributions, which is exactly what has been pursued in the original Cantor search investigations of normal distributions by Wei (1993), Nakada (1993), and Badiru (1993).

Preliminary investigations have been conducted using the restricted 1/3 interval sectioning regardless of the distribution peakedness and shape. A research agenda is to conduct investigations and experimentations with alternate sectioning approaches (i.e., different values for n) with different non-normally distributed sets of data. The results of such investigations will help establish conditions and parameters under which Cantor search technique will be more effective than the binary search for a range of data distributions. The intelligence aspect of the research comes from the fact that the set sectioning methodology will be inherently adaptive to data distribution shapes and sizes. A new approach for effectively estimating the modal value of any sorted set or distribution of data is anticipated. The new methodology is incorporated into a tri-sectioning search technique originally defined as Cantor Set search. Preliminary analyses and results have shown that for a sorted array ranging from 150 to 1500 data points, Cantor search is strictly superior to the binary search for a range between 10% and 20% of data points closest to the modal concentration, regardless of the type of distribution and its skewness. The above percentage is a function of the database size and decreases as the number of data point increases. Cantor search is further competitive for the range of data points between 30% and 54% around the modal concentration, this percentage being inversely proportional to the size of the database. As the distance of the search key increases further beyond these percentage values for a fixed database size, the performance of the Cantor search severely deteriorates, and the binary search becomes more effective. Further research is needed to formulate ways to exploit the potentials that have been identified for the Cantor search methodology. The cantor research has intrinsic merit because the result could contribute to a practical taxonomy for the selection of search techniques for large data sets. The benefit of a new methodology comes from its being adaptive to the specific distribution being searched. The intelligent neural network adaptive search technique has several potential application domains including the following:

- Detection and categorization of elements in large data sets.
- Taxonomical organization of training sets in large-scale neural network applications.
- Exploratory research in information generation for industrial applications.
- Analysis and effective utilization of massive data sets in data mining applications.
- Pruning of large data sets collected with modern data gathering tools.
- Segmentation of large information bases for transmission.

The practical implementation potential for the proposed Cantor Set search is quite significant because we can do things with AI software implementations now that were not possible previously. This creates an environment that is conducive for practical applications of AI in diverse areas.

REFERENCES

Ahlswede, R. and Ingo Wegener (1987). *Search Problems*. John Wiley & Sons, New York.
Baase, S. (1988). *Computer Algorithms: Introduction to Design and Analysis*. Addison-Wesley, Reading, MA.
Badiru, A. B. (1992). *Expert Systems Applications in Engineering and Manufacturing*. Prentice-Hall, Hoboken, NJ.
Badiru, A. B. (1993). "A New Computational Search Technique for AI Based on the Cantor Set". *Applied Mathematics and Computation* 57: 255–274.
Badiru, A. B. and J. Cheung (2002). *Fuzzy Engineering Expert Systems with Neural Network Applications*. John Wiley & Sons, New York.
Badiru, A. B., L. Gruenwald, and T. Trafalis (1992). "A New Search Technique for Artificial Intelligence Systems". *Proceedings of 1992 Oklahoma Symposium on Artificial Intelligence*, University of Tulsa, November 11–12.
Folk, R. L. and W. C. Ward (1957). "Brazos River Bar: A Study in the Significance of Grain Size Parameters". *Journal of Sedimentary Petrology*, 27(1): 3–26.
Gardiner, V. and G. Gardiner (1978). *Analysis of Frequency Distributions. Concepts and Techniques in Modern Geography, No. 19*. Leicester University and Formerly North Staffordshire Polytechnic.
Gonnet, G. H. (1984). *Handbook of Algorithms and Data Structures*. Addison-Wesley, London, 23–36.
Goodman, S. E. and S. T. Hedetniemi (1977). *Introduction to the Design and Analysis of Algorithms*. McGraw-Hill, New York, 225–238.
Inman, D. L. (1952). "Measures for Describing the Size Distribution of Sediments". *Journal of Sedimentary Petrology* 22(3): 125–145.

Knuth, D. (1980). *The Art of Computer Programming*, Vol. 3. Addison-Wesley, Reading, MA.

McCammon, R. B. (1962). "Efficiencies of Percentile Measures for Describing the Mean and Sorting of Sedimentary Particles". *Journal of Geology* 70: 453–465.

Mehlhorn, K. (1984). *Data Structures and Algorithms 1: Sorting and Searching.* Springer-Verlag, Reading, MA.

Milatovic, M. (1996). Development of Mode Estimating Technique for Cantor Search of Sorted Data in Artificial Intelligence Systems and Design Manufacturing. MS Thesis, University of Oklahoma.

Milatovic, M. and A. B. Badiru (1998). "Fast Estimation of the Modal Position for Unimodally Distributed Data". *Intelligent Data Analysis* 2(1). On Line Journal, http://www-east. elsevier.com/ida/

Nakada, M. (1993). Experimental Investigation of Alternate Search Key Distributions for Cantor Set Search Algorithm. MS Thesis, University of Oklahoma.

Otto, G. H. (1939). "A Modified Logarithmic Probability Graph for the Interpretation of Mechanical Analyses of Sediments". *Journal of Sedimentary Petrology* 9: 62–76.

Sedgewick, R. (1992). *Algorithms in C++.* Addison-Wesley, Reading, MA, 193–230, 373–388.

Stone, L. D. and D. H. Wagner (1980). *Brief Overview of Search Theory and Applications. Search Theory and Applications.* Edited by K. Brian Haley and L. D. Stone. Plenum Press, New York.

Trask, P. D. (1932). *Origin and Environment of Source Sediments of Petroleum.* Gulf Pub. Co., Houston, 323.

Wei, H. (1993). Improving Cantor Set Search for Applications in Database and Artificial Intelligence. MS Thesis, University of Oklahoma.

3 Set-theoretic Systems for AI Applications

3.1 SET SYSTEMS IN PROBLEM DOMAINS

Following the treatment of Cantor Set in the preceding chapter, we now consider a more general set-theoretic foundation for AI applications in this chapter. Everything is a system in its own right. Every system comprises a collection of sets. The sets may include qualitative, quantitative, or conceptual elements. A set-theoretic approach may help us to have a better grasp of how sets fit together to form systems, thereby leading to a higher potential for systems integration. In spite of the much touted need to achieve systems integration, only limited successes have been accomplished in practice for the purpose of actualizing innovation. This is often so because there is no analytical basis for identifying and modeling the parameters that the subsystems have in common. Consequently, attempts to achieve systems integration degenerate to mere conceptual formulations. Sometimes the conceptual formulations work, but most times they do not. The premise of this chapter is that a rigorous set-theoretic formulation can help identify the dominant common elements of subsystems, thereby facilitating practical implementation that is amenable to quantitative optimization. The chapter presents fundamental set theory and principles for modeling systems interactions for the purpose of advancing innovation, particularly in technical systems, as elucidated by Badiru and Ibidapo-Obe (2008), Badiru et al. (2012), and Badiru and Cheung (2002).

In order for two systems to be integrated, they must overlap in scope or performance with respect to time, space, or physics; subsequently, this requires mathematical representations for manipulating the integrated parameters of the systems. In spite of the much touted need to achieve systems integration, very little has been accomplished in practice. This is often because there is limited analytical basis for modeling the parameters that the subsystems have in common. Consequently, attempts to achieve systems integration degenerates to mere conceptual formulations. Sometimes the conceptual formulations work, but most times they do not. Omicini and Papadopoulos (2001) present systems integration as the coordination and management of interactions among entities of a system. The premise of this chapter is that a rigorous set-theoretic formulation can help identify the dominant common elements of subsystems, thereby facilitating practical implementation that is amenable to quantitative optimization. The chapter explores fundamental set theory and principles for modeling system interactions. A system is often defined as a collection of interrelated elements working together in synergy to produce a composite output that is greater than the sum of the individual outputs of the components. Systems integration is a desired state whereby subsystems of the system are in proper consonance of operation and input-output interactions. Methodologies pertaining to

DOI: 10.1201/9781003247395-3

the various characteristics of integrated systems are available in the literature. Yang et al. (2006) present techniques for assessing reliability of integrated systems under time-varying tasks. Chen and Liu (2004) address the time-varying aspect by incorporating fuzzy logic reasoning. Heymann et al. (2005) and Zhang et al. (2001) considered integration complexity of hybrid systems. As a practical application example, Balkhi (1999) addressed integrated inventory systems with time-varying demand. The approach presented in this chapter differs from those existing in the literature by focusing on set-theoretic modeling. See Aokl (1968), Golub and Reinsch (1970), Golub and Van Loan (1979), Ibidapo-Obe (1982a), Ibidapo-Obe (1982b), Itô (1951), Laintiotis (1971), Rashwan and Ahmed (1982), Soong (1977), and (Stewart, 1973).

3.2   SETS AND SYSTEMS IN INNOVATION

There are soft sets, as in qualitative systems. There are hard sets, as in quantitative systems. Set theory is applicable in both cases, albeit on different mathematical platforms. Every pursuit of innovation has some underlying basis in sets. Without a rigorous understanding of sets, management of innovation in a technical environment will be more difficult or less achievable. Set theory is the mathematical concept that governs sets, which represent collections of objects. Sets are defined by the elements contained in the collection and the respective memberships of the elements in the sets. Set theory provides the formalism in which mathematical objects are described. This makes set theory suitable for defining systems integration. In this conception, sets and set membership are fundamental concepts describing points and lines in Euclidean geometry. The properties of sets essential for modeling systems integration include:

1. Ordered Pairs
2. Relations
3. Functions
4. Set Cardinality

3.3   ORDERED PAIRS ON SETS

The notion of *ordered pair* suggests that if a and b are sets, then the *unordered pair* $\{a, b\}$ is a set whose elements are exactly a and b. The "order" in which a and b are put together plays no role. Thus, $\{a, b\} = \{b, a\}$. For many applications, we need to pair a and b in a way making possible to "read off" which set comes "first" and which set comes "second." We denote this *ordered pair* of a and b by (a, b); a is the *first coordinate* of the pair (a, b), b is the *second coordinate*. To assess systems integration properties, the ordered pair has to be a set. It should be defined in such a way that two ordered pairs are equal if and only if their first coordinates are equal and their second coordinates are equal. This guarantees in particular that $(a, b) \neq (b, a)$ if $a \neq b$. With this understanding, we can distinguish systems integration from the perspective of "push system" or "pull system" because they may not necessarily be identical.

Definition. $(a, b) = \{\{a\}, \{a, b\}\}$.

If $a \neq b$, (a, b) has two elements, a singleton $\{a\}$ and an unordered pair $\{a, b\}$. We find the first coordinate by looking at the element of $\{a\}$. The second coordinate is then the other element of $\{a, b\}$. If $a = b$, then $(a, a) = \{\{a\}, \{a, a\}\} = \{\{a\}\}$ has only one element. In any case, it seems obvious that both coordinates can be uniquely "read off" from the set (a, b).

Ordered Pair Theorem. $(a, b) = (a', b')$ if and only if $a = a'$ and $b = b'$.

Proof. If $a = a'$ and $b = b'$, then, of course, $(a, b) = \{\{a\}, \{a, b\}\} = \{\{a'\}, \{a', b'\}\} = (a', b')$. The other implication is more intricate. Let us assume that $\{\{a\}, \{a, b\}\} = \{\{a'\}, \{a', b'\}\}$. If $a \neq b$, $\{a\} = \{a'\}$ and $\{a, b\} = \{a', b'\}$. So, first, $a = a'$ and then $\{a, b\} = \{a, b'\}$ implies $b = b'$. If $a = b$, $\{\{a\}, \{a, a\}\} = \{\{a\}\}$. So $\{a\} = \{a'\}, \{a\} = \{a', b'\}$, and we get $a = a' = b'$, so $a = a'$ and $b = b'$ holds in this case also.

Ordered pairs can be extended to define *ordered triples*: $(a, b, c) = ((a, b), c)$. Similarly, we can define *ordered quadruples*: $(a, b, c, d) = ((a, b, c), d)$, and so on. These extensions are useful for modeling subsystems that number more than two.

3.4 SET RELATIONS IN INNOVATION SYSTEMS

Set relation is defined in terms of a binary relation, which is determined by specifying all ordered pairs of objects in that relation.

Definition. A set R is a *binary relation* if all elements of R are ordered pairs, that is, if for any z in R there exist x and y such that $z = (x, y)$.

It is customary to write xRy instead of $(x, y) \in R$. We say that *x is in relation R with y* if xRy holds.

The set of all x which are in relation R with some y is called the *domain* of R and denoted by "dom R." So, dom $R = \{x \mid \text{there exists } y \text{ such that } xRy\}$. dom R is the set of all first coordinates of ordered pairs in R.

The set of all y such that, for some x, x is in relation R with y is called the *range* of R, denoted by "ran R." So, ran $R = \{y \mid \text{there exists } x \text{ such that } xRy\}$. In modeling systems integration parameters, the relations mapped from one subsystem to another become defining characteristics of the integration.

3.5 FUNCTIONS ON SETS

A function represents a special type of relation. A function is a rule that assigns to any object a, from the domain of the function, a unique object b, the value of the function at "a." The function implies a relation where every object "a" from the domain is related to precisely one object in the range, namely, to the value of the function at a. This allows one-to-one mapping for systems integration purposes.

Definition. A binary relation F is called a *function* (or *mapping, correspondence*) if aFb_1 and aFb_2 imply $b_1 = b_2$ for any a, b_1, and b_2. In other words, a binary relation F is a function if and only if for every a from dom F there is exactly one b

such that aFb. This unique b is called the *value of F at a* and is denoted $F(a)$ or F_a. Note that $F(a)$ is not defined if $a \in$ dom F. If F is a function with dom $F = A$ and range $F \in B$, we use the notations $F: A \in B$, $<F(a) \mid a \in A>$, $<F_a \mid a \in A>$, $<F_a>_a \in A$ for the function F. The range of the function F can then be denoted $\{F(a) \mid a \in A\}$ or $\{F_a\}_{a \in A}$.

The Axiom of Extensionality can be applied to functions as follows.

Lemma. Let F and G be functions. $F = G$ if and only if dom $F =$ dom G and $F(x) = G(x)$ for all $x \in$ dom F.

A function f is called *one-to-one* or *injective* if $a_1 \in$ dom f, $a_2 \in$ dom f, and $a_1 \neq a_2$ implies $f(a_1) \neq f(a_2)$. In other words if $a_1 \in$ dom f, $a_2 \in$ dom f, and $f(a_1) = f(a_2)$, then $a_1 = a_2$.

3.6 CARDINALITY OF SETS

The most obvious property of a set relates to how many elements it contains. It is a fundamental observation that we can define the statement "sets A and B have the same number of elements" without knowing anything about numbers.

Definition. Sets A and B have *the same cardinality* if there is a one-to-one function f with domain A and range B. We denote this by $|A| = |B|$.

Definition. The cardinality of A is less than or equal to the cardinality of B (notation: $|A| \leq |B|$) if there is a one-to-one mapping of A into B.

Notice that $|A| \leq |B|$ means that $|A| = |C|$ for some subset C of B. We also write $|A| < |B|$ to mean that $|A| \leq |B|$ and not $|A| = |B|$, that is, that there is a one-to-one mapping of A onto a subset of B, but there is no one-to-one mapping of A onto B.

Equation Chapter 1 Section 1.1 **Lemma**.

1. If $|A| \leq |B|$ and $|A| = |C|$, then $|C| \leq |B|$.
2. If $|A| \leq |B|$ and $|B| = |C|$, then $|A| \leq |C|$.
3. $|A| \leq |A|$.
4. If $|A| \leq |B|$ and $|B| \leq |C|$, then $|A| \leq |C|$.

Cantor-Bernstein Theorem. If $|X| \leq |Y|$ and $|Y| \leq |X|$, then $|X| = |Y|$.

3.7 RELATIONSHIPS OF SET-TO-SYSTEM AND SUBSET-TO-SUBSYSTEM

Based on the underlying principles of sets and relations, we can now take a system as a set and subsets as subsystems. Each system is represented by n-th tuples as an extension of an ordered pair. For systems that are describable by a matrix, the eigenmodes, in order of magnitude, will be the n-th tuple and we take the dominant modes. This will generate a reduced order representation using singular value decomposition, which is addressed in the next section.

Consider the domain of set relations such that if **R** is a relation from set A to set **B**, then the domain of **R** is the set of all elements "*a*" belonging to A such that *aRb* for some element "*b*" belonging to **B**. That is,

$$Dom(\mathbf{R}) = \{a \in A | (a,b) \in \mathbf{R} \text{ for some } b \in \mathbf{B}\} \quad (3.1)$$

Consider the two framed sets in Figure 3.1 (Badiru and Cheung, 2002). Let Frame **A** contain the subsets *a*1, *a*2, *a*3, *a*4, and *a*5, while Frame **B** contains the subsets *b*1, *b*2, *b*3, *b*4, *b*5, *b*6, *b*7, and *b*8. Define a *integrating* relation **Z** from **A** to **B** such that an element a belonging to **A** is related to an element *b* belonging to **B**, if and only if there is a rule in **B** that has *a* as a premise and *b* as a complement.

From Figure 3.1, we have the following:

$$\mathbf{Z} = \text{Relation of relevant subset correspondence}$$
$$\text{Domain of } \mathbf{Z} = \{a1, \ a2, \ a3\} \quad (3.2)$$
$$\text{Image of } \mathbf{Z} = \{b2, \ b5, \ b6\}$$

Now suppose only the following relationships exist in Frame **B**:

$$a1 \rightleftharpoons b2$$
$$a2 \rightleftharpoons b5 \quad (3.3)$$
$$a3 \rightleftharpoons b6,$$

where $\rightleftharpoons$ represents reciprocity of parameter correspondence. The domain of **Z** is then given by the set:

$$Dom(\mathbf{Z}) = \{a1, a2, a3\}, \quad (3.4)$$

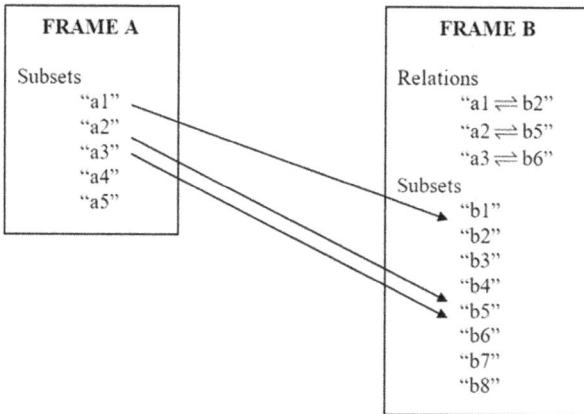

FIGURE 3.1 Relations on system subsets.

Since the elements $a1$, $a2$, and $a3$ are the only elements of **A** that can successfully integrate with elements in **B**. The image of the relation, **R**, is then defined as:

$$Im(\mathbf{R}) = \{b \in \mathbf{B} | (a,b) \in \mathbf{R} \text{ for some } a \in A\} \tag{3.5}$$

Thus, the image of the relation **Z** (measure of system integration) is:

$$Im(\mathbf{Z}) = \{b2, b5, b6\}, \tag{3.6}$$

which corresponds to the set of integrating relations that are effected in Frame **B**. A special kind of relation on integrating sets is parameter mapping in which there is one-to-one correspondence between elements in subsets (subsystems). An identity relation in a subsystem is a relation that integrates every element with itself such that:

$$\mathbf{R} = \{(a,a) | a \in A\}, \tag{3.7}$$

which is *reflective* property indicating that every subsystem perfectly integrates with itself. In a similar vein, we can define a binary operation on a subsystem, **S**, whereby the Cartesian square **S×S** produces another element of **S**. For example, let **S** be the set of words in a dictionary. An operation may be defined for the process of forming natural language statements (or new groups of words) in the set. The new subset of words belongs to the original set of words, and can possibly be used to covey a measure of systems integration across systems.

3.8   INTEGRATION MAPPING OF SUBSETS

If the set **C** is a subset of the Cartesian product, **A×B**, of the sets **A** and **B**, then **C** is an integration mapping from **A** to **B** such that for each a belonging to **A** there is exactly one b belonging to **B** for which (a, b) belong to **C**. Thus, the mapping, **C**, is the collection of the elements of **A×B** that have one-to-one correspondence or integration with the elements of **A** and **B**. The elements of the sets **A**, **B**, and **C** are specific parameters of objects contained in the main set of interest. In the word-set example mentioned earlier, **C** = **A×B** since each element a of **A**, in integration with each element b of **B**, yields a unique (distinct) ordered pair of words. The ordered pair formation can be extended to n-tuple to represent a finite sequence (or ordered list) of elements pulled from multiple systems to form an integrated subset. Thus, n-tuples can be used to mathematically describe integrated objects that consist of specified elements of cooperating systems of dimension n.

3.9   MODEL REDUCTION APPROACH

One approach to modeling system of systems is to develop representational equations of model reduction of integrated systems. For example, we consider canonical representation based upon a singular value decomposition approach, which is more efficient computationally.

Consider an nth-order linear system S_1 defined by

$$S_1 : \dot{x} = Ax + Bu \tag{3.8}$$

where x is the n-dimensional state vector, A is an $n \times n$ system matrix, and u is a p-dimensional input vector. Let z be an m-vector $(m < n)$ related to x by

$$z = Cx. \tag{3.9}$$

In model reduction, it is desirable to fine an mth-order system S_2 described by

$$S_2 : \dot{z} = Fz + Gu. \tag{3.10}$$

The $m \times n$ matrix C in Eq. (3.2) is the aggregation matrix and S_2 is the aggregated system or the reduced model. It is easy to show that $G = CD$ and that F must satisfy the matrix equation

$$FC = CA. \tag{3.11}$$

Equation (3.4) defines an over-specified system of equation for the unknown matrix F, and hence F must be approximated. In [1], a multivariate linear regression scheme is used to yield a "best" approximation for F in the form

$$\hat{F} = CAC^T \left(CC^T\right)^{-1}, \tag{3.12}$$

where T and -1 denote matrix transpose and matrix inverse, respectively. The rank of C is assumed to be m. The result given by Eq. (3.5) is interpreted as a linear, unbiased, minimum-variance estimate of F and its form agrees with that given by Aoki [2], following an ad hoc procedure.

In addition, the covariance of $\hat{F}$ is found to be

$$con\left(vec\hat{F}\right) = \sigma^2 \left[\left(CC^T\right)^{-1} \otimes I_m\right] \tag{3.13}$$

and it is shown in [1] that this covariance matrix can be used for model reduction error assessment. In Eq. (3.6), the Kronecker product $\otimes$ and the "vec" operator are defined as

$$P \otimes Q = \begin{bmatrix} Pu & Q \end{bmatrix} \tag{3.14}$$

$$vec(P) = \begin{bmatrix} P_1 & P_2 \dots \end{bmatrix}^T, \tag{3.15}$$

where P and Q are matrices of arbitrary dimensions and P_k is the kth column of P.

3.10   SINGULAR VALUE DECOMPOSITION IN INNOVATION SYSTEMS

From the computational point of view, it is desirable to circumvent the use of matrix inverses in Equations (3.5) and (3.6), particularly for systems having large aggregation matrices. In what follows, this is accomplished through the use of matrix singular value decomposition (SVD), which has found useful application in linear least squares problems. The SVD concept gives the Moore-Penrose pseudo-inverse of C as

$$C^+ = VAU^T \tag{3.16}$$

where U and V are unitary matrices whose columns are the eigenvectors of matrices DD^T, and D^TD, respectively, and matrix $\mathbf{A}$ is n × n in dimension. Embedded within $\mathbf{A}$ is an m × m sub-matrix with elements $\sigma_1 \geq \sigma_2 \geq \cdots \geq \sigma_m \geq 0$, called singular values, which are the nonnegative square roots of the eigenvalues of D^TD. A discussion of this decomposition and its properties can be found in Stewart (1973). Now, Eq. (3.4) gives

$$\hat{F} = CAC^+ \tag{3.17}$$

And, using SVD, we can write

$$\hat{F} = CA\left(VAU^T\right) \tag{3.18}$$

The matrix C can also be written in the form

$$D = U\Sigma V^T \tag{3.19}$$

$$\hat{F} = U\Sigma V^T AVAU^T \tag{3.20}$$

Compared with Eq. (3.5), either Eq. (3.12) or Eq. (3.15) provides a more efficient method of computation for $\hat{F}$ due to elimination of the matrix inverse. Similarly, advantages are realized in the calculation of $con\left(vec\hat{F}\right)$. Following the SVD scheme,

$$\left(DD^T\right)^{-1} = \left(D^T\right)^+ D^+$$
$$= \left(UAV^T\right)\left(VAU^T\right) \tag{3.21}$$
$$= \left(UA^2U^T\right)$$

Equation (3.6) now takes the form

$$con\left(vec\ \hat{F}\right) = \sigma^2\left[UA^2U^T \otimes I_m\right], \tag{3.22}$$

which is clearly of a simpler structure that Eq. (3.6).

3.11   SUBSYSTEM SURFACE PROJECTION INTEGRALS

In this idea, we consider the projection of one subsystem onto the plane of another subsystem to identify common region, area, or parameters for integration. The idea is to identify sub-sets that represent segments (subsystems) of the systems to be integrated. Fischer-Cripps (2005) presents formulations of line integrals that are adapted for the systems modeling in this section. Consider a surface $z = f(x,y)$ in Figure 3.2. Let a line increment on the curve C in the xy plane be given by Δs. The area of the band formed from the line increment to the surface is given by the product expressed as shown below:

$$f(x,y)\Delta s$$

The total area under the surface along the curve from P1 to P2 represents the sum as Δs goes to zero. That is,

$$\lim_{n \to \infty; \Delta s \to 0} \Sigma f(x,y)\Delta s \qquad (3.23)$$

This limit is called the line integral of $f(x, y)$ along the curve C from P1 to P2. Note that C is used to represent an interval or group of intervals along the curve.

Referring to the Figure, the line integral is written as:

$$\int_{P1}^{P2} f(x,y)ds = \int_C f(x,y)ds \qquad (3.24)$$

If the curve C can be represented as a **vector function** in terms of the path or arc length parameter s, such that:

$$\mathbf{r}(s) = x(s)\mathbf{i} + y(s)\mathbf{j}$$

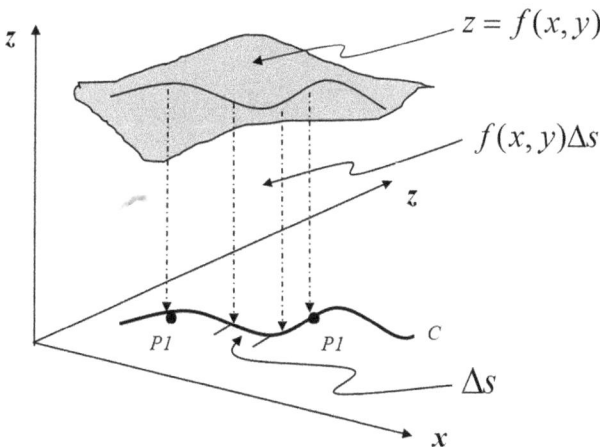

FIGURE 3.2   Line integral of surface projection.

Then the line integral becomes:

$$\int_{P1}^{P2} f\big(x(s), y(s)\big) ds \qquad (3.25)$$

If the curve C is given in terms of the parameter t such that:

$$\mathbf{r}(t) = x(t)\mathbf{i} + y(t)\mathbf{j}$$

Then, we have:

$$\int_{C} f(x, y) ds = \int_{P1}^{P2} f\big(x(t), y(t)\big) \frac{ds}{dt} dt \qquad (3.26)$$

where

$$\frac{ds}{dt} = \sqrt{\frac{dr}{dt} \cdot \frac{dr}{dt}} = \sqrt{\left(\frac{dx}{dt}\right)^2 + \left(\frac{dy}{dt}\right)^2} \qquad (3.27)$$

The line integral can be extended to three dimensions where $F = f(x, y, z)$. Then, the line integral is represented as:

$$\int_{C} f(x, y, z) ds \qquad (3.28)$$

It is possible to express the line integral of $F(x, y)$ along C with respect to x (or y) instead of the arc length s. This gives the projected area on the xz plane (for Δx) or the yz plane (for Δy). That is, the curve segment Δs projects onto x-axis as Δx and onto y-axis as Δy. Consequently, in this case, the surface z may be represented by the function $f(x, y)$ in the x direction and $g(x, y)$ in the y direction. The line integral then becomes:

$$\int_{C} F(x, y) ds = \int_{C} f(x, y) dx + \int_{C} g(x, y) dy$$
$$= \int_{C} \big(f(x, y, z) dx + g(x, y, z) dy\big) \qquad (3.29)$$

For the general case for three dimensions, the line integral is given by the sum of the projected line integrals and the functions, f, g, and h. Then, we have:

$$\int_{C} F(x, y, z) ds = \int_{C} f(x, y, z) dx + \int_{C} g(x, y, z) dy + \int_{C} h(x, y, z) dz$$
$$= \int_{C} \big(f(x, y, z) dx + g(x, y, z) dy + h(x, y, z) dz\big) \qquad (3.30)$$

The functions f, g, and h may be components of a vector function $\mathbf{F}$, such that:

$$
\begin{aligned}
\mathbf{F} &= f(x,y,z)\mathbf{i} + g(x,y,z)\mathbf{j} + h(x,y,z)\mathbf{k} \\
&= F_1\mathbf{i} + F_2\mathbf{j} + F_3\mathbf{k}
\end{aligned}
\tag{3.31}
$$

Now, we have:

$$
\begin{aligned}
F_1 dx + F_2 dy + F_3 dz &= \left(F_1 \frac{dx}{ds} + F_2 \frac{dy}{ds} + F_3 \frac{dz}{ds} \right) ds \\
&= \left(\mathbf{F} \cdot \frac{d\mathbf{r}}{ds} \right) ds
\end{aligned}
\tag{3.32}
$$

where

$$
d\mathbf{r} = dx\mathbf{i} + dy\mathbf{j} + dz\mathbf{k}
$$

and

$$
\frac{d\mathbf{r}}{ds} = \frac{dx}{ds}\mathbf{i} + \frac{dy}{ds}\mathbf{j} + \frac{dz}{ds}\mathbf{k}
\tag{3.33}
$$

Consequently, the line integral $\displaystyle\int_C \mathbf{F} \cdot \frac{d\mathbf{r}}{ds}\, ds$ can be written as $\displaystyle\int_C \mathbf{F}.d\mathbf{r}$, which is referred to as the circulation of $\mathbf{F}$ around C. In general, the value of the line integral depends upon the path of integration. If, however, the vector field $\mathbf{F}$ can be represented by:

$$
\mathbf{F} = \nabla\phi,
\tag{3.34}
$$

Then the line integral is independent of the path. In this case, ϕ is called the scalar potential. The field F is said to be a conservative field and $d\phi$ is called an exact differential. Now, we have

$$
\begin{aligned}
\int_{P1}^{P2} \mathbf{F}.d\mathbf{r} &= \int_{P1}^{P2} \nabla\phi.d\mathbf{r} \\
&= \int_{P1}^{P2} \nabla\phi.\left(dx\mathbf{i} + dy\mathbf{j} + dz\mathbf{k} \right) \\
&= \int_{P1}^{P2} \left(\frac{\partial\phi}{\partial x}\mathbf{i} + \frac{\partial\phi}{\partial y}\mathbf{j} + \frac{\partial\phi}{\partial z}\mathbf{k} \right).\left(dx\mathbf{i} + dy\mathbf{j} + dz\mathbf{k} \right) \\
&= \int_{P1}^{P2} \left(\frac{\partial\phi}{\partial x}dx + \frac{\partial\phi}{\partial y}dy + \frac{\partial\phi}{\partial z}dz \right) \\
&= \int_{P1}^{P2} d\phi \\
&= \phi(x_2, y_2, z_2) - \phi(x_1, y_1, z_1)
\end{aligned}
\tag{3.35}
$$

The integral depends only on the end points $P1$ and $P2$ and not on the path joining them. This is verified by the fact that the form or equation of the curve C is not defined in the above formulations. If the path of integration of C is a closed curve, and if **F** is conservative, then we have:

$$\oint_C \mathbf{F}.d\mathbf{r} = 0, \tag{3.36}$$

where the circle in the integral sign denotes integration around a closed curve. For systems integration application, the area of integration can be used to represent the magnitude of overlap between subsystems.

3.12 SET PROJECTION AND SYSTEM OVERLAP

Systems overlap for integration purposes can further be represented as projection integrals by considering areas bounded by the common elements of subsystems. In Figure 3.3, the projection of a flat plane onto the first quadrant is represented as area A, while Figure 3.4 shows the projection on an inclined plane as area B.

$$A = \int_{A_y} \int_{A_x} z(x,y)dydx \tag{3.37}$$

$$B = \int_{B_y} \int_{B_x} z(x,y)dydx \tag{3.38}$$

The net projection encompassing the overlap of A and B is represented as area C, shown in Figure 3.5, and computed as:

$$C = \int_{C_y} \int_{C_x} z(x,y)dydx \tag{3.39}$$

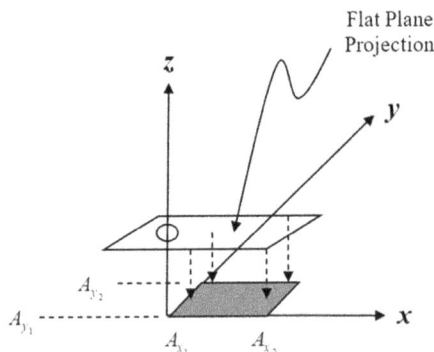

FIGURE 3.3 Flat plane projection.

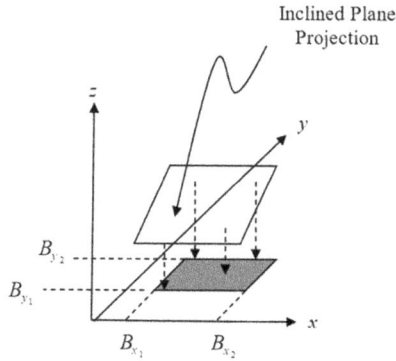

FIGURE 3.4   Inclined plane projection.

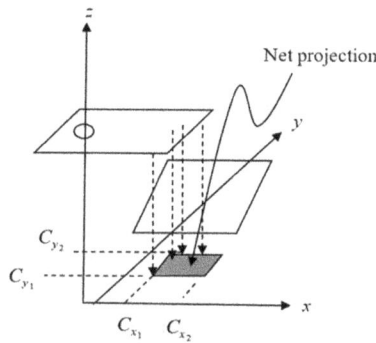

FIGURE 3.5   Net projection area.

3.13 TIME-VARIANT SYSTEMS INTEGRATION

Time-variant systems pose additional complexity to systems integration. This can be addressed by stochastic techniques. The problem of stochastic estimation in non-linear systems has been formulated using different mathematical models and techniques such as the stochastic differential equation approach and function-space methods. Although some of the expositions demonstrate organic generality of the proposed non-linear structures, the techniques available for the analysis and synthesis of these structures for stochastic processes do not yield solutions for the generalized models. An enhanced technique for optimal estimation of system parameters in a stochastic environment with non-linear sensors is presented in this chapter. It is based on the eigenmode representation of generalized displacements and orthogonal expansion of the conditional joint probability distribution. The algorithm proposed in this chapter for realization of the innovations and system state constitutes an initial platform for sustainable systems integration. A linear dynamical system is used to illustrate the correspondence between the innovations process, the observations process and the integrated system response. The proposed algorithm may be applied to several civil engineering and space structures, best described by the distributed parameter equation

Equation Section (Next)

$$m(x)u_{tt}(x,t) + Bu_t(x,t) + Cu(x,t) = F_0(x,t) \tag{3.40}$$

with the initial conditions

$$u(x,t_0) = u^0 \quad \text{and} \quad u_t(x,t_0) = u_t^0 \tag{3.41}$$

where $m(x)$ is the mass per unit length (assumed unity); $u(x, t)$ the vector of generalized displacements; B and C are in general spatial differential operators representing the damping and restoring forces respectively; and $F_0(x, t)$ is the external impressed forces (wind loads, earthquake excitations, etc.).

3.14   MODAL CANONICAL REPRESENTATION

Let the displacement be expressed in terms of structural dominant mode shapes $\{\phi_i(x)\}_1^N$ with associated frequencies $\{w_i\}_1^N$

$$u(x,t) = \sum_1^N u_i(t)\phi_i(x); \quad F_0(x,t) = \sum_1^N F_1(t)\phi_i(t) \tag{3.42}$$

so that
and

$$\left.\begin{array}{c} u(x,t) = U^T(t)\Phi(x) \\[2mm] F_0(x,t) = F_0^T(t)\Phi(x) \end{array}\right\} \tag{3.43}$$

where

$$U = (u_1,u_2\dots,u_N)^T, \quad F_0 = (F_1^0,F_2^0,\dots,F_N^0)^T \tag{3.44}$$

and

$$\Phi = (\phi_1,\phi_2,\dots,\phi_N)^T \tag{3.45}$$

Substituting Eq. (3.2) to Eq. (3.1) gives

$$U^T(t)\Phi(x) + BU^T(t)\Phi(x) + CU^T(t)\Phi(x) = F_0^T(t)\Phi(x) \tag{3.46}$$

thus obtaining the ith mode amplitude equation

$$\ddot{u}_i(t) + 2\xi_i\omega_i\dot{u}_i(t) + \omega_i^2 u_i(t) = F_i^0(t) \tag{3.47}$$

where $C\phi_i = \omega_i^2\phi_i$ and $B\phi_i = 2\xi_i\omega_i\phi_i$ with appropriate initial conditions.
Equation (3.3) can now be put in the state space form

$$\dot{q}(t) = Aq(t) + F \tag{3.48}$$

such that

$$q = (q_1, q_2)^T \tag{3.49}$$

with

$$q_1 = u_i, \quad q_2 = \dot{u}_i$$
$$A = \begin{pmatrix} 0 & 1 \\ -\omega_i^2 & -2\xi_i\omega_i \end{pmatrix} \tag{3.50}$$

and

$$F = (0, F_i^0)^T \tag{3.51}$$

The sensors are assumed to have the form

$$y(t) = z(t) + v(t); \quad z(t) = h[q(s); \; s \le t] \tag{3.52}$$

where

$$E\{v(t)\} = 0; \quad E\{v(t)v^T(s)\} = R(t)\delta(t-s);$$
$$E\{z(t)\} = 0; \quad [z(t)] \le M < \infty \text{ for all } t;$$

$$\int_0^T E\{z(t)z^T(t)\} \, dt < \infty; \quad E\{v(t)F^T\} = 0 \quad \text{and } h \quad \text{is a functional of } q(s).$$

It is further assumed that the forcing function F is additively (preferably Gaussian) random.

3.15   CANONICAL ESTIMATION PROCEDURE

The least squares estimate of $q(t)$ is

$$\hat{q}(t|\tau) = E\{q(\tau)|y(s), \; 0 \le s \le \tau\} \tag{3.53}$$

Assuming that the property of "causal equivalence" (Clark, 1969) holds, the sensor equations may be transformed into a white Gaussian process $v(t)$, called the strict sense innovations process with a simpler structure so that Eq. (3.6) can now be put in the form

$$\hat{q}(t|\tau) = \int_0^\tau E\{q(t)v^T(s)|v(\sigma), \ 0 \le \sigma < s\}v(s)ds \qquad (3.54)$$

where $v(t) = y(t) - \hat{z}(t|t)$, $0 \le t < \tau$ and $\int$ is the Itô (1951) integral. A differential structure for Eq. (3.7) is obtained as

$$\dot{\hat{q}}(t|t) + A\hat{q}(t|t) + K(t)v(t) \qquad (3.55)$$

where

$$K(t) = E\{\hat{q}(t)v^T(t)|v(\sigma), \ 0 \le \sigma < t\} \qquad (3.56)$$

Conditional joint probability distribution functions are required to obtain explicit solution to the stochastic differential Eq. (3.8). A method for the evaluation of the probability densities is obtained in the form of a partition theorem (Laintiotis, 1971).

$$p(q|f') = A(t|q)p(q_0)|\int P(q_0)A(t|q)dq \qquad (3.57)$$

with $f' = \{y(\tau)\}_0^t$ and

$$A(t|q) = \exp\left\{ \int_0^t \dot{\hat{h}}(s,q)R^{-1}(s)y(s)ds - \frac{1}{2} \int_0^t \left\| \hat{h}(s,q) \right\|_{R^{-1}(s)}^2 ds \right\} \qquad (3.58)$$

where

$$\hat{h}(s,q) = E\{h(s,q)|f^s\} \qquad (3.59)$$

and $p(q_0)$ is the priori density function. Equation (3.9) can be approximated by expanding the distributions $A(t|q)$ in a Volterra (Barrett and Lampard, 1955; Biglieri, 1973) power series with functional terms

$$A(t|q) = \sum_{i=0}^{\infty} A_i(t|q) \qquad (3.60)$$

where

$$A_i\left(t|q\right)=\frac{1}{i!}\int_0^\infty d_{\tau_1}\cdots\int_0^\infty d_{\tau_i}g_i\left(\tau_1,\tau_2,\ldots,\tau_i\right)\prod_i^{k-1}v\left(t-\tau_k\right)$$ (3.61)

such that $g_i(\ldots)$, $i=1, \ldots, n$ are the integral kernels describing the system and $v(\cdot)$ is the innovations process. It has been assumed that in order to satisfy the requirements of physical realizability the kernels are zero for any argument less than zero. The kernels can be identified in the following manner.

(i) First order kernel: $g_i(t)$

Let $v(t) = A\delta_0(t)$ be an impulse of strength A so that

$$A\left(t|q\right)=Ag_1\left(t\right)+A^2g_2\left(t,t\right)+A^3g_3\left(t,t,t\right)+\ldots$$ (3.62)

and

$$g_1\left(t_1\right)=\left.\frac{\delta A\left(t_1|q\right)}{\delta A}\right|_{A=0}$$ (3.63)

(ii) Second order kernel: $g_2(t, t)$

Let $P(t) = A\delta_0(t) + B\delta_0(t + \tau)$ be two impulses at t and $t + \tau$ of strength A and B respectively, where A and B are specified constants as above; then

$$A\left(t|\theta\right)=Ag_1\left(t\right)+Bg_1\left(t+\tau\right)+A^2g_2\left(t,t\right)$$
$$+2ABg_2\left(t,t+\tau\right)+B^2g_2\left(t+\tau,t+\tau\right)+\ldots$$

and the second-order kernel

$$g_2\left(t,t+\tau\right)=\frac{1}{2}\left.\frac{\delta^2 A\left(t_1|q\right)}{\delta A\delta B}\right|_{\substack{A=0\\B=0}}$$ (3.64)

Higher order kernels can be computed in similar manners.
The system state estimates can now be obtained as

$$\hat{q}\left(t|t\right)=J\left(t|t_0\right)\hat{q}\left(t|t_0\right)+\int_{t_0}^i J\left(t|\tau\right)k\left(\tau\right)v\left(\tau\right)d\tau$$ (3.65)

where

$$J\left(t|t_0\right)=\exp\left\{A\left(t-t_0\right)\right\}$$ (3.66)

Is the transition/fundamental matrix; $k(\tau)$ having been approximated using the Equations (3.9) and (3.10).

3.16   HYPOTHETICAL EXAMPLE

A hypothetical linear dynamical system is solved below, using the Kalman filtering equations, to illustrate the close correlation (in a non-statistical sense) between the observations process, the innovations process and the systems state. It has been assumed that the distributed parameter system is reducible to a linear ordinary differential equation system as follows:

System dynamics:

$$\dot{x}(t) = 0.5x(t) + u(t), \quad x(0) = 0 \tag{3.67}$$

Observation process:

$$y(t) = x(t) + v(t) \tag{3.68}$$

It will be interesting to run a simulation of Equations (3.67) and (3.68) to generate observations to develop plots of observation magnitudes against time, where the time axis represents sequential observation processes. Similarly, a plot of the dynamic introduction of innovations can be plotted against time.

3.17   CONCLUSION

Chapter 3 presents advanced set-theoretic modeling of innovation processes. Every innovation process, inherently, consists of sets. These sets are interrelated either in quantitative terms or qualitative relationships. Mapping of set elements is essential for a full understanding of the innovation environment for the purpose of systems integration. A well-integrated system is expected to drive innovation more successfully.

REFERENCES

Aokl, M. (1968). "Control of Large Scale Dynamic Systems B Aggregation". *IEEE Transactions on Automatic Control* AC-13: 246–253.

Badiru, A. B. and J. Cheung (2002). *Fuzzy Engineering Expert Systems with Neural Network Applications.* John Wiley & Sons, New York.

Badiru, A. B. and O. Ibidapo-Obe (2008, December). "On Set-Theoretic Fundamentals of Systems Integration". *African Journal of Computing & ICT (Journal of the IEEE Nigeria Computer Chapter)* 1(2): 3–25.

Badiru, A. B., O. Ibidapo-Obe and B. J. Ayeni (2012). *Industrial Control Systems: Mathematical and Statistical Models and Techniques.* Taylor & Francis Group / CRC Press, Boca Raton, FL.

Balkhi, Z. T. (1999). "On the Optimal Production Stopping and Restarting Times for an EOQ Model with Deteriorating Items". *The Journal of the Operational Research Society* 51(8): 999–1001.

Barrett, J. F. and D. G. Lampard (1955). "An Expansion for Some Second-Order Distributions and Its Application to Noise Problems". IRE Transactions on Information Theory 1: 10.

Biglieri, E. (1973). "Power Series with Functional Terms". *Proceedings of the Institution of Electrical Engineers* 61: 251.

Chen, Bing and Xiaoping Liu (2004). "Reliable Control Design of Fuzzy Dynamic Systems with Time-Varying Delay". *Fuzzy Sets and Systems* 146: 349–374. DOI: 10.1016/S0165-0114(03)00326-9

Clark, J. M. C. (1969). Technical Report. Equivalence of Sensor Equations, Department of Computing and Control, Imperial College, London, UK.

Fischer-Cripps, A. C. (2005). *The Mathematics Companion*. Institute of Physics Publishing, Philadelphia, PA.

Golub, G. H. and C. Reinsch (1970). "Singular Value Decomposition and Least Squares Solutions". *Numerische Mathematik* 14: 403–420.

Golub, G. H. and C. Van Loan (1979). "Total Least Squares," in *Smoothing Techniques for Curve Estimation* (T. Gasser and M. Rosenblatt, Eds.), pp. 62–76, Springer Verlag, New York.

Heymann, M., F. Lin and S. Resmerita (2005) "Analysis of Zeno Behaviors in a Class of Hybrid Systems". *IEEE Transactions on Automatic Control* 40(10): 16–24. DOI:10.1109/TAC.2005.843874

Ibidapo-Obe, O. (1982a). "A Modal Approach for Estimation in Distributed Parameter Systems". *International Journal of Systems Science* 13(10): 1165–1170.

Ibidapo-Obe, O. (1982b). "A Note on Model Reduction of Large Scale Systems". *Journal of Mathematical Analysis and Applications* 90: 480–483.

Itô, K. (1951). "On Stochastic Differential Equations". *Memoirs of the American Mathematical Society* 4.

Laintiotis, D. G. (1971). "Bayes-Optimal Recursive Supervised Learning Structure and Parameter Adaptive Pattern Recognition Systems". *International Journal of Control* 14: 1137.

Omicini, A. and G. A. Papadopoulos (2001). "Why Coordination Models and Languages in AI?" *Applied Artificial Intelligence* 15(3): 1–10.

Rashwan, A. H. and A. S. Ahmed (1982). *IFAC Proceedings*, India.

Soong, T. T. (1977). "On Model Reduction of Large Scale Systems". *Journal of Mathematical Analysis and Applications* 60: 477–482.

Stewart, G. W. (1973). *Introduction to Matrix Computations*. Academic Press, New York.

Yang, E., D. M. Cunnold, R. J. Salawitch, M. P. McCormick, J. Russell, J. M. Zawodny, S. Oltmans and M. J. Newchurch (2006). "Attribution of Recovery in Lower-Stratospheric Ozone". *Journal of Geophysical Research*. DOI: 10.1029/2005JD006371

Zhang L, A. Ohta, H. Horiuchi, M. Takagi and R. Imai (2001). "Multiple Mechanisms Regulate Expression of Low Temperature Responsive (LOT) Genes in Saccharomyces Cerevisiae". *Biochem Biophys Res Commun* 283(2): 531–535.

4 AI Mathematical Modeling for Product Design

4.1 INTRODUCTION

Artificial intelligence (AI) can be productively applied to the design of new products, whereby multiple options are available for product characteristics. The contents present here are adapted from Wu et al. (2011). Product design has normally been performed by teams, each with expertise in a specific discipline such as material, structural, electrical systems. Traditionally, each team would use its member's experience and knowledge to develop the design sequentially. Collaborative design decisions explore the use of optimization methods to solve the design problem incorporating a number of disciplines simultaneously. It is known that the optimum of the product design is superior to the design found by optimizing each discipline sequentially due to the fact that it is enabled to exploit the interactions between the disciplines. In this chapter, a bi-level decentralized framework based on Memetic Algorithm (MA) is proposed for collaborative design decision using forearm crutch as the case. Two major decisions are considered: the weight and the strength. In this chapter, we introduce two design agents for each of the decisions (Wu et al., 2011). At the system level, one additional agent termed facilitator agent is created. Its main function is to locate the optimal solution for the system objective function, which is derived from the Pareto concepts; thus, Pareto optimum for both weight and strength is obtained. It is demonstrated that the proposed model can converge to Pareto solutions.

Under collaborative design paradigm, the first common topic is Multidisciplinary Design Optimization (MDO), which is defined as "an area of research concerned with developing systematic approaches to the design of complex engineering artifacts and systems governed by interacting physical phenomena" (Alexandrov, 2005). Researchers agree that interdisciplinary coupling in the engineering systems presents challenges in formulating and solving the MDO problems. The interaction between design analysis and optimization modules and multitudes of users is complicated by departmental and organizational divisions. According to Braun and Kroo (1997), there are numerous design problems where the product is so complex that a coupled analysis driven by a single design optimizer is not practical, as the method becomes too time consuming either because of the lead time needed to integrate the analysis or because of the lag introduced by disciplinary sequencing. Some researchers have taken the project management as a means to facilitate and coordinate the design among multiple disciplines (Badiru and Theodoracatos, 1995, Thal et al., 2007) the product design.

DOI: 10.1201/9781003247395-4

Early advances in MDO involve problem formulations that circumvent the organizational challenges, one of which is to protect disciplinary privacy by not sharing full information among the disciplines. It is assumed that a single analyst has complete knowledge of all the disciplines. As indicated by Sobieszczanski-Sobieski and Haftka (1997), most of the work at this phase aims to tackle the problems by a single group of designers within one single enterprise environment where the group of designers share a common goal and require less disciplinary optimum. The next phase of MDO gives birth to two major techniques: Optimization by Linear Decomposition (OLD) and Collaborative Optimization (CO). These techniques involve decomposition along disciplinary lines and global sensitivity methods that undertake overall system optimization with minimal changes to disciplinary design and analysis. However, Alexandrov and Lewis (2000) explore the analytical and computational properties of these techniques and conclude that disciplinary autonomy often causes computational and analytical difficulties, which result in severe *convergence* problems.

Parallel to these MDO developments, there also evolves the field of Decision-Based Design (Huang, 2004, Li et al., 2004, Simpson et al., 2001, Hernandez and Seepersad, 2002, Choi et al., 2003, Wassenaar and Chen, 2003), which provides a means to model the decisions encountered in design and aims at finding "satisfying" solutions (Wassenaar et al., 2005, Nikolaidis, 2007). Research in Decision-Based Design includes the use of adaptive programming in design optimization (Simon, 1955), the use of discrete choice analysis for demand modeling (Hernandez and Seepersad, 2002, Choi et al., 2003). In addition, there has been extensive research ranging from single-objective Decision-Based Design (Hernandez and Seepersad, 2002) to multi-objective models (Lewis and Mistree, 1998, 1999). It combines Game Theory, Utility Theory, and Decision Sciences for collaborative design, which can be conducted among a group of designers from different enterprises. This technique has several computational difficulties in calculating the "best reply correspondence" and the rational reaction sets especially when the designs are very complex. Besides, several approximations like using response surfaces within these techniques make them prone to errors (Fernandez et al., 2005).

Note most methods reviewed above have strict assumption on the utility functions and/or constraints (e.g., convexity and quasi-linear of the functions), which limits the application to product design. In this research, we explore the use of a heuristic method, MA, a combination of Local Search (LS) and Genetic Algorithm (GA) to forearm crutch design, which has non-convex objective for one of the decisions. Forearm crutches had been exclusively used by people with permanent disability. Nowadays, it is beginning to serve for some shorter-term purposes as well. The design of forearm crutch needs to consider multidisciplinary decisions. For example, the structure designer wants to ensure that the design is lightweight. The material engineer wants composite material to have the right number of layered at right angles to make the product durable. The outsourcing engineer wants the supplier to provide low cost, high reliable, and light weight parts. Another important factor impacting the design is cost. Here, we introduce the design agent for each disciplinary decision problem and one system agent facilitating the communication among the design agents and guiding the design to convergence. To achieve this, the overall decision space is partitioned into two sets: one for coupled variables (the ones shared by at least two designers), one for local variables (the ones can be fully controlled by each designer). Next, an iterative process between design agent decisions on local variables and facilitator agent decisions on the whole design space launches. It is demonstrated that a converged Pareto optimum is achieved after a number of iterations for the forearm crutch design, which has nonlinear form decision functions.

This chapter is organized as follows: the related literature is briefly reviewed in Section 4.2, followed by the detailed explanation on the proposed bi-level decentralized framework in Section 4.3. The forearm crutch case is explained in Section 4.4 with the conclusions being drawn in Section 4.5.

4.2 PRODUCT DESIGN BACKGROUND

Collaborative Optimization (CO), introduced by Braun and Kroo (1997), is a bi-level optimization approach where a complex problem is hierarchically decomposed along disciplinary boundaries into a number of subproblems which are brought into multidisciplinary agreement by a system level coordination process. With the use of local subspace optimizers, each discipline is given complete control over its local design variables subject to its own disciplinary constraints. The system-level problem sets up target values for variables from each discipline. Each discipline sets the objectives to minimize the discrepancy between the disciplinary variable values and the target values. The system-level optimization problem is formulated as minimizing a global objective subject to interdisciplinary consistency constraints. The interdisciplinary consistency constraints are equality constraints that match the system-level variables with the disciplinary variables. In OLD (Sobieszczanski-Sobieski et al, 1982, 1985, 1988), the disciplines are given the autonomous task of minimizing disciplinary design infeasibility while maintaining system-level consistency. The system-level problem is to drive design infeasibility to zero. At the local-level problem, the disciplines use their local degrees of freedom to minimize the violation of the disciplinary design constraints, subject to matching the target value for the disciplinary output that is fed into the discipline. Balling et al. (1995) introduced a combination of CO and OLD where the disciplinary subproblems minimize the discrepancy in the system-level targets as well as the disciplinary design infeasibility given the disciplinary design constraints.

Both CO and OLD depend on a design problem's amenability to hierarchical decomposition with the system objective explicitly defined. On the other hand, Concurrent Sub-Space Optimization (CSSO) (Sobieszczanski-Sobieski, 1988) is a non-hierarchic system optimization algorithm that optimizes decomposed subspaces concurrently, followed by a coordination procedure for directing system problem convergence and resolving subspace conflicts. In CSSO, each subspace optimization is a system-level problem with respect to the subset of the total system design variables. Within the subspace optimization, the non-local variables that are required to evaluate the objective and the constraint functions are approximated using Global Sensitivity Equations (GSE). Interested readers please refer to Sobieszczanski-Sobieski, 1988, for detailed description of GSEs.

The Bi-Level Integrated Synthesis (BLISS) (Sobieszczanski-Sobieski and Kodiyalam, 1998) method uses a gradient-guided path to reach the improved system design, alternating between the set of modular design spaces (the disciplinary problems) and the system-level design space. BLISS is an all-in-one like method in that the complete system analysis performed to maintain multidisciplinary feasibility at the beginning of each cycle of the path. The overall problem is decomposed such that a set of local optimization problems deal with the detailed local variables, which are large in number and one system-level optimization problem that deals with a small number of global variables.

Decision-Based Design (Fernandez et al., 2002; Hernandez and Seepersad, 2002; Simpson et al., 2001; Choi et al. 2003) is a paradigm focusing on distributed and collaborative design efforts. For the cases where continuous variables are used, adaptive linear programming (Lewis and Mistree, 1999) is employed; in case of mixed discrete and continuous variables, Foraging-Directed Adaptive Linear Programming has been used (Lewis and Mistree, 1999). In a non-cooperative environment, game theoretic principles are used to arrive at the best overall design (Lewis and Mistree, 1998, 1999). Recently, Design-for-Market systems grow out of the Decision-Based Design and emerge as an area focusing on establishing a solid basis in decision theory, by taking microeconomics into account, to support engineering design. Kumar et al. (2007) propose a hierarchical choice model based on discrete choice analysis to manage and analyze customer preference data in setting design targets. Azarm's group studies new product designs that are robust from two perspectives – from the engineering perspective in terms of accounting for uncertain parameters and from the market perspective in terms of accounting for variability in customer preferences measurement (Besharati et al., 2006). They conclude that incorporating consumer heterogeneity in considering the variability in customer preferences may have a significant impact on the ultimate design. Research led by Michalek explores the use of game-theoretic approach to finding market equilibrium under various regulation scenarios (Shiau and Michalek, 2007). A metric for agility measurement is introduced by Sieger et al. (2000) to explore the product development for mass customization.

In general, some common criticisms and/or challenges facing collaborative design decisions are the convergence and information sharing issues:

- *Will the decision model converge? If yes, under what condition (assumptions on the function form and design spaces) will it converge? How fast will it converge?*
- *Most models (CO, OLD, BLISS, etc.) take a top-down approach with the full knowledge of the design space (e.g., the form of utility functions, constraints) being available. For the cases when the design information is partially known, what decision model is appropriate?*

To address these challenges, we propose a general decision framework based on MA that allows distributed design teams to arrive at Pareto solutions, which is explained in detail in the next section.

4.3 MEMETIC ALGORITHM AND ITS APPLICATION TO COLLABORATIVE DESIGN

Memetic Algorithm is one of the emerging areas in evolutionary computation. It integrates GA with LS to improve the efficiency of searching complex spaces. In MA, GA is used for global exploration, while LS is employed for local exploitation. The complementary nature of GA and LS makes MA an attractive approach for large-scale, complex problems, for example, collaborative design.

4.4 A FRAMEWORK FOR COLLABORATIVE DESIGN

Let us consider a general collaborative design with z design teams. The problem can be represented as:

$$\min \mathbf{J}\left(\mathbf{x},\mathbf{y}\right)$$

$$\mathrm{St.}\mathbf{g}\left(\mathbf{x}\right)\leq 0$$

$$\mathbf{h}\left(\mathbf{x},\mathbf{y}\right)\leq 0$$

$$x_i^{LB} \leq x_i \leq x_i^{UB} \left(i=1,\ldots,n_1\right)$$

$$y_i^{LB} \leq y_i \leq y_i^{UB} \left(i=1,\ldots,n_2\right)$$

where $\mathbf{J} = [J_1\,(\mathbf{x},\mathbf{y})\ldots J_Z\,(\mathbf{x},\mathbf{y})]^T$, $\mathbf{x} = [x_1\ldots x_{n1}]^T$, $\mathbf{y} = [y_1\ldots y_{n2}]^T$, $\mathbf{g} = [g_1\,(\mathbf{x})\ldots g_{m1}\,(\mathbf{x})]^T$, $\mathbf{h} = [h_1\,(\mathbf{x},\mathbf{y})\ldots h_{m2}\,(\mathbf{x},\mathbf{y})]^T$, $\mathbf{x}$ is the set of n_1 local variables, $\mathbf{y}$ is the set of n_2 coupled variables, $\mathbf{g}$ is the set of m_1 local constraints, and $\mathbf{h}$ is the set of m_2 coupled constraints.

Figure 4.1 illustrates the iterative decision process between system facilitator agent and disciplinary design agents. First, the facilitator initializes the global

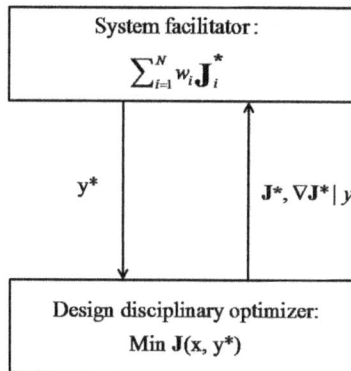

FIGURE 4.1 Design optimizer framework.

solution space over both local and coupled variables. For any solution, for example, [**x***, **y***], each design agent will execute local optimizer over the sub-design space, which consists of **x** only, that is, Min **J** (**x**, **y***). The results fed back to the facilitator are the value of objective function and the gradient of objective function over coupled variables. The facilitator will (1) employ LS for the recent results updated by each designer using the related gradient information for the improved design; (2) next, traditional GA operators, crossover, and mutation are applied to introduce new candidates to the solution space.

4.5 THE PSEUDO CODE

This section shows the layout of the Pseudo Code for the proposed methodology as illustrated in Figure 4.2.

Parameters:

N: no. of disciplinary design agents;
w_i: weight for the objective function of ith disciplinary design agent,
where $i = 1, \ldots, N$;
$1/W$: weight step size;
P: population size;

Pseudo Code Lines:

As shown in the above Pseudo Code, there exist three loops, from outer to inner in the proposed method: weight enumeration (lines 11–37), GA loop (lines 13–37), and local search loop (lines 24–28). That is, given a weights combination (e.g., $w1 = 0.3$, $w2 = 0.7$ for two agents), GA is triggered, which applies crossover and mutation operators and selection mechanism (in this case study, elitism selection is employed) for the population update. In addition, for the updated population, local search is further employed to identify improved solutions within the neighborhood. This is achieved by having sub-gradient information from each designer on the coupled variables fed back to the facilitator. Specifically, given any chromosome from the population, each design agent assumes the coupled variables are set and thus conducts optimization on the local variables only. Each design agent would also study the gradients on the coupled variables. Thus, given the values of the coupled variables, both the optimal design on local variables and the sub-gradient on the coupled variables are sent back to the facilitator. Since the priorities of the objective functions reflected by the weight assignments are enumerated exhaustively, all the possible Pareto solutions are located forming the Pareto frontier. In some cases where the priority is known, the weight loop can be removed. Please note that the Pareto filter operation (lines 29–36) is triggered by the facilitator within each weight combination. That is, it is possible that some Pareto solutions given a specific weight may be dominated by the Pareto solutions obtained with other weights.

One distinguishable feature of this proposed approach from other existing methods is the information exchanged iteratively between the facilitator and the design agent is values instead of function forms. For example, passing from the facilitator to the design agent (top-down) is the values of the coupled variable; passing from the

```
(1)    //Initialization
(2)    The set of final Pareto solutions FP = ∅;
(3)    The set of GA population PS = ∅;
(4)    The set of weights combination WS = ∅;
(5)    Given N objective functions, we have ∑ᵢ₌₁ᴺ wᵢJᵢ(xᵢ, y)
(6)    Begin (at facilitator agent level)
(7)       //Enumerate weights combination
(8)       Set w₁ = w₂ = ⋯ = w_{N−1} = 0;
(9)       Given weight step size 1/W;
(10)      Let each wᵢ (i = 1, …, N − 1) increases 1/W, w_N = 1 − w₁ − ⋯ w_{N−1},
             and add (w₁, w₂, …, w_N) to WS;
(11)      //Weights Loop
(12)      For each weights combination (w₁, w₂, …, w_N) in WS, ∑ᵢ₌₁ᴺ wᵢJᵢ(xᵢ, y) is
             constructed;
(13)         //GA loop
(14)         //Initialization
(15)         Generate random population of P solutions and add them to PS;
(16)         For n = 1 to maximum # of generations for GA loop;
(17)            //Crossover and Mutation
(18)            Random select two parents p_a and p_b from PS;
(19)            Generate two offspring p'_a and p'_b by crossover operator;
(20)               if p'_a and/or p'_b are not feasible, generate new feasible offspring
(21)               p''_a and/or p''_b using mutation operator;
(22)            //Selection
(23)            Using fitness function (∑ᵢ₌₁ᴺ wᵢJᵢ(xᵢ, y)) to evaluate the solution, update PS
                  with improved solutions;
(24)            //Local Search Loop
(25)            For each chromosome pⱼ in PS;
(26)               Call each Design Agent for local optimization on x (note different
                     optimization engines can be employed based on the design
                     disciplines);
(27)               Given updates from Design Agent on x, Facilitator agent employs
                     sub-gradient algorithm [19] as local search algorithm to
                     iteratively locate improved solution p'ⱼ with respect to y;
(28)            Next pⱼ;
(29)            //Pareto Filter:
(30)            For each chromosome pⱼ in the set PS;
(31)               If pⱼ is not dominated by all the solutions in the set FP;
(32)                  Add pⱼ to the set FP;
(33)               Else If there are solutions in the set FP are dominated by pⱼ;
(34)                  ← Remove those solutions in the set FP;
(35)               End If;
(36)            Next pⱼ;
(37)         Next n;
(38)   End;
```

FIGURE 4.2 Pseudo code 1.

design agent back to the facilitator (bottom-up) is the values of the objective function and associated gradient values, passing from the facilitator to the design agents (top-down) is the values of the coupled variables. The main advantage of this approach is a "black box" disciplinary optimizer can be easily plugged in. Second, since the facilitator explores the solution space based on the knowledge of the solution candidates (x^*, y^*), the candidate performance (J^*) instead of the function formulation, a truly decentralized decision without the full knowledge of the design space can be implemented. An industry case is explored to demonstrate the applicability of the proposed framework.

4.6 CASE EXAMPLE OF FOREARM CRUTCH DESIGN

Crutches are mobility aids used to counter mobility impairment or an injury that limits walking ability. Forearm crutches are used by slipping the arm into a cuff and holding the grip (Figure 4.3). It has been increasingly used for patients with shorter-term needs. Earlier study conducted by National Medical Expenditure Survey (NMHS) in 1987 indicates that an estimated 9.5 million (4%) non-institutionalized U.S. civilians experience difficulty in performing basic life activities; some need crutches for leg support for walking. This number increases due to the baby boomer effect.

Typical forearm crutches are made of aluminum and are criticized by customers for being heavy, noisy, and less durable. Customers suggest that a small reduction in the weight of forearm crutches would significantly reduce the fatigue experienced by crutch users. However, the reduction in weight should not be accompanied by a strength reduction. Most crutches on the market are designed for temporary use and

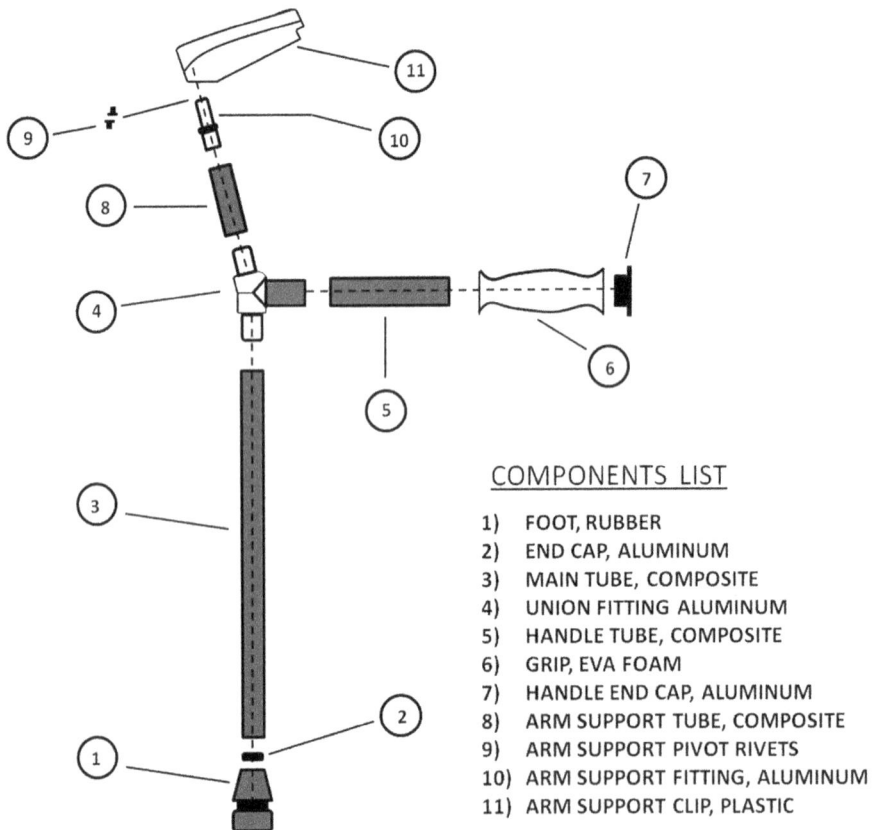

COMPONENTS LIST

1) FOOT, RUBBER
2) END CAP, ALUMINUM
3) MAIN TUBE, COMPOSITE
4) UNION FITTING ALUMINUM
5) HANDLE TUBE, COMPOSITE
6) GRIP, EVA FOAM
7) HANDLE END CAP, ALUMINUM
8) ARM SUPPORT TUBE, COMPOSITE
9) ARM SUPPORT PIVOT RIVETS
10) ARM SUPPORT FITTING, ALUMINUM
11) ARM SUPPORT CLIP, PLASTIC

FIGURE 4.3 Exploded view of a forearm crutch.

wear out quickly. Crutch users commonly have to replace their crutches two to three times a year. This drives the need to redesign forearm crutches which are robust, appropriate for a wide range of users from lighter weight adults to users weighing up to 250 pounds with considerable upper body strength and who may use them aggressively on a continuous basis.

One solution is to use composite material for crutch, which is *light weight* with good performance in *strength*. However, it comes with relatively expensive cost. After in-depth marketing survey, the design team decides to outsource the aluminum union fitting (component #4 in Figure 4.3), use appropriate composite tube, and apply adhesive from Hysol to bond the tubes with union fitting.

Aluminum Union: The design team first develops a computer model based on finite element method to determine the necessary wall thickness and to calculate the load on the handle necessary to produce yielding. An aluminum union, which costs $150 and stands ≥630 lbs, is used. The use of Hysol adhesive to bond the union with the tube needs to be tested to ensure the strength requirement is satisfied.

Composite Tube: A typical composite tube is 39" in length. The tube can be cut into smaller pieces for the forearm crutch assembly. Approximately 2 ½ tubes are needed to make a pair of crutches. Here, three smaller tubes are used as handle (component 5 in Figure 4.3), which is fixed as 4.75 inch, arm support tube (component 8 in Figure 4.3), which usually ranges from 6.5 to 7.8 inch, main tube (component 3 in Figure 4.3), which ranges from 30.69 to 34.25 inch. The inner diameter of the tube is critical to maintain the proper bondline thickness for each adhesive joint. It ranges from 0.7605 to 0.7735 inch. The outer diameter is determined by the number of plies and it ranges from 0.922 to 0.935 inch. Usually, the arm support tube is less concerned with strength, and the main tube needs to be tested for the strength consideration. Thus, we have two decision problems constructed: weight and strength.

4.7 DESIGN PROBLEM FORMULATION

In this research, we focus on the weights of the tubes (arm support and main tubes). A minimization problem is introduced as:

$$\min : W = W_u + W_L$$

$$\text{St}: \quad W_u = \rho\pi\left[\left(\frac{D_o}{2}\right)^2 - \left(\frac{D_i}{2}\right)^2\right] \times L_u$$

$$W_L = \rho\pi\left[\left(\frac{D_o}{2}\right)^2 - \left(\frac{D_i}{2}\right)^2\right] \times L_L$$

$$30.6875 \le L_L \le 34.25$$

$$6.5 \le L_u \le 7.8$$

$$0.922 \le D_\text{o} \le 0.935$$

$$0.7605 \le D_\text{i} + 2T/1000 \le 0.7735$$

where W_u (inch) is the weight of arm support tube, W_L (inch) is the weight of main tube, ρ (lbs/inch^3) is the density of the composite tube, which is 0.08 in this chapter, L_u (inch) is the length of the arm support tube, L_L (inch) is the length of the main tube, D_o (inch) is the outer diameter, D_i (inch) is the inner diameter, and T (mils, 1 mils = 0.001 inch) is the bondline adhesive material thickness.

4.8   DESIGN AGENT FOR STRENGTH DECISION

Since the strength from aluminum fitting is satisfactory from Finite Element Analysis (FEM), the strength model will consider two potential failures: the adhesive applied joint and the strength of the main tube. Thus, the problem is constructed as:

$$\max : S = \min\left(S_\text{L}, S_\text{A}\right)$$

St:

$$S_\text{L} = \frac{\pi EI}{L_\text{L}^2}$$

$$I = \pi\left(D_\text{o}^4 - D_\text{i}^4\right)/64$$

$$12 \le E \le 16$$

$$S_\text{A} = \left(-6.0386T^2 + 7.7811T + 4644.5\right)\times\frac{\pi}{4}\times\left(D_\text{o}^2 - D_\text{i}^2\right)$$

$$0.922 \le D_\text{o} \le 0.935$$

$$0.7605 \le D_\text{i} + 2T/1000 \le 0.7735$$

$$30.6875 \le L_\text{L} \le 34.25$$

$$0 \le T \le 17$$

where S_L (lbs) is the strength of the bottom of the lower tube, E (msi, 1 msi = 10^6 psi) is the modulus of elasticity, I (inch^4) is the area moment of inertia, and S_A (lbs) is the strength of the joint after applying adhesive.

4.9   SYSTEM IMPLEMENTATION

For the decision problems explained above, optimization code written in MATLAB®
is executed. Here, we provide detailed explanation about how the system problem is
constructed and how the facilitator agent guides the design agents to converge to the
solution using MA.

Step 1 Initialization: Given w_1, w_2, construct system search space as $w_1 W^*$
  $- w_2 S^*$ (W^* and S^* are the values of the objectives from each design
  agent, $w_1 + w_2 = 1$).

Step 2 Real Code Genetic Algorithm: The chromosome is represented with
  real numbers, that is, $(L_u, L_L, D_o, D_i, T, E)$. Note L_L, D_o, D_i, T are coupled
  variables, L_u is the local variable for weight agent, and E is the local
  variable for strength agent.

  Step 2.1 (Initial population): For $(L_u, L_L, D_o, D_i, T, E)$, without loss
    of generalization, let's assume a and b represent the lower
    bound and upper bound of one of the variable, r be a random
    number $r \in [0, 1]$, we get $(b - a)r + a$. Thus, a new chromo-
    some is generated as for the initial population. A pool of 40
    chromosomes is created.

  Step 2.2 (Selection of parents): To ensure all chromosomes have the
    chances to be selected, solutions are classified into three
    groups according to their fitness: high fitness level, medium
    fitness level, and low fitness level. The fitness is assessed
    based on $w_1 W^* - w_2 S^*$, the lower, the better.

  Step 2.3 (Crossover): Given two chromosome $C_1 = (L^1_u, L^1_L, D^1_o, D^1_i,$
    $T^1, E^1)$ and $C_2 = (L^2_u, L^2_L, D^2_o, D^2_i, T^2, E^2)$, the offspring are
    generated as:

$$C'_1 = \theta C_1 + (1-\theta) C_2$$

$$C'_2 = (1-\theta) C_1 + \theta C_2$$

    where $\theta \in [0, 1]$

  Step 2.4 (Mutation): Mutation is applied by simply generating a new
    feasible solution to replace the infeasible one.

Step 3 (Local Search): The facilitator agent applies sub-gradient method-based
  LS over coupled variables to improve the solutions. First, each design
  agent evaluates the gradients of the design decision problems (disciplinary) wrt the coupled variables. For example, given the coupled variables $L_L = L^*_L, D_o = D^*_o, D_i = D^*_i, T = T^*$, each decision problem is solved
  independently for W^* and S^*. The gradients are obtained as follows:

$$\lambda_{W,L_L} = \left.\frac{\partial W}{\partial L_L}\right|_{L_L = L^*_L, D_o = D^*_o, D_i = D^*_i, T = T^*}, \quad \lambda_{S,L_L} = \left.\frac{\partial S}{\partial L_L}\right|_{L_L = L^*_L, D_o = D^*_o, D_i = D^*_i, T = T^*}$$

$$\lambda_{w,D_i} = \frac{\partial W}{\partial L_i}\bigg|_{L_L = L_L^*, D_o = D_o^*, D_i = D_i^*, T = T^*}, \quad \lambda_{S,D_i} = \frac{\partial S}{\partial D_i}\bigg|_{L_L = L_L^*, D_o = D_o^*, D_i = D_i^*, T = T^*}$$

$$\lambda_{w,D_i} = \frac{\partial W}{\partial D_i}\bigg|_{L_L = L_L^*, D_o = D_o^*, D_i = D_i^*, T = T^*}, \quad \lambda_{S,D_i} = \frac{\partial S}{\partial D_i}\bigg|_{L_L = L_L^*, D_o = D_o^*, D_i = D_i^*, T = T^*}$$

$$\lambda_{w,T} = \frac{\partial W}{\partial T}\bigg|_{L_L = L_L^*, D_o = D_o^*, D_i = D_i^*, T = T^*}, \quad \lambda_{S,T} = \frac{\partial S}{\partial T}\bigg|_{L_L = L_L^*, D_o = D_o^*, D_i = D_i^*, T = T^*}$$

The gradients of the system problem are then calculated as:

$$\lambda_{L_L} = w_1 \lambda_{w,L_L} - w_2 \lambda_{S,L_L}$$

$$\lambda_{D_o} = w_1 \lambda_{w,D_o} - w_2 \lambda_{S,D_o}$$

$$\lambda_{D_i} = w_1 \lambda_{w,D_i} - w_2 \lambda_{S,D_i}$$

$$\lambda_T = w_1 \lambda_{w,T} - w_2 \lambda_{S,T}$$

Based on $\lambda = [\lambda_{LL}, \lambda_{Do}, \lambda_{Di}, \lambda_T]$, the facilitator agent non-summable diminishing method to update the coupled variables. That is, at iteration $k + 1$,

$$\begin{bmatrix} L_L \\ D_o \\ D_i \\ T \end{bmatrix}^{(k+1)} = \begin{bmatrix} L_L \\ D_o \\ D_i \\ T \end{bmatrix}^{(k)} - \alpha_{k+1} \begin{bmatrix} \lambda_{L_L} \\ \lambda_{D_o} \\ \lambda_{D_i} \\ \lambda_T \end{bmatrix}^{(k)}$$

where step size α_k satisfies:

$$\begin{cases} \lim_{k \to \infty} \alpha_k = 0 \\ \sum_{k=1}^{\infty} \alpha_k = \infty \end{cases}$$

The coupled variables are updated based on the above sub-gradient method until no further improvement of the weighted system problem is required.

4.10   SYSTEM RESULTS AND ANALYSIS

The Pareto frontier obtained by the proposed decentralized framework is shown in Figure 4.4. Note that the problem has Min-Max structure. Since this project focuses

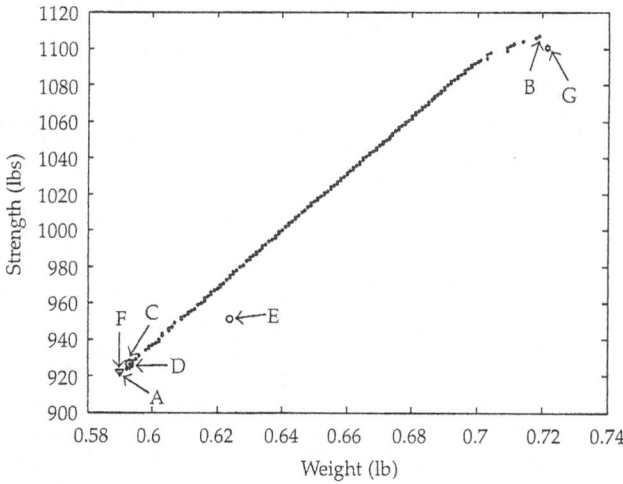

FIGURE 4.4   Pareto frontier in performance space for the crutch design.

on the composite tube design (main tube and handle tube), the weight for the handle tube (component #5) is computed as

$$\rho\pi\left[\left(\frac{D_o}{2}\right)^2 - \left(\frac{D_i}{2}\right)^2\right] \times 4.75$$

Other components in Figure 4.3 are outsourced with the weights summarized in Table 4.1.

We choose Pareto solution (A and B) to compare with the composite crutch from Ergonomics and the Invacare crutch, which are two commercial products in Table 4.2. Apparently, most composite crutches outperform Ergonomics and Invacare for both weight and strength except Design B outweighs Ergonomics by 0.09 lb. However, Design B is much durable with strength being 1105 lb compared to 715 lb of Ergonomics.

It is expected that the cost of the composite crutch will be high. In this case, it is around $460 in total (tube and other components shown in Figure 4.3). The crutch

TABLE 4.1
Weight for Each Component of the Crutch

Components (Figure 4.3)	Weight (lb)
#2	0.006
#4	0.05
#7	0.0074
#10	0.025
Others (#1, #6, #9, #11)	0.2

TABLE 4.2
Comparison of Crutch Weight and Strength

Crutch Design	Weight (lbs.)	Strength (lbs.)
Invacare	2.3	630
Ergonomics	1	715
Pareto design (A)	0.9498	921
Pareto design (B)	1.0945	1107
Nash equilibrium (C)	0.9532	926
Weight leader strength follower (D)	0.9532	926
Strength leader weight follower (E)	0.9879	951
Weight complete control (F)	0.9499	922
Strength complete control (G)	1.0952	1100

produced by Invacare and Ergonomics price range is $60 to $250. Although the composite crutch is several times expensive, it lasts much longer. Instead of having replacement two to three times a year, it can be used for a number of years since the lighter composite crutch could sustain greater than 1100 pounds load.

4.11 CONCLUSION

Collaborative design decisions involve designers from different disciplines with different specific domain knowledge. The decision process is a sequence of phases or activities where mathematical modeling can be employed. In this chapter, a bi-level distributed framework based on MA is proposed. Since the information communicated is neither the form of the decision function nor the decision space, private information is protected. In addition, in the cases where the information is not complete, the proposed framework can still guarantee the convergence to Pareto solutions. To demonstrate the applicability of the framework, a forearm clutch design is studied in detail. The results confirm converged Pareto set can be obtained for any form of decision function. While promising, the decision problem constructed are deterministic, our next step is to explore the use of this framework for design decisions under uncertainty. Computational efficient approach in the area of reliability-based design optimization would be explored.

REFERENCES

Alexandrov, N. M. (2005). "Editorial – Multidisciplinary Design Optimization". *Optimization and Engineering* 6(1): 5–7.

Alexandrov, N. M. and R. M. Lewis (2000, June). An Analysis of Some Bilevel Approaches to Multidisciplinary Design Optimization" Technical Report, Institute for Computer Applications in Science and Engineering, Mail Stop 132C, NASA Langley Research Center, Hampton, VA 23681-2199.

Badiru, A. and V. Theodoracatos (1995). "Analytical and Integrative Expert System Model for Design Project Management". *Journal of Design and Manufacturing* 4: 195–213.

Balling, R. and J. Sobieszczanski-Sobieski (1995). Optimization of Coupled Systems: A Critical Overview. *AIAA-94-4330, AIAA/NASA/USAF/ISSMO 5th Symposium on Multidisciplinary Analysis and Optimization*, Panama City Beach, FL, September 1994, publication in AIAA J.

Besharati, B., L. Luo, S. Azarm and P. K. Kannan (2006). "Multi-Objective Single Product Robust Optimization: An Integrated Design and Marketing Approach". *Journal of Mechanical Design* 128(4): 884–892.

Braun, R. D. and I. M. Kroo (1997). "Development and Application of the Collaborative Optimization Architecture in a Multidisciplinary Design Environment," in *Multidisciplinary Design Optimization: State of the Art*, pp. 98–116. SIAM, Boston, MA.

Choi, H., J. H. Panchal, K. J. Allen, W. D. Rosen and F. Mistree (2003).Towards a Standardized Engineering Framework for Distributed Collaborative Product Realization. *Proceedings of Design Engineering Technical Conferences and Computers and Information in Engineering Conference*, Chicago, IL.

Fernandez, G. M., W. D. Rosen, K. J. Allen and F. Mistree (2002). A Decision Support Framework for Distributed Collaborative Design and Manufacture. *9th AIAA/ISSMO Symposium on Multidisciplinary Analysis and Optimization*, Atlanta, GA, September 4–6.

Fernandez, M. G., J. H. Panchal, J. K. Allen and F. Mistree (2005). An Interactions Protocol for Collaborative Decision Making – Concise Interactions and Effective Management of Shared Design Spaces. *ASME Design Engineering Technical Conferences and Computer and Information in Engineering Conference*, Long Beach, CA. Paper No. DETC2005-85381.

Hernandez, G. and C. C. Seepersad (2002). "Design for Maintenance: A Game Theoretic Approach". *Engineering Optimization* 34(6): 561–577.

Huang, C. C. (2004). "A Multi-Agent Approach to Collaborative Design of Modular Products". *Concurrent Engineering: Research and Applications* 12(2): 39–47.

Kumar, D., C. Hoyle, W. Chen, N. Wang, G. Gomez-Levi and F. Koppelman (2007, September). Incorporating Customer Preferences and Market Trends in Vehicle Packaging Design. *Proceedings of the DETC 2007, ASME 2007 International Design Engineering Technical Conferences and Computers and Information in Engineering Conference*, Las Vegas, NV.

Lewis, K. and F. Mistree (1998). "Collaborative, Sequential and Isolated Decisions in Design". *ASME Journal of Mechanical Design* 120(4): 643–652.

Lewis, K. and F. Mistree (1999). "FALP: Foraging Directed Adaptive Linear Programming. A Hybrid Algorithm for Discrete/Continuous Problems". *Engineering Optimization* 32(2): 191–218.

Li, W. D., S. K. Ong, J. Y. H. Fuh, Y. S. Wong, Y. Q. Lu and A. Nee (2004). "Feature-Based Design in a Distributed and Collaborative Environment". *Computer-Aided Design* 36: 775–797.

Nikolaidis, E. (2007). "Decision-Based Approach for Reliability Design". *ASME Journal of Mechanical Design* 129(5): 466–475.

Shiau, C. S. and J. J. Michalek (2007). A Game-Theoretic Approach to Finding Market Equilibria for Automotive Design Under Environmental Regulation. *Proceedings of the ASME International Design Engineering Technical Conferences*, Las Vegas, NV.

Sieger, D., A. Badiru and M. Milatovic (2000). "A Metric for Agility Measurement in Product Development". *IIE Transactions* 32: 637–645.

Simon, H. (1955). "A Behavioral Model of Rational Choice". *Quarterly Journal of Economics* 6: 99–118.

Simpson, T. W., C. C. Seepersad and F. Mistree (2001). "Balancing Commonality and Performance within the Concurrent Design of Multiple Products in a Product Family". *Concurrent Engineering Research and Applications* 9(3): 177–190.

Sobieszczanski-Sobieski, J. (1982). A Linear Decomposition Method for Large Optimization Problems Blueprint for Development, National Aeronautics and Space Administration, NASA/TM-83248 1982, Washington, DC.

Sobieszczanski-Sobieski, J., Benjamin James and Augustine Dovi (1985). "Structural Optimization by Multilevel Decomposition". *AIAA Journal* 23: 1775–1782.

Sobieszczanski-Sobieski, J. (1988, September). Optimization by Decomposition: A Step from Hierarchic to Nonhierarchic Systems, Technical Report TM 101494, NASA, Hampton, VA, Washington, DC.

Sobieszczanski-Sobieski, J. and R. T. Haftka (1997, August). "Multidisciplinary Aerospace Design Optimization: Survey of Recent Developments". *Structural Optimization* 14(1): 1–23.

Sobieszczanski-Sobieski, J. and S. Kodiyalam (1998). "BLISS/S: ANew Method for Two-Level Structural Optimization". *Structural Multidisciplinary Optimization* 21: 1–13.

Thal, A. E., A. Badiru and R. Sawhney (2007). "Distributed Project Management for New Product Development". *International Journal of Electronic Business Management* 5(2): 93–104.

Wassenaar, H. and W. Chen (2003). "An Approach to Decision-Based Design with Discrete Choice Analysis for Demand Modeling". *ASME Journal of Mechanical Design* 125(3): 480–497.

Wassenaar, H., W. Chen, J. Cheng and A. Sudjianto (2005). "Enhancing Discrete Choice Demand Modeling for Decision-Based Design". *ASME Journal of Mechanical Design* 127(4): 514–523.

Wu, T., S. Soni, M. Hu, F. Li and A. Badiru (2011). "The Application of Memetic Algorithms for Forearm Crutch Design: A Case Study". *Mathematical Problems in Engineering* 3(4): 1–15.

5 Mathematical Formulation of the Pursuit Problem for AI Gaming

5.1 INTRODUCTION TO THE PURSUIT PROBLEM

Ibidapo-Obe et al. (2001) present a mathematical modeling for the pursuit problem. Pursuit in the generalized sense is not limited to physical pursuits. It may very well be focused on database searches and signal detections. A rigorous mathematical formulation of the pursuit in E^3 space is presented. The classical solution is extended, followed by the formulation of a new method of solution defined as the intercept approach. This new technique involves the specification of an arbitrary 'escape function' having real zeros. The method has been used to model the path of a dog in pursuit of a fish in a newly developed computer game named FISHER. A simple numerical problem is solved to illustrate the ease of use of the new approach over the classical method. The methodology is useful for modern gaming techniques based on artificial intelligence (Ibidapo-Obe et al., 2001).

The pursuit problem is an ancient and practical problem that is of immense importance in the study of exterior ballistics, prey—predator escapades, etc. (Bender & Orszag, 1987). The problem description is as follows.

Let a body B, termed the target or evader move along a space curve $B(t)$ in an Euclidean Cartesian frame of reference. At time $t - T1$, another body A, termed the pursuer, takes off from the origin such that its instantaneous velocity is directed at the instantaneous position of B.

$$\text{At } t = T, B \text{ is at the point} \left(x_0, y_0, z_0 \right) \text{with } y_0 \neq 0.$$

The problem is to determine the path of the pursuer and to find the necessary conditions for interception, given that the target would have crossed into safety at $t - TF$

One motivation for this research is to aid the development of more efficient auto-tracking defense systems. For example, TIMES International Magazine (TIMES International Magazine special issue on the Gulf War, 1991) reported that the US Patriot missiles achieved a 75% success during the Gulf War. Also, the PENTAGON disclosed (The Oklahoma Daily, 1999) that a new High-Altitude Area Defense missile designed to intercept enemy missiles failed its sixth test. Deterministic

DOI: 10.1201/9781003247395-5

formulations of the problem have been given by Sokolnikoff (Sokolnikoff & Redheffer, 1966) and Green et al. (Green & Shimar, 1989). For the case where the pursuer has only partial observation of the evader's state, Yavin (1991) solved the problem using stochastic optimal control theory. A partial solution has been provided by Shenichnyi (1995) when yo is unknown and the evader is stationary.

5.2   INTRODUCTION

The pursuit problem is an ancient and practical problem that is of immense importance in the study of exterior ballistics, prey—predator escapades, etc. (Bender & Orszag, 1987). This chapter is based on Ibidapo-Obe et al. (2001). The mathematical basis for the pursuit problem provides an excellent backdrop for the design of modern AI-based games. In the example of this chapter, a rigorous formulation of the pursuit in 3-D Euclidean Space is presented. The classical solution is extended, and we then proceed to formulate a new method of solution defined as the intercept approach. This new technique involves the specification of an arbitrary 'escape function' having real zeroes. The method has been used to model the path of a dog in pursuit of a fish in a newly developed computer game—FISHER. A simple numerical problem is solved to illustrate the ease of use of the new approach over the classical method. The problem description is as follows.

Let a body B, termed the target or evader move along a space curve $B(t)$ in an Euclidean Cartesian frame of reference. At time $t = T_1$, another body A, termed the pursuer, takes off from the origin such that its instantaneous velocity is directed at the instantaneous position of B. At $t = T_1$, B is at the point (x_0, y_0, z_0) with $y_0 \neq 0$.

The problem is to determine the path of the pursuer, and to find the necessary conditions for interception, given that the target would have crossed into safety at $t = T_F$.

One motivation for this research is to aid the development of more eminent auto-tracking defense systems. For example, TIMES International Magazine (TIMES International Magazine special issue on the Gulf War, 1991) reported that the US Patriot missiles achieved a 75% success during the Gulf War. Also, the PENTAGON disclosed (The Oklahoma Daily, 1999) that a new High-Altitude Area Defense missile designed to intercept enemy missiles failed its sixth test. Deterministic formulations of the problem have been given by Sokolnikoff (Sokolnikoff & Redheffer, 1966) and Green et al. (Green & Shimar, 1989). For the case where the pursuer has only partial observation of the evader's state, Yavin (1991) solved the problem using stochastic optimal control theory. A partial solution has been provided by Shenichnyi (1995) when y_0 is unknown and the evader is stationary.

5.3   THE CLASSICAL APPROACH

The instantaneous pursuit of target implies that the desired path of the pursuer $A(t) = (x(t), y(t), z(t))$ must satisfy the following system of ordinary differential equations:

Let

$$Y = y_0 - y. \tag{5.1}$$

Then

$$\frac{dY}{dx} = \frac{Y_B - Y}{x_B - x} \tag{5.2}$$

$$\frac{dZ}{dx} = \frac{z_B - z}{x_B - x} \tag{5.3}$$

$$\frac{dY}{dx} = \frac{z_B - z}{Y_B - Y} \tag{5.4}$$

In the classical approach to the solution, it is assumed (Bender & Orszag, 1987; Sokolnikoff & Redheffer, 1966; Green & Shimar, 1989; Yavin, 1991) that

1. The pursuit motion is planar, that is, $z_B = z_0 = 0$.
   The instantaneous prey—pursuer speed ratio is a known constant: $m = v_B/u_A$.
   For this case, if in addition $Y_B = 0$ (which without loss of generality can be achieved by co-ordinate transformation whenever $B(t)$ is a linear space curve), the solution as indicated in (Sokolnikoff & Redheffer, 1966) is

$$x = \frac{y_0}{2} \left[\frac{Q}{1+m} \left(\left[\frac{Y}{y_0} \right]^{1+m} - 1 \right) - \frac{1}{Q(1-m)} \left(\left[\frac{Y}{y_0} \right]^{1-m} - 1 \right) \right], \tag{5.5}$$

where

$$Q = \frac{x_0}{y_0} + \sqrt{1 + \left(\frac{x_0}{y_0} \right)^2} \tag{5.6}$$

and as an implicit function of time

$$\frac{y_0}{2} \left[\frac{Q}{1+m} \left(\left[\frac{Y}{y_0} \right]^{1+m} - 1 \right) - \frac{1}{Q(1-m)} \left(\left[\frac{Y}{y_0} \right]^{1-m} - 1 \right) \right] = \frac{x_0 - x_B}{m} \tag{5.7}$$

Interception occurs at time $t = c$ (solving (5.5) and (5.7) with $Y = 0$)

$$\text{if } c < T_F. \tag{5.8}$$

This implies that m must be less than some fractional value (Smith & Smith, 1977).

Consider the case where $Y_B \neq 0$, identically. This may be tackled using the method developed herein tagged.

2. The extended classical approach
   From (5.2) and (5.4)

$$x_B - x = (Y_B - Y)\frac{dx}{dY} \tag{5.9}$$

$$z_B - z = (Y_B - Y)\frac{dz}{dY} \tag{5.10}$$

Let

$$\dot{x}_B = -m\sqrt{\dot{x}^2 \pm \beta^2 Y^2}, \tag{5.11}$$

$$\dot{Y}_B = -m\dot{Y}\alpha, \tag{5.12}$$

$$\dot{z}_B = -m\sqrt{\dot{z}^2 \pm \lambda^2 Y^2}, \tag{5.13}$$

where

$$\alpha^2 + \beta^2 \pm \lambda^2 = 1 \tag{5.14}$$

with

$$0 < \alpha < 1, 0 < \beta < 1, 0 < \lambda < 1.$$

Also $-\beta^2$, $-\lambda^2$ must be used if $\dot{z}_B = 0, Y_B = 0$ identically, respectively.
   Differentiating (5.9) and (5.10) with respect to y and using (5.11) and (5.13) gives

$$= -m\sqrt{\left(\frac{dx}{dY}\right)^2 \pm \beta^2} = -mx\frac{dx}{dY} + (Y_B - Y)\frac{d^2x}{dY^2}, \tag{5.15}$$

$$= -m\sqrt{\left(\frac{dz}{dY}\right)^2 \pm \lambda^2} = -mx\frac{dz}{dY} + (Y_B - Y)\frac{d^2z}{dY^2}. \tag{5.16}$$

Let

$$p_1 = \frac{dx}{dY}, p_2 = \frac{dz}{dY}, \tag{5.17}$$

From (5.12), $Y_B = m\alpha(y_0 - Y)$,and hence (5.15) and (5.16) yields

$$m\alpha p_1 - m\sqrt{p_1^2 \pm \beta^2} = \left(m\alpha y_0 - Y(1 + m\alpha)\right)\frac{dp_1}{dY}$$

$$m\alpha p_2 - m\sqrt{p_2^2 \pm \lambda^2} = \left(m\alpha y_0 - Y(1+m\alpha)\right)\frac{dp_2}{dY} \tag{5.18}$$

It is obvious that (5.1) can only be solved numerically and not in closed form. The extended classical approach is fraught with encumbrances. Besides, an assumed or projected constant prey—pursuer ratio may not accurately model a situation. For example, a dog pursuing a vehicle that is moving with a uniform velocity is bound to get tired and lose speed! An entirely new approach is now formulated.

5.4 THE INTERCEPT APPROACH

Let $q = q(t)$ represent any of x, y and z. Eqs. (5.2)–(5.4) can be combined.
Let

$$\frac{\dot{x}}{x_B - x} = \frac{\dot{Y}}{Y_B - Y} = \frac{\dot{z}}{z_B - z} = w(t). \tag{5.19}$$

Define

$$E(t) = \frac{1}{w(t)}, \tag{5.20}$$

$w(t)$ is an arbitrarily prescribed time varying function and shall be defined as the *interception function*. We also define its reciprocal $E(t)$ as the *escape function*.
The square of the speed of the pursuer is given by

$$u_A^2 = \sum \dot{q}^2. \tag{5.21}$$

Equations (5.23) and (5.21) implies

$$u_A^2 = w^2 \sum \left(q_B - q\right)^2.$$

Hence

$$r_{AB} = u_A / w. \tag{5.22}$$

Proposition 1.

The instantaneous distance between the pursuer and the prey is given by the instantaneous product of the pursuer's speed and the value of an arbitrarily pre-scribed escape function.

Theorem 1.

Interception may possibly occur only for values of $t < TF$ for which the escape function vanishes.

Proof. From (5.20) and (5.22)

$$r_{AB} = 0 \Leftrightarrow E(t) = 0$$

for u_A not necessarily zero.

Let the minimum positive root of $E(t) = 0$ be $t = c$. Then, $E(t)$ must be prescribed such that $c < T_F$ is a necessary condition for interception. This however is not sufficient because the pursuer is inevitably constrained by practical speed limit considerations.

Analysis: Now

$$\dot{q} = w(q_B - q). \tag{5.23}$$

Differentiating (5.23) with respect to t yields

$$\ddot{q} + (w - \dot{w}/w)\dot{q} = w\dot{q}_B. \tag{5.24}$$

When the $q_B s$ S are piecewise continuous, by carefully selecting w and its parameters not only can a feasible interception time be obtained, (5.23) can always be solved in closed form. A fundamental step towards this is to employ the integrating factor $R = e^{\int w dt}$ The solution is given by

$$\int_{T_1}^{t} \frac{d}{dt}(qR)\,dt = \int_{T_1}^{t} \frac{d}{dt} wRq_B dt. \tag{5.25}$$

Consider the case for which the coefficient of $\dot{q}$ vanishes in (5.24). This implies

$$w = w^* = \frac{1}{t - (T_1 + 1/w(T_1))}. \tag{5.26}$$

(Hence $w(T_1)$ is the reciprocal of the expected flight time of pursuer.) This is a special case of

$$w = \frac{-k}{t - c}, k > 0.$$

The form of w as given by (5.26) is suitable when each q_B has a Taylor series expansion in t, that is

$$q_B = \sum_{n=0}^{N} a_n t^n, \tag{5.27}$$

then Eq. (5.23) becomes

$$\dot{q} - \frac{k}{(t-c)}q = -\frac{k}{(t-c)}q_B.$$

(5.28)

By using an integrating factor $(t-c)^{-k}$, binomial expansion and the substitutions $\rho = T_1 - c$, $T = (t-c)/\rho$ if k is an integer and $k \le N$, then

$$q = q(T_1)T^k + \sum_{n=0}^{N} a_n \left[-kC_k^n c^{n-k} \rho^k T^k \ln T + \sum_{r=0,r\ne k}^{n} -kC_r^n c^{n-r} \rho^r \frac{(T^r - T^k)}{k-r} \right]$$

(5.29)

else

$$q = q(T_1)T^k + \sum_{n=0}^{N} a_n \left[\sum_{r=0}^{n} kC_r^n c^{n-r} \rho^r \frac{(T^r - T^k)}{k-r} \right],$$

(5.30)

where the C_r^n are the binomial coefficients and ln denotes natural logarithms.

When the q_B are periodic functions, each has a unique Fourier series representation of the form

$$q_B = \frac{a_0}{2} + \sum_{n=1}^{N} a_n \cos(nt) + b_n \sin(nt),$$

(5.31)

where N is the number of series terms taken. A possible choice of w is now

$$w = \frac{a\cos(at-b)}{\sin(at-b)}$$

(5.32)

$$a = \pi / (4k+c), b = ac$$

(5.33)

The solution of (5.23) becomes (assuming that a is not an integer)

$$q = \frac{1}{2\sin(at-b)} \left[\begin{array}{l} 2q(T_1)\sin(aT_1 - b) + a_0 \sin(at - b) \\ -\sin(aT_1 - b) + \Omega_1 + \Omega_2 \end{array} \right],$$

(5.34)

$$\Omega_1 = -\sum_{n=1}^{N} aa_n \left(\frac{\sin(b-t(a+n)) - \sin(b - T_1(a+n))}{a+n} \right)$$

$$+ \sum_{n=1}^{N} ab_n \left(\frac{\cos(t(a-n)-b) - \cos(T_1(a-n)-b)}{a-n} \right),$$

$$\Omega_2 = -\sum_{n=1}^{N} aa_n \left(\frac{\sin\left(t(a-n)-b\right)-\sin\left(T_1(a-n)-b\right)}{a-n} \right)$$
$$- \sum_{n=1}^{N} ab_n \left(\frac{\cos\left(b-t(a+n)\right)-\cos\left(b-T_1(a+n)\right)}{a+n} \right),$$

$$(5.35)$$

with $q(c) = qB(c)$.

Generalized solutions: For the general form of the $q_B S$ (e.g., when its terms contain powers of t and transcendental functions), we may use any w of the form

$$w = \frac{f(t,c,k)}{f(t,c,k)}.$$

$$(5.36)$$

For the particular choice

$$w = \frac{2t-(c-k)}{(t-c)(t+k)}$$

$$(5.37)$$

$$q = \frac{\left(2t-(c-k)\right)\int_{T_1}^{t} q_B dt - 2\int_{T_1}^{t}\int_{T_1}^{t} q_B dt\, dt + q(T_1)(T_1-c)(T_1+k)}{(t-c)(t+k)}$$

$$(5.38)$$

Also, (5.35) holds.

In general, the interception function w shall be an explicit function of time. It has parameters k and c. We may prescribe c and use the actual take-off speed of pursuer at $t = T_1$ in (5.22) to determine k. Alternatively, it may be expedient to prescribe k (as in simulations). This value will then be used to compute the necessary take-off speed. A more realistic approach is to satisfy speed constraints; thus, let the possible maximum value of the initial and final speeds of the pursuer be u_{A1}, u_{A2}, respectively. The simultaneous inequalities below must be solved for c and k, such that $T_1 < c < T_F$ for an admissible solution.

$$w(T_1)r_{AB}(T_1) \le u_{A1},$$

$$(5.39a)$$

$$\sum \dot{q}^2(c) \le u_{A2}^2.$$

$$(5.39b)$$

5.5   EXAMPLE

A cat commenced the pursuit of a rat dashing with a uniform speed along a straight path in a field at the instant when the rat is closest to it. Given that the initial distance between them is unity, and that the cat—rat speed ratio is 1.618, determine a path for the cat and probable time of interception.

Solution: $xo = 0, z - 0, yo = 1, T_1 = 0$, Without loss of generality, let

$$V_B \Rightarrow 1, x_B = t, y_B = 1, z_B = 0.$$

For the *classical approach*: Assuming a constant rat-cat speed ratio $m = 1/1.618 = 0.618$ gives

$$z = 0, Y = 1 - y,$$

$$x = 0.5 \left[\frac{Y^{1.618} - 1}{1.618} - \frac{Y^{0.382} - 1}{0.382} \right],$$

$$3.226t - 0.618 \left(Y^{1.618} - 1 \right) + 2.618 \left(Y^{070382} - 1 \right) = 0.$$

Interception occurs when $Y = 0$, that is at $x = l, t = c = 1$.

For the *intercept approach*: We use the solution for polynomial $q_B S$, $w(t)$ given by (5.26)

Assuming that $k \neq 1$ implies

$$T = 1 - \frac{t}{c},$$

$$Y = 1 - T^k, x = c \left(1 - T^k \right) - \frac{kc \left(T - T^k \right)}{k - 1}, z = 0,$$

$$u = \sqrt{\left(\frac{kT^{k-1}}{c} \right)^2 + \left(\frac{k(1 - T^{k-1})}{k - 1} \right)^2}$$

Using $u_{A1} = u_{A2}$.

Case 1: Let the minimum Cat speed be 1.618, (5.39) becomes

$$k = c + 1, \sqrt{\left(\frac{k(0.5)^{k-1}}{c} \right)^2 + \left(\frac{k(1 - (0.5)^{k-1})}{k - 1} \right)^2} = 1.618.$$

The approximate solution of which are (obtained by using the Solver tool of Microsoft Excel 7.0 software application) $k = 1.793, c = 0.793$, $u_{max} = 2.261$.

Case 2: Case 2: Let the maximum Cat speed be 1.618, (5.39) becomes

$$k = c + 1, \frac{k}{k - 1} = 1.618.$$

The approximate roots of which are: $k = 2.618, c = 1.618, u_{min} = 1.212$.

Case 3: To obtain an interception time of $c = 1.000$
The values are found to $k = 2$, $c = 1$, $u_{min} = 1.414$, $u_{max} = 2$ (Tables 5.1 and 5.2; Figures 5.1 and 5.2).

The instantaneous distance and speed values were plotted against percent of interception time as shown. The graphs show that the new approach initially gives comparatively shorter distances to the classical approach, but this trend is reversed after about 45% of the flight time. The greater the interception time, the lower the maximum speed attained. It is obvious that Case 3 seems ideal, that is, a situation whereby the constant Cat speed of the classical approach is close to the average of the varying speed of the new approach. Figures 5.1 and 5.2 illustrate the search paths in the pursuit problem.

TABLE 5.1
Instantaneous Distance for Various % Interception Time

% C	r (Classical)	r (Case I)	r (Case 2)	r (Case 3)
0	1	1	1	1
0.1	0.843	0.831	0.772	0.815
0.2	0.699	0.683	0.608	0.66
0.3	0.567	0.555	0.499	0.533
0.4	0.451	0.447	0.428	0.433
0.5	0.351	0.358	0.374	0.354
0.6	0.265	0.283	0.322	0.288
0.7	0.191	0.218	0.261	0.228
0.8	0.125	0.155	0.186	0.165
0.9	0.062	0.085	0.098	0.091
1	0	0	0	0

TABLE 5.2
Instantaneous Speed for Various % Interception Time

% C	u (Classical)	u (Case I)	u (Case 2)	u (Case 3)
0	1.618	2.261	1.618	2
0.1	1.618	1.932	1.579	1.811
0.2	1.618	1.748	1.503	1.649
0.3	1.618	1.641	1.406	1.523
0.4	1.618	1.6	1.303	1.442
0.5	1.618	1.618	1.121	1.414
0.6	1.618	1.6	1.303	1.442
0.7	1.618	1.641	1.406	1.523
0.8	1.618	1.748	1.503	1.649
0.9	1.618	1.932	1.579	1.811
1	1.618	2.261	1.618	2

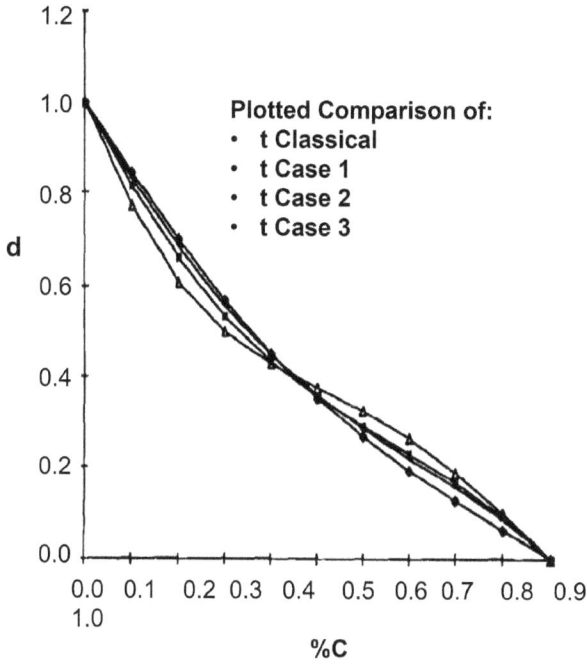

FIGURE 5.1   Rat -< Cat instantaneous distance d against % interception time.

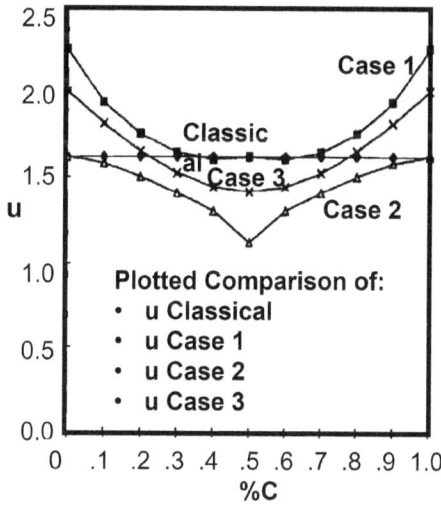

FIGURE 5.2   Cat's speed against % interception time.

5.6   CONCLUSION

Although the extended classical approach may be further investigated, the intercept approach as developed herein gives generalized solutions to the pursuit in three-dimensional Euclidean space. In fact, extensions to n-dimensional space follow immediately. The solutions obtained thereby are explicit functions of time and can be manipulated to give feasible interception times. Further work is being done on the choice of interception functions and the determination of the optimal parameters. The results of this research are not only applicable to PC games and natural processes but also to high-technological warfare such as in the programming of the path of anti-missile missiles. Furthermore, the concepts of this work and the tracking approach of Tsoularis (Tsoularis & Kambhapati, 1998) may lead to a solution of the robot path-planning and obstacle avoidance problem, formulated as a constrained pursuit problem.

REFERENCES

Bender, C. M. and S. A. Orszag (1987). *Advanced Mathematical Methods for Scientist and Engineers*, p. 33. McGraw Hill, New York.

Green, J. and M. Guelman Shimar (1989). The Effect of Pursuer's Final Velocity Constraint in a Planar Pursuit Evasion Game of a Kind. *The Collection of Papers of the 30th Israel. A Conference on Aviation and Astronautics*, Israel, pp. 226–237.

Ibidapo-Obe, O., O. S. Asaolu and A. B. Badiru (2001). "Generalized Solutions of the Pursuit Problem in Three-Dimensional Euclidean Space". *Applied Mathematics and Computation* 119(1): 35–45.

Shenichnyi, P. (1995). "Special Pursuit-And-Evasion Problem with Incomplete Information". *Cybernetics and Systems Analysis* 31(2): 246.

Smith, A. and D. Smith (1977). "Mechanics". *English Libr. Book Soc.* 35(10): 125–126.

Sokolnikoff, I. S. and R. M. Redheffer (1966). *Mathematics of Physics and Modern Engineering*, pp. 110–111. McGraw Hill, New York.

The Oklahoma Daily. (1999, March 29 Monday). Missile defense system fails direct-hit attempt.

TIMES International Magazine special issue on the Gulf War. (New Yor, Dec 1991).

Tsoularis, A. and C. Kambhapati (1998). "On-Line Planning for Collision Avoidance on the Nominal Path". *Journal of Intelligent and Robotic Systems* 21: 327–371.

Yavin, Y. (1991). "Open-Loop Evasion Strategies in a Pursuit-Evasion Problem in A Reduced State Space". *Computers & Mathematics with Applications* 22(1): 3–13.

6 AI Framework for the Financial Sector

6.1 INTRODUCTION

This chapter is based on Asaolu (2016), as a contribution to the increasing roles of artificial intelligence (AI) in financial transactions. Given the expanding mobile phone subscribers of national populations and the potentials of deploying financial transactions on this platform, mobile payment services and mobile-banking have surfaced in several continents but have met with slow success relative to the initial expectations. We believe that there is a great need for collaboration, consensus, standard protocols and an application service to facilitate the delivery of mobile payments/banking. It is equally expedient if not apposite to integrate this with the myriad of other existing payment services in a complimentary fashion. Worldwide, the present cacophonic situation in which each phone vendor, telecom provider, payment processor and/or bank introduces its own proprietary and closed mobile payment or banking solution militates against widespread adoption and successful implementation of the technology. This work presents a universal framework for multi-modal authenticated and secured SMS-based mobile application for financial transactions. It also specifies how this could be easily integrated with existing online payment systems to constitute an application service for mobile phone subscribers irrespective of whether such have bank accounts or not. We show that by harnessing existing structures optimally, it is feasible to empower mobile phone users with the capacity to seamlessly send or receive payments anywhere, anytime, at their own convenience and to convert mobile payments into raw cash.

The use of the mobile phone is ubiquitous and is greatly transforming the way we live and do business in both the developed and developing nations. Given the expanding mobile phone subscribers of national populations and the potentials of deploying financial transactions on this platform, mobile payment services and m-banking have surfaced in several continents but have met with slow success relative to the initial expectations (Is the humble handset the handmaiden of the future of banking in Africa?, 2009). Combined market for all types of mobile payments was expected to reach more than $600 B globally by 2013 (Juniper Research Forecasts Total Mobile Payments to Grow Nearly Ten Fold by 2013, 2012). Nigeria has a population of about 170 million people with 20 million bank accounts and 90 million mobile phone lines. Although several mobile payments licenses have been awarded by the Nigerian Central Bank (Regulating Mobile Money in Nigeria, 2013), none has enjoyed wide acceptance. The Flash Me Cash service pioneered by a defunct bank was moribund, while the other succeeding models seem quite inadequate. We believe that there is a great need for collaboration, consensus, standard protocols and an application service to facilitate the delivery of mobile payments/banking. It is equally expedient if not

DOI: 10.1201/9781003247395-6

opposite to integrate this with the myriad of other existing payment services in a complimentary fashion. Worldwide, the present cacophonic situation in which each phone vendor, telecom provider, payment processor and/or bank introduces its own proprietary and closed mobile payment solution militates against widespread adoption and successful implementation of the technology (Why Mobile e-Commerce Is Struggling, 2013). Consider the SMS as a standard way of sending *text* messages; the public expects this service to be supported on their mobile phones and by their telecom provider and takes it for granted that a message can be sent from *a familiar application interface* to another subscriber on any other network. It is not too complicated to compose, receive or read such messages especially in this age of smart phones with qwerty like keyboards (Balakrishnan et al., 2011)! In like manner, divers payment systems can be integrated and supported on our proposed framework and integrated into an application service for mobile phone subscribers irrespective of whether such have bank accounts or not. That mobile phone manufacturers have agreed with the European Commission to standardize and introduce a universal charger that fits all new phones is a welcomed development that will save consumers a lot of hassles (Universal phone charger deal done, 2013). Similarly, it is time to empower mobile phone users with the capacity to seamlessly send or receive payments anywhere, anytime, at their own convenience and to convert mobile payments into raw cash.

6.2 METHODOLOGY

There are advantages and disadvantages of each of the four primary models for mobile payments (Mobile Payments, 2009): Premium SMS-based transactional payments, Direct Mobile Billing, Mobile web payments (WAP) and Contactless NFC (Near Field Communication). The proposed scheme is termed the CashMobile Service (CMS). Its engine will be a web application server interconnected with a secure mobile phone application. The framework could be adopted by collaborating industry stakeholders. It is envisaged to leverage on existing platforms/providers and harness their competencies and deliverables. The proposed solution is an innovation using appropriate protocols that will be of interest and benefit to the Nigerian populace and could be extended to become a global service.

The framework involves:

- The adoption of secure SMS by telecoms service providers (Toorani & Shirazi, 2008; Secure SMS, 2013).
- The creation of a special purpose phone application for mobile financial transactions.
- The integration of various payment services in the background by means of server inter-connectivity.
  1. There is need to provide secure text messaging (SSMS) <u>directly</u> as an optional commercial service by telecoms providers. It should be specifically deployed to support financial transactions. This will obviate the complex and confusing SMS formats currently being used for vulnerable and unsecured mobile payments in many parts of the world. We could facilitate the phone manufacturing companies and telecoms operators association to agree to minimal standards.

2. A phone application with a consistent user-friendly interface will be created to handle secure mobile financial transactions and will have many features including but not limited to:

- Ability to Send or Receive Payments (with daily limits)
- Selection of a number directly from the Phone Contacts list
- Specification of Amount to be sent in local currency (based on country telephone calling code)
- Inclusion of text message or Memo for the recipient
- Transaction validation by phone owner by inclusion of a 4-letter PIN or alphanumeric Password (which he may reset as desired)
- Ability to fund the CashMobile account (to be administered by a special company) via the subscriber's GSM credit or airtime balance with telecom provider, received payments, a Credit Card, a Debit Card or their combination
- Ability to check the account Balance among other Options, accessible as indicated in Figure 6.1.
- Ability to set currency for transaction and interfacing with a globally acceptable real-time currency converter.
- Ability to withdraw into the airtime balance, a Debit Card, or by sending fund (at a token fee) to a local collection center, for example, Western Union

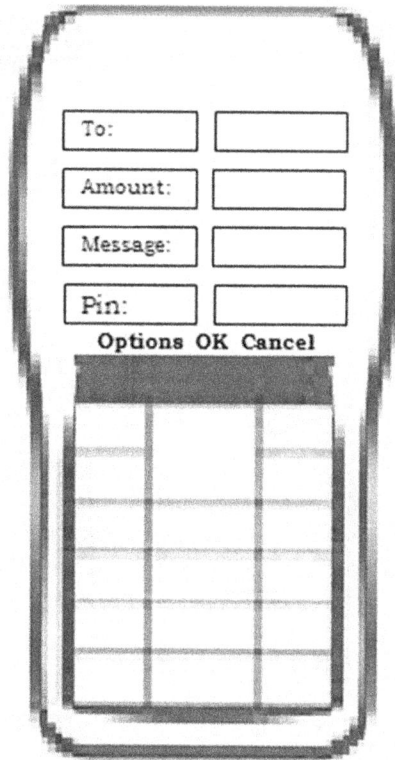

FIGURE 6.1 Proposed phone application interface.

to be cashed over the counter with one's phone as alternative ID (since Transaction details, e.g., Control Number, Question & Answer, Time-stamp, etc., are on it) or an ATM machine.

A subscriber will only need to walk into his telecoms provider's customer care office (or bank) to affirm personal details (Name, Address, Sex, Photo, Date of Birth, Nationality, Next of Kin & optionally an E-mail) in order to activate the service. This fulfils verification/national financial regulation compliance requirements of Know Your Customer (KYC). The recent adoption of the Bank Verification Number (BVN) in Nigeria in 2014 is a good development. This number serves as a unique ID within the financial system and is biometric based; it also links all bank accounts of an individual for easy tracking and cross referencing.

3. This product need not complete with existing web-based services or struggle to 're-invent the wheel' rather it could partner with such and utilize them for internet payments! For example, PayPal though popular worldwide is non-existent in many developing countries including Nigeria. This could be selected by CashMobile subscribers for online purchase payments simply by entering a masking and virtual CashMobile email and their CashMobile account password (e.g., fulltelephonenumber@cashmobile.com, password) at e-commerce site checkout pages. The PayPal server could make API calls to CashMobile server to auto-complete the transaction. This is because CashMobile would be the PayPal Client and the initiating user account would appear as a sub or delegated account with limits or assigned roles.

6.3   DISCUSSION

For implementation in Nigeria, for example, the major players would be

- National Communications Commission,
- The Central Bank of Nigeria (CBN)
- Mobile Phone Manufacturers
- Mobile Telecoms Service Providers, for example MTN, Zain, Globacom, Etisalat, Visafone
- Commercial Banks
- Card Payment Processors, for example VISA, MasterCard, Interswitch, ValuCard, E-Transact.
- Internet Payment Services, for example PayPal, ValuGold, etc.
- Traditional Money Transfer Operators, for example Western Union Money Transfer, MoneyGram, NIPOST

The following action plan would need to be carried out:

i Develop a business plan around resulting framework for the CashMobile enterprise, seek investors, incorporate firm and obtain operational license from the CBN.

ii Refine model via industry consultations with potential partners/stakeholders and promotion of standards/protocols, adoption of secure SMS on national scale.

iii Develop generic phone application for financial services with telecom providers buy-in and integration of Card Payment Processors, Online and Traditional Cash Transfer Services

iv Systems Deployment and Testing

v   Project Rollout

6.4 CONCLUSION

We have conceived and presented a unified framework for initiating, processing and managing financial transactions via the mobile phone and allied devices. The proposed system requires a synergy of operators and regulators in the telecoms and banking industries as well as the general ICT community. It could facilitate seamless adoption and usage compared to the present schemes where each operator introduces a propriety concept that suffers from non-interoperability and ease of usage. Although exploratory, the project if practically pursued could yield the needed solution for a global standard and workable approach to mitigate the issues identified.

REFERENCES

Asaolu, O. S. (2016). "A Universal Framework for Secured Financial Transactions on Mobile Phones", in *Building Entrepreneurial Universities in a Developing Economy: Issues, Challenges and Prospects* (J. Eyisi and A. Ichu-Ituma, eds.), FUNAI Press, Ebonyi State, Nigeria.

Balakrishnan, V., S. F. Guan and R. G. Raj (2011). "A One-Mode-For-All Predictor for Text Messaging". *Maejo International Journal of Science and Technology* 5(02): 266–278.

Editorial Staff. (2009). "Is the Humble Handset the Handmaiden of the Future of Banking in Africa?" (2009, March). *Africa Investor Magazine*, pp. 31–32.

Juniper Research Forecasts Total Mobile Payments to Grow Nearly Ten Fold by 2013. Retrieved September 3, 2012 from http://www.juniperresearch.com/shop/viewpressrelease.php?pr=106

Mobile Payments. Retrieved September 3, 2009 from http://en.wikipedia.org/wiki/Mobile_payment

Regulating Mobile Money in Nigeria. Retrieved March 3, 2013 from http://community.zdnet.co.uk/blog/0,1000000567,10012315o-2000587089b,00.htm

Secure SMS. Retrieved October 6, 2013 from http://www.celltrust.com/products/sms-gateway/celltrust-smsgatewaysuite-securesms.html

Toorani, M. and Beheshti Shirazi (2008). SSMS – A Secure SMS Messaging Protocol for the m-Payment Systems. *A IEEE Symposium on Computers and Communications, 2008. ISCC 2008.*

Universal phone charger deal done. *BBC NEWS*. Retrieved January 13, 2013 from http://news.bbc.co.uk/newsbeat/hi/technology/newsid_8124000/8124293.stm

Why Mobile e-Commerce Is Struggling. Retrieved October 6, 2013 from http://www.readwriteweb.com/archives/mobile_e-commerce_is_struggling.php

7 AI Neuro-Fuzzy Model for Healthcare Prediction

7.1 INTRODUCTION

This chapter is based on the work presented by Orieke et al. (2019) for an application in the healthcare industry. Diabetes mellitus (DM) is a metabolic disorder that affects the ability of the human body to properly utilize and regulate glucose. It is pervasive worldwide yet tenuous and costly to manage. DM is also difficult to model because it is nonlinear, dynamic and laden with mostly patient-specific uncertainties. A neuro-fuzzy model for the prediction of blood glucose level (BGL) in Type 1 diabetic patients using coupled insulin and meal effects is developed. This study establishes that the necessary and sufficient conditions to predict BGL in a Type 1 diabetes mellitus patient are knowledge of the patient's insulin effects and meal effects under diverse metabolic scenarios and the transparent coupling of the insulin and meal effects. The neuro-fuzzy models were trained with data collected from a single Type 1 diabetic patient covering a period of two months. Clarke's Error Grid Analysis (CEGA) of the model shows that 87.5% of the predictions fall into region A, while the remaining 12.5% of the predictions fall into region B within a four (4) hour prediction window. The model reveals significant variation in insulin and glucose responses as the body mass index (BMI) of the patient changes.

Diabetes is described as a chronic disease that occurs either when the pancreas does not produce enough insulin or when the body cannot effectively use the insulin it produces (World Health Organization, 2013). DM is estimated to affect about 415 million people worldwide (that is 1 in 11 adults) (International Diabetes Federation, 2016). In Africa, Nigeria is the worst hit, with over 4 million people with diabetes (International Diabetes Federation, 2013). This alarming number is on the fast increase daily and in 2012, an estimated 1.5 million deaths were directly caused by diabetes (World Health Organization, 2014). A projection on the global mortality and burden of DM reveals that 80% of the deaths associated with the disease occur in the middle-income and low-income countries (Mathers and Loncar, 2006). WHO projects that DM will be the seventh leading cause of death in 2030 (2011).

Diabetes mellitus is of different types; however, the most common are Type 1, Type 2 and Gestational DM. In Type 1 DM (T1DM), the pancreas undergoes an autoimmune attack by the body itself and is rendered incapable of making insulin (Melissa, 2012). Type 2 DM is characterized by insulin resistance, which may be combined with relatively reduced insulin secretion (Shoback, 2011). Gestational DM is detected during pregnancy and occurs in about 2–5% of all pregnancies. Poorly controlled BGL inflicts damaging effects to major organs of the human body. Direct medical implications of overly raised BGL include diabetic ketoacidosis and nonketotic hyperosmolar coma. Long-term damages include retinopathy, neuropathy,

DOI: 10.1201/9781003247395-7

nephropathy and heart diseases, which are associated with diseased small and large blood vessels as the case may be.

Presently, there is no cure for diabetes mellitus. This metabolic disorder can only be managed/controlled using oral medication or insulin, depending on the type of diabetes. Proper management of BGL in T1DM cannot be achieved without regular measurement. The invasive method of measuring blood glucose is the clinically acceptable means of measuring blood glucose. The finger strip meter and the continuous glucose meter are the most common and they require access to the blood or interstitial body fluid, respectively. BGL in T1DM is affected by exogenous insulin, meal, exercise, stress, etc. The amount of insulin required in T1DM to correct elevated BGL depends on the type of insulin used and on the sensitivity of the individual to the insulin. The amount of carbohydrate required for one unit of insulin is given by the Insulin to Carbohydrate Ratio (ICR). The ICR of a patient may change due to the type of food consumed, body mass index (BMI), hormonal balance and fitness level. Exercise is important in improving the lives of T1DM patients, as it increases the permeability of glucose in peripheral tissues (Charles, 2002; American Council of Exercise, 2013). Although regular activity is beneficial for all patients, vigorous exercise can cause major disturbances in blood glucose. The glycemic response depends on the type, intensity and duration of the activity, as well as the circulating insulin and glucose counter-regulatory hormone concentrations (Michael and Bruce, 2006).

T1DM patient does not produce insulin; hence, normal BGL cannot be maintained. Most patients have a vague idea of how they expect their BGL to vary after a meal, exercise or insulin injection. Although this is not to be relied upon as patients have slumped into episodes of hypoglycemia or hyperglycemia, it paints a bright picture of a possibility to predict BGL in DM. Hence, the need to use a neuro-fuzzy network (NFN) to capture the trend in BGL variation and patient-specific responses relates to specific activities (e.g., exercise, insulin, and meal) that affect BGL.

The idea of predicting BGL based on past blood glucose values was suggested by Bremer and Gough (1999). They identified the statistical dependence of glycemic data in both diabetic and non-diabetic individuals. Sparacino et al. (2007) predicted BGL using first-order AR model with time-varying parameters, identified by recursive least squares (RLS) with a constant forgetting factor. They achieved 20–25 minutes prediction of hypoglycemic threshold ahead in time. Shanthi et al. (2010) BGL prediction model combines the use of AR as a linear aggregation of previous glucose values and a moving average model that considers previous variations in blood glucose. They obtained maximum RMSE of 0.9, 2.7 and 4.2 mg/dl within the prediction horizon of 10, 20 and 30 minutes, respectively. Stahl and Johansson (2009) presented a log-normalized linear model based on subspace-based identification and the GTFM-Wiener model for BGL prediction. However, they did not meet the 9 mg/dl accuracy target within the two-hour prediction window. Ahmed and Mahmud (2013) developed a PID controller as an artificial pancreas whose outputs are based on the history and rate of change of the error signal. Their results revealed that the response of the PID controller is not acceptable since the BGL has an oscillation, causing a drop in BG level below the basal level.

Zarita et al. (2009) combined principal component analysis (PCA) and Wavelet neural network (WNN) with Gaussian wavelet in predicting BGL in a single diabetic patient. The Gaussian WNN predictive model resulted in a RMSE of 0.0450, 0.0348,

0.0330, 0.0170 mmol/dl for the morning, afternoon, evening and night predictive period, respectively. Scott et al. (2010) proposed BGL predictive model using feed forward three-layer neural network with a predictive horizon of 75 minutes, and back propagation training algorithm. The overall error of the prediction model using the patient-specific model and the general neural network model are 7.9% and 15.9%, respectively. Clarke's Error Grid Analysis (CEGA) show that 95.1% and 69.8% of the prediction falls in region A for the patient-specific and general neural network model, respectively. Neuro-fuzzy networks have seen more application in the prediction of the onset of diabetes and in fuzzy logic controllers for artificial pancreas. Deutsch et al. (1990) suggested that due to the high variability and uncertainty of the observed blood glucose data, a qualitative means of pattern recognition will be more suitable for analysis and pattern recognition. Ghevondian et al. (2001) developed a Fuzzy Neural Network Estimator (FNNE) to predict the onset of hypoglycemia using heart rate and skin impedance as system inputs. The model revealed that hypoglycemia leads to increased heartbeat by approximately 21 bpm and reduced skin impedance by approximately 111 ohms for T1DM patients. Moshe et al. (2013) investigated logic-based artificial pancreas revealing superior performance in the reduction of nocturnal hypoglycemia and tighter blood glucose control. Ahmed and Mahmud (2013) developed a fuzzy logic algorithm for implementing an artificial pancreas. In their result, the response of this controller in steady state kept blood glucose concentration almost at the basal level, although it had a little overshoot before the steady state.

Juan and Chandima (2016) proposed a personalized blood glucose predictor based on time-series model of historical glucose measurements, pooled panel data regression and pre-clustered personalized regression. Their best performance was a 43-cluster personalized regression that achieved a Root Mean Square Error (RMSE) of 27.458 mg/dl and a correlation coefficient (R^2) of 0.8883. Hidalgo et al. (2017) proposed the use of a variant of grammatical evolution model and a tree-based genetic programming model that uses a three-compartment model for carbohydrate and insulin dynamics. The model achieved 90% of the prediction within regions A and B of CEGA, with 5–10% falling into regions D (serious error) and 0.5% in region E (very serious error). Kyriaki et al. (2018) explored the combination of autoregressive models with exogenous inputs and pharmacokinetic compartment models to predict blood glucose. Clarke's Error Grid Analysis of the model showed that 53.84% of the predictions were in region A in the 4-hour prediction window.

7.2 COUPLED INSULIN AND MEAL EFFECT NEURO-FUZZY NETWORK MODEL

7.2.1 MODEL ASSUMPTIONS

The following assumptions are made in formulating the model

1) There is no production of insulin by the T1DM patient.
2) The stress level, activity level, hormonal balance, fat mass index and lean body mass of the patient are fixed and do not affect the NFN training data collected.
3) The effect of the change in age during the period of investigation is insignificant.

7.2.2 Model Formulation

In other to create patients' awareness to their unique body metabolism, a neuro-fuzzy model is formulated. This is based on the proposition that the necessary and sufficient conditions to predicting BGL in a T1DM patient are knowledge of the patient's: 1) insulin effects, 2) meal effects and 3) transparent coupling of the insulin and meal effects. Insulin effect refers to the change in BGL for specific metabolic indices because of injected insulin defined by neuro-fuzzy weights (feature extracts). In the same vein, meal effect refers to the change in BGL for specific metabolic indices because of ingested meal defined by neuro-fuzzy weights (feature extracts). Combinations of physiological states that affect BGL variations in T1DM patients define metabolic scenarios. Metabolic index defines specific metabolic values of the physiological states that make up the metabolic scenario, for example BMI of 22.12.

The control variables for the model are:

1) State Variable: Measured BGL $x(k)$ in mg/dl
2) Input Variables: Insulin u_1, Meal u_2, in units (IU) and serving spoon (SS), respectively.
3) Sampling Time: The sampling time (Δt) is the time between successive measurements of the BGL, which is one (1) hour.
4) Output: Predicted BGL $x(k + 1)$ mg/dl
5) k is the discrete time index
6) Patient's specifics (Δp): Injection site, Body/Fat Mass Indices, Lean Body Mass, Other Illness, etc.

7.2.3 Blood Glucose Level (BGL), $x(k)$

The wide range of variation of BGL in a diabetic patient poses a major bottle neck to NFN training. Hence, it is necessary to normalize the BGL. For this research, the range of BGL is between 40 and 600 mg/dl.

$$\bar{x}(k) = \frac{x(k)}{\max(x(k))} \tag{7.1}$$

7.2.4 Insulin Injection, u_1

Insulin effect is modeled as a function of insulin injection u_1, current BGL, sampling time and patient's specifics. This captures the uniqueness of the patient's reaction to injected insulin. The injected insulin u_1 is a scalar function g of the insulin type T and quantity Q injected.

$$u_1 = g(Q, T) \tag{7.2}$$

Given that u_1 is injected, the insulin effect on the BGL $I(k)_{u_1}$ given by the function f_1 is

$$I(k+1)_{u_1} \propto f_1\left(u_1, x(k), \Delta t, \Delta p\right)$$

$$I(k+1)_{u_1} = K_1 f_1\left(u_1, x(k), \Delta t, \Delta p\right) \tag{7.3}$$

$$\bar{I}(k+1)_{u_1} = \frac{I(k+1)_{u_1}}{\max\left(I(k+1)_{u_1}\right)} \tag{7.4}$$

7.2.5 MEAL INTAKE, U_2

Meal effect is modeled as a function of meal intake, current BGL, sampling time and patient's specifics. The ingested meal u_2 is a scalar function m of the meal type M and quantity Q ingested

$$u_2 = m(Q, M) \tag{7.5}$$

Given that u_2 is consumed, the meal effect on BGL $I(k)_{u_2}$ given by the function f_2 is

$$I(k+1)_{u_2} \propto f_2\left(u_2, x(k), \Delta t, \Delta p\right)$$

$$I(k+1)_{u_2} = K_2 f_2\left(u_2, x(k), \Delta t, \Delta p\right) \tag{7.6}$$

$$\bar{I}(k+1)_{u_2} = \frac{I(k+1)_{u_2}}{\max\left(I(k+1)_{u_2}\right)} \tag{7.7}$$

7.3 FUZZIFICATION OF STATE AND INPUT VARIABLES

BGL variation is fuzzified using the Trapezoidal Membership Function (MF), while insulin, insulin effect, meal and meal effect are fuzzified using the Triangular MF as given in Equations (7.8) and (7.9), respectively.

$$\mu_F\left(\bar{u}; \alpha, \beta, \gamma, z\right) = \begin{cases} 0, & \bar{u} \leq \alpha \\ \dfrac{\bar{u} - \alpha}{\beta - \alpha}, & \alpha < \bar{u} \leq \beta \\ 1, & \beta < \bar{u} \leq \gamma \\ \dfrac{\gamma - \bar{u}}{\beta - \bar{u}}, & \gamma < \bar{u} \leq z \\ 0, & \bar{u} > z \end{cases} \tag{7.8}$$

Where $\bar{u}$ is the in situ variable of interest, α, β, γ, z are projections of the vertices of a trapezium to its base (representing the in situ values of $\bar{u}$), from left to right.

$$\mu_F\left(\bar{u};\alpha,\beta,\gamma\right) = \begin{cases} 0, & \bar{u} \leq \alpha \\ \dfrac{\bar{u}-\alpha}{\beta-\alpha}, & \alpha < \bar{u} \leq \beta \\ \dfrac{\gamma-\bar{u}}{\beta-\bar{u}}, & \beta < \bar{u} \leq \gamma \\ 0, & \bar{u} > \gamma \end{cases} \tag{7.9}$$

Where $\bar{u}$ is the in situ variable of interest, α, β, γ are projections of the vertices of a triangle to its base (representing the in situ values of $\bar{u}$), from left to right.

The MFs for the state and input variables are given in Table 7.1

$$F = \left\{\bar{x} \mid \mu_F\left(\bar{x}\right)\right\} \text{ for } \bar{x} \in X \tag{7.10}$$

Where X is the universe of discourse and $\mu_F\left(\bar{x}\right)$ is the degree of membership of object $\bar{x}$ in the fuzzy set F and it is a real number that lies between [0, 1]. Each of the control variables represents a universe of discourse. The control variables are each fuzzified to have a three-member fuzzy set (set 1, set 2 and set 3). Hence, the fuzzification process is generically represented in Equation (7.11). The span of each fuzzy set is dependent on the nature of the universe of discourse and its members.

$$F \in \left\{\text{set1}, \text{set2}, \text{set3}\right\} \tag{7.11}$$

TABLE 7.1
In situ State and Input Variables Fuzzification Ranges

State Variable	Min	Max	Hypoglycemic	Normal	Hyperglycemic	
BGL (mg/dl)	40[a]	600[b]	40, 40, 66, 72	66, 72, 120, 138	120, 180, 600, 600	
Input Variable	**Min**	**Max**	**Type**	**Low**	**Medium**	**High**
Insulin (u_1) (IU)	0	40	Biphasic Human Insulin	0, 0, 10	0, 10, 20	10, 40, 40
Insulin Effect $(I(k)_{u_1})$ $k = 1, 2, 3, 4$	0	-120	Biphasic Human Insulin	0, 0, -60	0, -60, -120	-60, -120, -120
Meal (u_2)	**Min**	**Max**	**Type**	**Small**	**Medium**	**Large**
Meal (u_2) (SS)	0	5	Rice	0, 1, 3	1, 3, 5	3, 5, 5
Meal Effect $(I(k)_{u_2})$ $k = 1, 2, 3, 4$	0	180	Rice	0, 0, 90	0, 90, 180	90, 180, 180

N/B: unit of meal size = serving spoon (SS); Units of Insulin and Meal Effect = mg/dl
[a] Hypoglycemic BGL set to avoid prolonged stay of the patient in the hypoglycemic range
[b] Maximum Hyperglycemic BGL that can be measured by the One Touch glucose meter used for the experiment

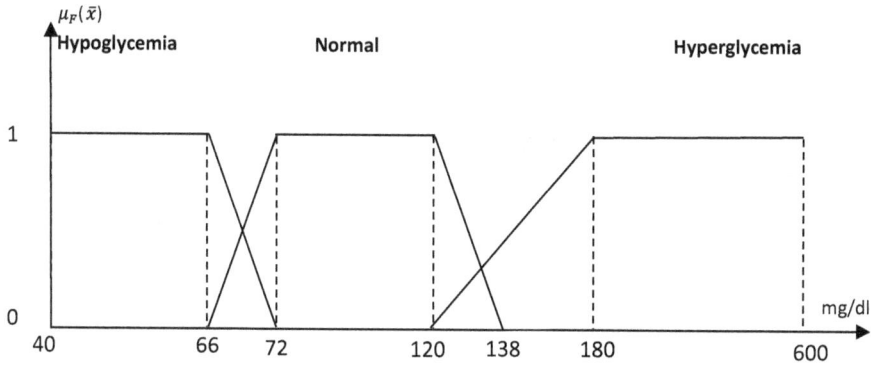

FIGURE 7.1 Fuzzification of in situ BGL.

The fuzzification of the BGL is given in Figure 7.1.

7.3.1 FORMULATION OF THE T-S MODEL RULE BASES

The Takagi-Sugeno Neuro-Fuzzy Network (T-S NFN) for the investigation of insulin, meal and insulin/meal effects is modeled as a discrete piece-wise linear system with discrete time variable k such that the rule base is modeled as a function of proportionality.

The insulin effect consequent part of the T-S model

$$\overline{I}\left(k+1\right)_{u_1} \propto f_1\left(u_1, \overline{x}\left(k\right), \Delta t, \Delta p\right) \tag{7.12}$$

The corresponding consequent part of the meal effect rule base is

$$\overline{I}\left(k+1\right)_{u_2} \propto f_2\left(u_2, \overline{x}\left(k\right), \Delta t, \Delta p\right) \tag{7.13}$$

The meal/insulin effect, which represents the coupling of the meal and insulin effects is

$$\overline{x}\left(k+1\right) \propto f\left(\overline{x}\left(k\right), \overline{I}\left(k+1\right)_{u_1}, \overline{I}\left(k+1\right)_{u_2}\right) \tag{7.14}$$

where $k = [1, 2, \ldots, N]$ and N is the number of samples taken.

A separate rule base is generated for each investigation; hence, there are three (3) rule bases, **RB1**, **RB2** and **RB3** for the insulin effect, meal effect and insulin/meal effect, respectively. **RB1** and **RB2** are composed of nine rules each, while **RB3** is composed of 27 rules. The rules represent different experiences of the T1DM patient. Given that the maximum discrete time of the input variables' effect on BGL variation is **z**, then the jth rule is of the form:

BGL VS Insulin Rule Base (RB1)
R_j**:IF**

$x(k)$ is F_1^j **AND** u_1 is F_2^j
THEN
FOR $k = 0, 1, \dots, z$

$$\bar{x}(k+1) = \bar{P}_A \bar{x}(k) + \bar{P}_B \, f_1\left(u_1, \bar{x}(k), \Delta t, \Delta p\right)$$

$$\bar{I}(k+1)_{u_1} = \bar{x}(k+1) - \bar{x}(k) \tag{7.15}$$

Where $\bar{P}_A, \bar{P}_B$ are state and input matrices, respectively, multipled by the output of
$RB1, j = 1, 2, \dots, 9$; F = fuzzy variable; $\bar{P}_A = \sum_{j=1}^{9} \sigma_j P_{A_j}$; $\bar{P}_B = \sum_{j=1}^{9} \sigma_j P_{B_j}$

BGL VS Meal Rule Base (RB2)
R$_j$:IF
$x(k)$ is F_1^j **AND** u_2 is F_3^j
THEN
FOR $k = 0, 1, \dots, z$

$$\bar{x}(k+1) = \bar{Q}_A \bar{x}(k) + \bar{Q}_B \, f_2\left(u_2, \bar{x}(k), \Delta t, \Delta p\right)$$

$$\bar{I}(k+1)_{u_2} = \bar{x}(k+1) - \bar{x}(k) \tag{7.16}$$

Where $\bar{Q}_A, \bar{Q}_B$ are state and input matrices respectively multiplied by the output of
$RB2, j = 1, 2, \dots, 9$; F = fuzzy variable; $\bar{Q}_A = \sum_{j=1}^{9} \sigma_j Q_{A_j}$; $\bar{Q}_B = \sum_{j=1}^{9} \sigma_j Q_{B_j}$

$$\sigma_j = \frac{\mu_j}{\sum_{j=1}^{9} \mu_j} ; \sum_{j=1}^{9} \sigma_j = 1.$$

BGL VS Insulin/Meal Rule Base (RB3)
R$_j$:IF
$x(k)$ is F_1^j **AND** $I(k+1)_{u_1}$ is F_4^j **AND** $I(k+1)_{u_2}$ is F_5^j
THEN
FOR $k = 0, 1, \dots, z$

$$\bar{x}(k+1) = \bar{A}\bar{x}(k) + \bar{B}\bar{I}(k+1)$$

$$\bar{y}(k) = \bar{C}\bar{x}(k) + \bar{D}\bar{I}(k+1) \tag{7.17}$$

Where A_j, B_j, C_j, D_j, are discrete time subsystem matrices; $\bar{I}(k+1) = \left[\bar{I}(k+1)_{u_1}, \bar{I}(k+1)_{u_2}\right]^T$, $j = 1, 2, ..., 27$, $F =$ fuzzy variable.

Given a current state vector $x(k)$ and an input vector u, the T-S fuzzy model infers $\bar{x}(k+1)$ as;

$$\bar{x}(k+1) = \sum_{j=1}^{27} \frac{\mu_j(A_j \bar{x}(k) + B_j \bar{I}(k+1))}{\sum_{j=1}^{27} \mu_j}$$

$$\bar{y}(k) = \sum_{j=1}^{27} \frac{\mu_j(C_j \bar{x}(k) + D_j \bar{I}(k+1))}{\sum_{j=1}^{27} \mu_j} \qquad (7.18)$$

$\mu_j = \min_{i=1}^{(n-1)} \mu_j^i(x, u_i)$ is the minimum of the MF for the fuzzy rule j, and $\mu_j^i(x)$ is the MF of the fuzzy term F_i^j for control variable x, $j = 1, 2, ..., 27$, n is the total number of input and state variables.

The overall fuzzy system model can be simplified as

$$\bar{x}(k+1) = \bar{A}\bar{x}(k) + \bar{BI}(k+1)$$

$$\bar{y}(k) = \bar{C}\bar{x}(k) + \bar{DI}(k+1) \qquad (7.19)$$

Where

$$\bar{A} = \sum_{j=1}^{27} \sigma_j A_j; \bar{B} = \sum_{j=1}^{27} \sigma_j B_j; \bar{C} = \sum_{j=1}^{27} \sigma_j C_j; \bar{D} = \sum_{j=1}^{27} \sigma_j D_j \sigma_j = \frac{\mu_j}{\sum_{j=1}^{27} \mu_j}; \sum_{j=1}^{27} \sigma_j = 1;$$

The overall system is nonlinear since $\bar{A}$ is a function of σ_j and σ_j is a function of $\bar{x}(k)$ the state variable.

Since the system output is the same as the future state (i.e. predicted BGL):

$$\bar{y}(k) = \bar{x}(k+1) \qquad (7.20)$$

Hence, the j-th rule for **RB3** simplifies to
R_j:IF
$x(k)$ is F_1^j **AND** $I(k + 1)_{u_1}$ is F_2^j **AND** $I(k + 1)_{u_2}$ is F_3^j
THEN
FOR $k = 0, 1, ... , z$

$$\bar{x}(k+1) = \bar{A}\bar{x}(k) + \bar{BI}(k+1) \qquad (7.21)$$

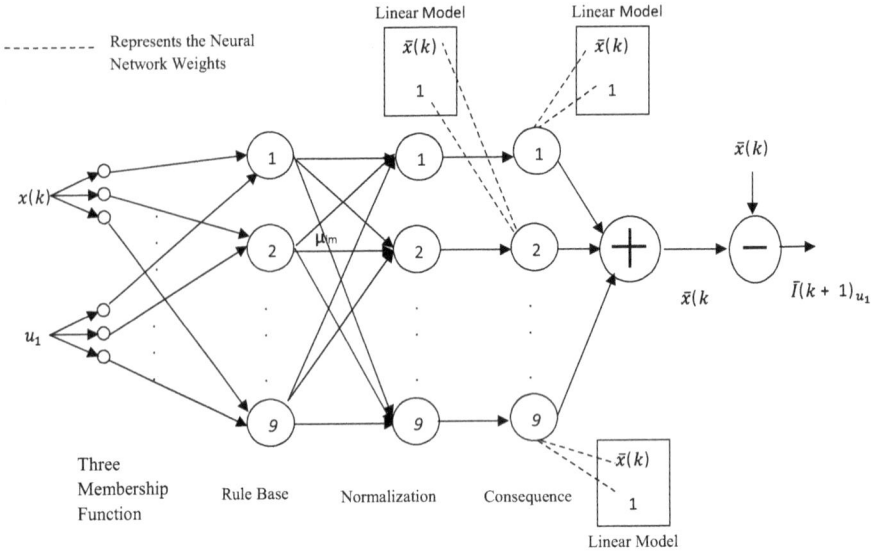

FIGURE 7.2   Insulin effect-based NFN model for the prediction of BGL in T1DM patients.

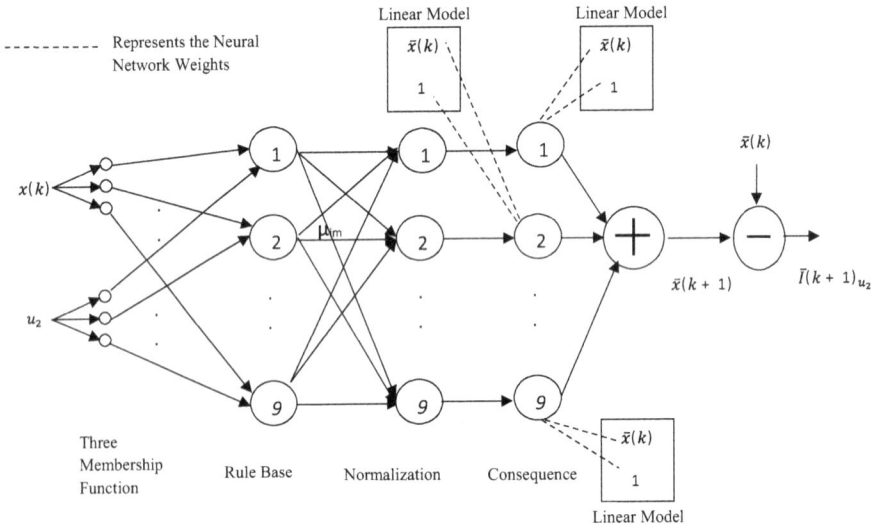

FIGURE 7.3   Meal effect-based NFN model for the prediction of BGL in T1DM patients.

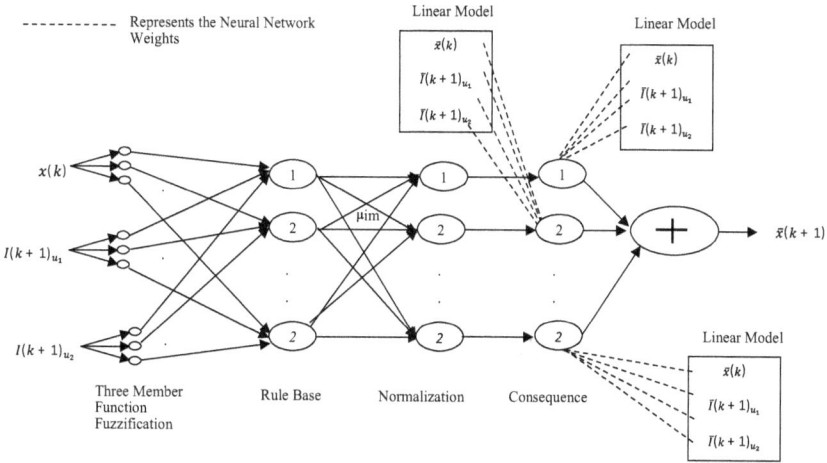

FIGURE 7.4 Couple insulin and meal effect-based NFN model for the prediction of BGL in T1DM patients.

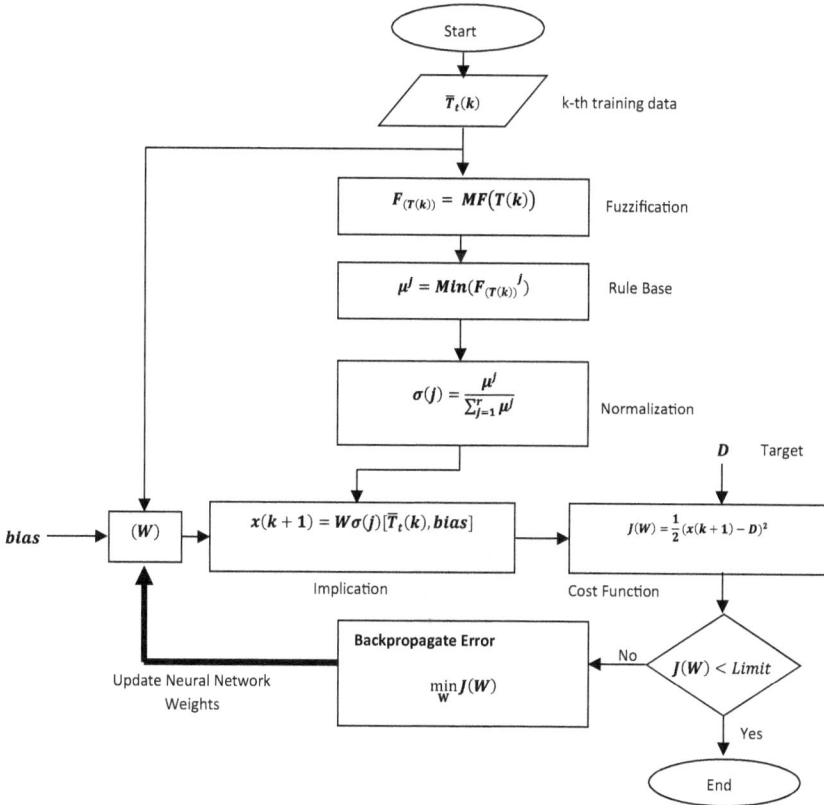

FIGURE 7.5 Flow chart of the forward/back propagation of training data set in T-S NFN.

7.4 PARAMETER IDENTIFICATION

In identifying the parameters $\bar{P}_A, \bar{P}_B, \bar{Q}_A, \bar{Q}_B, \bar{A}$ and $\bar{B}$, the fuzzy neural network is trained using least square error function and back propagation algorithm. Hence, the three (3) rule bases are formulated in the form:

BGL VS Insulin Rule Base (RB1)
R$_j$:IF
$x(k)$ is F_1^j **AND** u_1 is F_2^j
THEN
FOR $k = 0, 1, \ldots, z$

$$\bar{x}(k+1) = p_{1j}\bar{x}(k) + p_{2j}$$

$$\bar{I}(k+1)_{u_1} = \bar{x}(k+1) - \bar{x}(k) \tag{7.22}$$

BGL VS Meal Rule Base (RB2)
R$_j$:IF
$x(k)$ is F_1^j **AND** u_2 is F_3^j
THEN
FOR $k = 0, 1, \ldots, z$

$$\bar{x}(k+1) = q_{1j}\bar{x}(k) + q_{2j}$$

$$\bar{I}(k+1)_{u_2} = \bar{x}(k+1) - \bar{x}(k) \tag{7.23}$$

BGL VS Insulin/Meal Rule Base (RB3)
R$_j$:IF
$x(k)$ is F_1^j **AND** $I(k + 1)_{u_1}$ is F_4^j **AND** $I(k + 1)_{u_2}$ is F_5^j
THEN
FOR $k = 0, 1, \ldots, z$

$$\bar{x}(k+1) = w_{1j}\bar{x}(k) + w_{2j}\bar{I}(k+1)_{u_1} + w_{3j}\bar{I}(k+1)_{u_2} + w_{4j} \tag{7.24}$$

The weights $w_{1j}, \ldots, w_{4j}$ are weights of the NN, which is trained using the input-output data collected from the T1DM patient. They represent the vague experiences of the patient, as he/she carries on his/her day-to-day activities. More so, the trained weights p_{1j} and p_{2j} represent insulin effect on BGL, while the trained weights q_{1j} and q_{2j} represent meal effect on BGL.

7.5 PREDICTION OF BLOOD GLUCOSE LEVEL

Let the NFN trained weights for patient's specific meal/insulin effect feature extraction be W_T.

Let the inputs (BGL, insulin effect, meal effects and bias) on which prediction is based be

$$\left[\bar{p}_{11}, \bar{p}_{21}, \bar{p}_{31}, 1\right].$$

Given that there are 'l' rules, let the result of the triggered rules because of the input be

$$H_T = \begin{bmatrix} \sigma_{11}\bar{p}_{11} \\ \sigma_{11}\bar{p}_{21} \\ \sigma_{11}\bar{p}_{31} \\ \sigma_{11} \\ \vdots \\ \sigma_{l1}\bar{p}_{11} \\ \sigma_{l1}\bar{p}_{21} \\ \sigma_{l1}\bar{p}_{31} \\ \sigma_{l1} \end{bmatrix} \tag{7.25}$$

Hence, the predicted output Y_p is given as

$$Y_p = W_T H_T \tag{7.26}$$

7.6   CHOICE OF TRAINING SCENARIO

The metabolic activities of T1DM patients are affected by their general state of health. Thus, when defining the indices of the general state of health of the patient as metabolic scenarios, different training windows are built. These metabolic scenarios span a myriad of indices, which include stress, exercise, injection site, BMI, fat mass index, etc. The training windows are chosen to capture the contributory effect of the metabolic indices to the variation in BGL in T1DM.

Robustness and flexibility of the coupled insulin and meal effect BGL predictive model lies in its modularity. This is achieved using insulin and meal effect rule bases (**RB1** and **RB2**) that span the metabolic scenarios. Insulin/meal rule base (**RB3**) transparently couples **RB1** and **RB2** making **RB3** independent of the specifics of the insulin and meal variables, but on their apparent effects. Hence, **RB3** is quite robust and flexible and spans the metabolic scenarios that have been used in the training. As a result of this design, **RB3** learns the diverse experiences of the T1DM patient, while **RB1** and **RB2** are updated as new metabolic scenes arise.

7.7   RESULTS, OBSERVATIONS AND DISCUSSIONS

The insulin, meal and insulin/meal effect models were implemented using MATLAB® 2012a. A 31-year-old male T1DM patient, who has been diabetic for 17 years, was recruited for this study with the approval of Lagos University Teaching Hospital Research and Ethics Committee. Patient's BMI ranged between 22.42 and 23.55. The training dataset was gathered under normal living condition to capture patient's real-life experience. Blood glucose readings were taken using a One Touch blood glucose monitor. Patient's exogenous insulin injection site is the thigh. Polished rice was considered in this study for meal effect investigation because it is the most staple food in Nigeria. Every reference to rice in this study specifically refers to polished rice. Readers should refer to Orieke et al. (2019) for graphical details of this case application.

The absence of food and insulin in the blood stream results in a significant rise in BGL as given by the positive gradient of 45 mg/dl per hour. This can be attributed to internal glucose production. **Hence, it is not advisable for a T1DM patient to skip medication**. The average rise in BGL is 45 mg/dl per hour for the '*Three (3) SS of rice/BMI:23.32*' condition, while it is 60 mg/dl per hour for the '*Five (5) SS of rice/BMI:22.42*' condition. This shows that BGL response to meal has a positive gradient. It is obvious from the findings that insulin and meal effects change as the BMI of the investigated T1DM patient change.

Coupling the obtained insulin and meal effects to predicting BGL variations showed promising results in reproducing the same effect as seen in the training data. It also extrapolates future responses. The training data was collected for a period of two (2) months. A stratified holdout procedure with captured dataset randomly split into 70% for training and 30% for testing was used in this study. The NFN T-S model was trained for 100,000 epochs, with a neural network weight adjustment step size of 0.03. The performance of the NFN T-S model on BGL prediction was assessed using new sets of measurements that were not used during training. The performance of the NFN T-S model is summarized in Table 7.2.

The Clarke's Error Grid Analysis (CEGA) of the performance of the NFN T-S model in predicting BGL in the T1DM patient is given in Figure 7.6. 87.5% of the

---

TABLE 7.2

Performance Analysis of NFN T-S Model on BGL Prediction Under Different Conditions

Day 1		Day 2		Day 3		Day 4	
BMI: 23.32		BMI: 23.32		BMI: 22.42		BMI: 22.42	
Insulin: 30IU		Insulin: 10IU		Insulin: 30IU		Insulin: 30IU	
Meal: 3		Meal: 1		Meal: 1		Meal: 5	
IBGL: 123		IBGL: 210		IBGL: 301		IBGL: 231	
MBGL	PBGL	MBGL	PBGL	MBGL	PBGL	MBGL	PBGL
190	193.2	229	216.8	266	270.6	286	288
115	128.3	175	199.2	219	218.7	272	248.7
77	86.1	187	174.7	103	133.4	217	189.9
52	45.6	152	148.8	42	48.8	125	90.9

IBGL: Initial BGL (mg/dl) is BGL before administration of insulin/meal; MBGL: Measured BGL (mg/dl) is hourly measured BGL after administration of insulin/meal; PBGL: Predicted BGL (mg/dl) is NFN predicted BGL; Meal (Serving Spoons); 1 Serving spoon of rice = 50 g

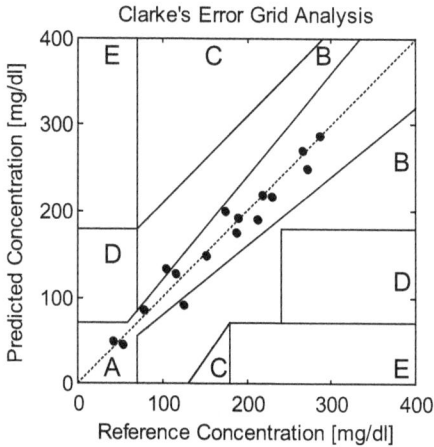

FIGURE 7.6 Clarke's error grid analysis of the performance of the coupled insulin/meal effect NFN BGL predictive model.

predictions fall into region A, while the remaining 12.5% of the predictions fall into region B within the 240 minutes prediction horizon. No predictions fall into regions C, D and E; hence, all the predictions are within the clinically acceptable range. Interested readers can refer to Orieke et al. (2019) to view plots of insulin and meal effects at different times.

Table 7.3 shows performance comparison of the coupled insulin and meal effect neuro-fuzzy model with previous predictive models.

7.8 CONCLUSION

The model developed establishes the fact that BGL variations in T1DM patients can be decoupled into manageable entities of feature extracts (NFN weights) describing the insulin and meal effects for specific metabolic scenarios. The transparent coupling of the insulin and meal effects can predict BGL variations for both known and uninvestigated scenarios. The feature extracts defining the insulin and meal effects depend on the prevailing metabolic indices, which include BMI, activity level, stress level, injection sites, etc. However, the transparent coupling of the insulin and meal effects is unaware of the prevailing metabolic indices.

The achievements made by the NFN T-S model in predicting BGL on limited training show very positive indication that predicting BGL variations for T1DM patients using NFN is viable. Its performance on limited training is comparable to the achievements seen in the Auto Regressive, Compartmental (State Space), Support Vector Regression and Artificial Neural Network Model. The introduction of the concept of insulin and meal effect for diverse metabolic scenarios provides a novel means to explaining the variability in BGL experienced by T1DM patients. This will further enhance patients' awareness of how their day-to-day activities affect their BGLs. Robustness and flexibility of the model lies in its modularity. This is achieved using **RB1** and **RB2** that span the metabolic scenarios. Hence, **RB3**, which is the

TABLE 7.3

Comparison of Couple Insulin/Meal Effect NFN with Previous Models

	Description	Coupled Insulin/Meal Effect Neuro-Fuzzy Model	Zarita et al. (2009)	Scott et al. (2010)	Eleni et al. (2011)
1	Model	Combination of Neuro-Fuzzy Network models	Combination of Principal Component Analysis and Wavelet Neural Network	Feed forward three layer neural network	Support Vector Regression and Compartmental Model
2	Participating Patients	One (1)	One (1)	One (1) patient specific, Five (5) general model	Seven (7)
3	Attributes	Current BGL, Insulin injection, Meal, Patient-specific attributes such as BMI, Injection Site, Stress and Exercise	Time, Insulin, Meal, Exercise, Stress	Time, CGM data, POC glucose test time/results, Insulin delivery type/dosage	Insulin/meal/exercise dynamics, CGM
4	Concept	Insulin Effect, Meal Effect, Coupled Insulin/Meal Effects all in mg/dl	No Concept Definition	No Concept Definition	Insulin/Meal/Exercise dynamics
5	Prediction Partition	Metabolic Indices and Time of the day	Time of the day	Not explicitly defined	Not explicitly defined
6	Accuracy	87.5% and 12.5% in regions A and B, respectively, of CEGA in 240 minutes prediction horizon	Measured in RMSE of 0.045 mmol/dl and no reference to CEGA	95.1% and 69.8% predictions in CEGA region A for patient-specific and general models, respectively	98.86%, 92.54%, 80.02%, 62.91% of predictions in CEGA region A for 15, 30, 60 and 120 minutes
7	Prediction Horizon	240 minutes	Not Specified Explicitly	75 minutes	15, 30, 60, 120 minutes
8	Multi-Scenario Application	Applicable in Multi-Scenario Environment	Restricted application	Highly Restricted application	Restricted application

coupling of **RB1** and **RB2**, is quite robust and flexible spanning the investigated metabolic scenarios. As a result of this design, **RB3** learns the diverse experiences of the T1DM patient, while **RB1** and **RB2** are updated as new metabolic scenes arise.

7.9 CONTRIBUTIONS TO KNOWLEDGE

1. This study established that glucose/insulin dynamics of a Type 1 diabetic patient can be decoupled into manageable entities of neuro-fuzzy feature extracts describing the insulin and meal effects for specific metabolic scenarios.
2. Furthermore, this study established that coupling of the insulin and meal effects can effectively predict blood glucose variations in T1DM patients.

For further research work, it is recommended that clinical investigations be carried out for more metabolic conditions, which include different meals, exercise levels, stress levels, sites of insulin injection, meal and insulin effect offset, etc. It is also recommended that more training data should be collected to enhance the performance of the NFN T-S-based BGL prediction model.

REFERENCES

Ahmed, Y. B. and A. E. Mahmud (2013). "A Fuzzy Controller for Blood Glucose-Insulin System". *Journal of Signal and Information Processing* 4(2): 111–117.

American Council of Exercise. (2013). "Exercise and Type I Diabetes". *Fit Facts.* Retrieved March 5, 2015 from http://wellnessproposals.com/fitness/handouts/health-challenges/exercise_diabetes.pdf

Bremer, T. and D. A. Gough (1999). "Is Blood Glucose Predictable from Previous Values?" *Diabetes* 48(3): 445–451.

Charles, D. C. (2002). *Pharmacology in Rehabilitation.* F.A. Davis, Pennsylvania.

Deutsch, T., E. R. Carson, E. Harvey, E. D. Lehmann, P. H. Sonksen, G. Tamas, K. Whitney and C. D. Williams (1990). "Computer Assisted Diabetes Management: A Complex Approach". *Computer Methods and Programs in Biomedicine* 32(3–4): 195–214.

Eleni, D., T. I. Athanasios, and D. P. Margaris (2011). "Evaluation of the Turbulence Models for the Simulation of the Flow". *Journal of Mechanical Engineering Research* 4(3). DOI: 10.5897/JMER11.074

Ghevondian, N., H. T. Nguyen and S. Colagiuri (2001). A Novel Fuzzy Neural Network Estimator for Predicting Hypoglycemia in Insulin-Induced Subjects. *Proceedings – 23rd Annual Conference – IEEE/EMVS*, 1657–1657.

Hidalgo, J. I., J. M. Colmenar, G. Kronberger, S. M. Winkler, O. Garnica and J. Lanchares (2017). "Data Based Prediction of Blood Glucose Concentrations Using Evolutionary Methods". *Journal of Medical Systems* 41(9): 142.

International Diabetes Federation. (2013). *IDF Diabetes Atlas* (6th ed.), Brussels, Belgium. Retrieved September 2, 2016 from http://www.idf.org/diabetesatlas/data-visualisations

International Diabetes Federation. (2016). *IDF Diabetes Atlas* (7th ed.). Retrieved September 2, 2016 from http://www.idf.org/diabetesatlas

Juan, Li and F. Chandima (2016). "Smartphone-Based Personalized Blood Glucose Prediction". *ICT Express* 2(4): 150–154.

Kyriaki, S., M. Martin, S. Katerina, P. Pavlína and L. Lenka (2018). "Predicting Blood Glucose Levels for a Type I Diabetes Patient by Combination of Autoregressive with One Compartment Open Model". *IFMBE Proceedings* 15(2): 771–774.

Mathers, C. D. and D. Loncar (2006). "Projections of Global Mortality and Burden of Disease from 2002 to 2030". *PLoS Medicine* 3(11): e442.

Melissa, C. S. (2012). "Diabetes (Type 1 and Type 2)". *Medicine Net*. Retrieved February 3, 2014 from http://www.medicinenet.com/diabetes_mellitus/page4.htm

Michael, C. R. and A. P. Bruce (2006). "Type 1 Diabetes and Vigorous Exercise: Applications of Exercise Physiology to Patient Management". *Canadian Journal of Diabetes* 30(1): 63–71.

Moshe, P., B. Tadej, A. Eran, K. Olga, B. Natasa, M. Shahar, B. Torben, A. Magdalena, M. D. Stefanija, M. Ido, N. Revital and D. Thomas (2013). "Artificial Pancreas for Nocturnal Glucose Control". *The New England Journal of Medicine* 368(9): 824–833.

Orieke, N. O., O. S. Asaolu, T. A. Fashanu and O. A. Fasanmade (2019). "A Coupled Insulin and Meal Effect Neuro-Fuzzy Model for the Prediction of Blood Glucose Level in Type 1 Diabetes Miletus Patients". *Annals of Science and Technology: An Official Journal of the Nigerian Young Academy* 4(1): 1–15.

Scott, M. P., J. B. Marilyn, D. C. Brent, E. B. Raymond, D. L. Jason, S. Desmond, C. Antonio and J. P. Thomas (2010). "Development of a Neural Network Model for Predicting Glucose Levels in a Surgical Critical Care Setting". *Patient Safety in Surgery Journal* 4(15): 1–5.

Shanthi, S., D. Kumar, S. Varatharaj and S. Santhana (2010). "Prediction of Hypo/Hyperglycemia through System Identification, Modelling and Regularization of Ill-Posed Data". *International Journal of Computer Science & Emerging Technologies* 1(4): 171–176.

Shoback, D. (2011). *Greenspan's Basic & Clinical Endocrinology*, 9th ed. McGraw-Hill Medical, New York.

Sparacino, G., F. Zanderigo, S. Corazza, A. Maran, A. Facchinetti and C. Cobelli (2007). "Glucose Concentration Can Be Predicted Ahead in Time from Continuous Glucose Monitoring Sensor Time-Series". *IEEE Transactions on Biomedical Engineering* 54(5): 931–937.

Stahl, F. and R. Johansson (2009). "Diabetes Mellitus Modelling and Short Term Prediction Based on Blood Glucose Measurements". *Mathematical Biosciences* 217(2): 101–117.

World Health Organization. (2011). *Global Status Report on Noncommunicable Diseases 2010*. Geneva. Retrieved September 2, 2016 from https://www.who.int/nmh/publications/ncd_report2010/en/

World Health Organization. (2013). *World Health Organization*. Retrieved March 5, 2014 from http://www.who.int/mediacentre/factsheets/fs312/en/

World Health Organization. (2014). *Global Health Estimates: Deaths by Cause, Age, Sex and Country, 2000-2012*. Geneva, Retrieved September 2, 2016 from https://www.who.int/healthinfo/global_burden_disease/estimates/en/index1.html

Zarita, Z., P. Ong and A. Cemal (2009). "A Neural Network Approach in Predicting the Blood Glucose Level for Diabetic Patients". *International Journal of Information and Mathematical Sciences* 5(1): 72–79.

8 Stochasticity in AI Mathematical Modeling

8.1 INTRODUCTION

Proper modeling of an artificial intelligence (AI) system is the first step to the formulation of an optimization strategy for practical applications. There are different types of models (Badiru et al. 2012). First, a **system** is a collection of **interrelated elements** brought together **to achieve a specified objective**. Normally, a system processes an input to yield an output or a response.

Next, models are needed to represent the system.

A model is an abstraction of a real-world system. It is a simplified representation of reality, which employs descriptive (linguistic, physical, or mathematical) concepts and/or symbols, that is, models mirror or approximate only some aspects of reality and evaluate others. Models provide a safe and cost-effective means for the study of entities and phenomena and their interactions. Two major types of models are material/iconic (including robotic/expert) models and mathematical models. Proper modeling of a system is the first step toward formulating an optimization strategy for the system. The criteria for objective/cost selection of a model would be to minimize the errors between the model and the actual system. The "goodness of fit" criteria can be evaluated when the model and the system are forced by sample inputs.

8.2 THE MATERIAL/ICONIC MODELS

The material/iconic models simulate the actual system as a prototype in the physical space. It could be a scaled model of an empirical system or a direct physical analogue. A study of its behavior under various conditions possible is undertaken, for example, wind-tunnel laboratories, test piloting, moot trials, etc. For example, there are robotic/expert models. This is a feedback control system, a device that can measure its own state and take actions based on it.

8.3 AI MATHEMATICAL MODELS

Mathematical modeling involves the application of mathematics/empirical knowledge to real-life problems. Stimulators to this approach include the advent of high-speed electronic computers (for large environmental problems – parallel computers), developments in information and communication technology, progress in applied mathematics (functional analysis and numerical methods), and progress in empirical knowledge (engineering). A mathematical model consists of a set of mathematical formulae giving the validity of certain fundamental "natural laws" and various hypotheses relating to physical processes.

DOI: 10.1201/9781003247395-8

The direct engineering problem is to find the output of a system given the input (see Figure 8.1) and the inverse problems are of three main divisions –design/synthesis, control/instrumentation, and modeling/identification (Figure 8.2).

Design/Synthesis: Given an input and output, find a system description which fits such a physically realizable relationship optimally.

Control/Instrumentation: Given a system description and a response, find the input which is responsible for the response (output).

Modeling/Identification: Given a *set* of inputs and corresponding outputs from a system, find a mathematical description (model) of the system.

The criteria for objective/cost function selection would be to minimize the errors between the model and actual system. The "goodness of fit" of the criteria can be evaluated when both the model and system are forced by sample inputs (see Figure 8.3).

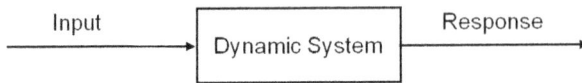

FIGURE 8.1   Input–response relationship in system modeling.

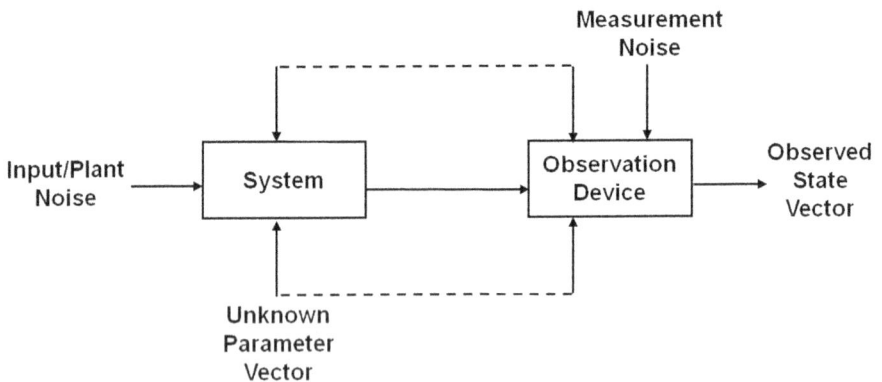

FIGURE 8.2   General system configuration.

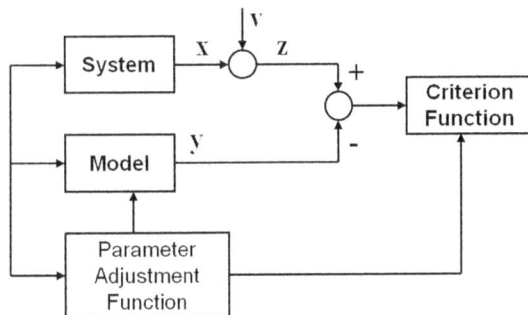

FIGURE 8.3   Parameter models.

General Problem Formulation.

$$\text{Let} \frac{dx(t)}{dt} = f\left(x(t), u(t), w(t), p(t), t\right) \quad \cdot \quad \text{(8.1)}$$

Be the system equation where

$x(t)$ is the system state vector;
$u(t)$ is the input signal/control;
$w(t)$ is the input disturbance/noise;
and $p(t)$ is the unknown parameters.

Assume that the observation is of the form

$$z(t) = h\left(x(t), u(t), w(t), p(t), v(t)t\right) \quad \text{(8.2)}$$

where $v(t)$ is the observation noise.

The identification/estimation problem is to determine $p(t)$ and perhaps $x(t)$ as well as the mean and variance coefficients of system noise $w(t)$ and observation noise $v(t)$.

System:

$$\frac{dx(t)}{dt} = f\left(x(t), u(t), w(t), p(t), t\right) \quad \text{(8.3)}$$

Observation:

$$z(t) = Dy + Eu, \; D, E \text{ are matrices} \quad \text{(8.4)}$$

Model:

$$\frac{dy(t)}{dt} = g\left(y(t), u(t), w(t), p', t\right) \quad \text{(8.5)}$$

Criterion Function:

$$J(T, p') = \int_0^T \left\| x(t) - y(t) \right\|_w dt \quad \text{(8.6)}$$

w is an appropriate weighing matrix

Problem: Seek an optimum set of parameters p^*, which minimizes J. That is,

$$J(T, p^*) = \min J(T, p') \quad \text{(8.7)}$$

Analytical expressions for p^* are possible, viz

$$\frac{dJ}{dp'} = 0 \text{ provided } \frac{d^2J}{dp'^2} > 0 \text{ in special cases.} \tag{8.8}$$

Search techniques are useful when the number of parameters is small. The technique consists of

(i) random selection of pre-selected grid pattern for parameters $p'_1, p'_2, \ldots$ and corresponding $J_1, J_2, \ldots$ and
(ii) simple comparison test for the determination of minimum J.

Gradient methods are based on finding the values of p' for which the gradient vector equals zero, namely,

$$\nabla_0 J = \left[\frac{\partial J}{\partial p_1}, \frac{\partial J}{\partial p_2}, \ldots \frac{\partial J}{\partial p_k} \right] = 0 \tag{8.9}$$

and

$$p^{(i+1)} = p^{(i)} - K\nabla_0 J\left(p^{(i)}\right) \tag{8.10}$$

where for

$$\text{Steepest Descent}, K = kJ, k = \text{constant} \tag{8.11}$$

$$\text{Newton Raphson}, K = \frac{J(p)}{\left\| \nabla J(p) \right\|^2} \tag{8.12}$$

$$\text{Newton}, K = H^{-1} = \left[\frac{\partial^2 J}{\partial p_j \partial p_k} \right]^{-1} \tag{8.13}$$

$$\text{Gauss-Newton}, K = \delta^{-1} = \left[\int_0^T 2\nabla y \nabla y' dt \right]^{-1} \tag{8.14}$$

It is desirable to have on-line or recursive identification so as to make optimum adaptation to the system goal possible in the face of environmental uncertainty and changing environmental conditions.

8.4 SYSTEMS FILTERING AND ESTIMATION

8.4.1 IDENTIFICATION

Identification problems can be categorized into two broad areas, namely, the total ignorance/"black box" identification and the grey box identification. In the grey box identification, the system equations may be known or deductible from the basic physics or chemistry of the process up to the coefficients or parameters of the equation. The methods of solution consist of the classical (deconvolution, correlation, etc.) and modern techniques (Kalman et al).

Given $u(t)$ and $y(t)$ for $0 \le t \le T$, determine $h(t)$.

(i) Observe input and output at N periodical sampled time intervals, say Δ secs apart in $[0,T]$ such that $N\Delta = T$

(ii) It is known that

$$y(t) = \int_0^t h(t-\tau)u(\tau)d\tau \tag{8.15}$$

called the Convolution Integral.

(iii) Assume that

$$u(t) = u(n\Delta) \text{ or } u(t) \approx \frac{1}{2}\{u(n\Delta) + u((n+1)\Delta)\} \tag{8.16}$$

for $n\Delta < t < (n+1)\Delta$

$$h(t) \approx h\left(\frac{2n+1}{2}\Delta\right), n\Delta <= t < (n+1)\Delta \tag{8.17}$$

(iv)

$$\text{If } y(n\Delta) = \Delta\sum_{i=0}^{n-1} h\left(\frac{2n-1}{2}\Delta - i\Delta\right)u(i\Delta) \tag{8.18}$$

(v) If

$$y(T) = \begin{bmatrix} y(\Delta) \\ y(2\Delta) \\ \cdot \\ y(N\Delta) \end{bmatrix}, \quad h(T) = \begin{bmatrix} h\left(\frac{\Delta}{2}\right) \\ h\left(\frac{3\Delta}{2}\right) \\ \cdot \\ h\left(\frac{(2N-1)\Delta}{2}\right) \end{bmatrix} \tag{8.19}$$

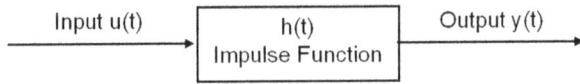

Input u(t) → | h(t) Impulse Function | → Output y(t)

FIGURE 8.4   Single input–output function.

$$\text{then } y(T) = \Delta U h(T) \tag{8.20}$$

$$\text{where } U = \begin{bmatrix} u(0) & 0 & 0 & 0 & 0 \\ u(\Delta) & u(0) & 0 & 0 & 0 \\ u(2\Delta) & u(\Delta) & u(0) & . & 0 \\ . & . & . & . & . \\ u((N-1)\Delta) & u((n-2)\Delta) & . & u(0) & (0) \end{bmatrix} \tag{8.21}$$

(vi) From Equation (8.20),

(vii) $h(T) = = U^{-1}\Delta^{-1}y(T)$ so that

$$h_n \cong h\left(\frac{2n-1}{2}\Delta\right), h_1 = \frac{y(\Delta)}{\Delta u(0)}$$

$$= \frac{1}{u(0)}\left\{\frac{y(n\Delta)}{\Delta} - \sum_{i=1}^{n-1}h_{n-i}u(i\Delta)\right\}$$

Advantages:

(i) Simple;

(ii) Quite effective for many identification problems;

(iii) FFT (Fast Fourier Transform) may be used to reduce the computational requirements;

(iv) Any input may be used (no need for special test inputs).

Disadvantages:

(i) Sequential/on-line use of algorithm impossible unless time interval of interest is short.

(ii) Numerical round-off errors make the technique inaccurate for $m \to \infty$.

8.5   CORRELATION TECHNIQUES

Correlation techniques use white noise test signal, $u(t)$; hence, it is necessary to have wide bandwidth to detect high-frequency components of $h(t)$. For zero error, $u(t)$ must be proper "white" (infinite bandwidth).

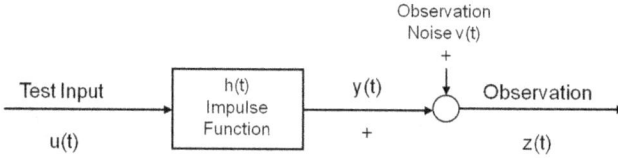

FIGURE 8.5   Configuration.

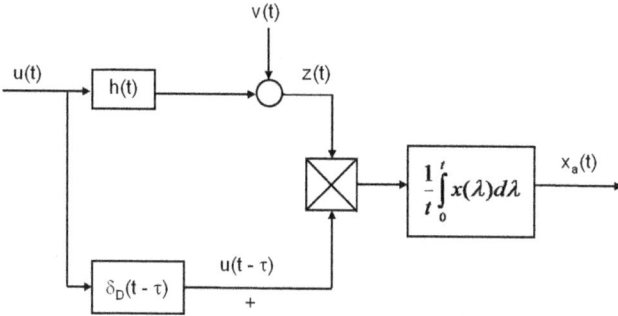

FIGURE 8.6   Identification correlator.

(i) It is assumed that steady state is reached.
(ii) Noise $u(t)$ and $v(t)$ ergodic and Gaussian distribution with zero mean

$$x_0(t) = \frac{1}{t}\int_0^t x(\lambda)d\lambda \tag{8.22}$$

$$x(t) = z(t)u(t-\tau) \tag{8.23}$$

$$z(t) = y(t) + v(t) \tag{8.24}$$

$$y(t) = \int_0^t h(\eta)u(t-\eta)d\eta \tag{8.25}$$

Now,

$$
\begin{aligned}
E\{x_a(t)\} &= \frac{1}{t}\int_0^t E\{x(\lambda)\}d\lambda \\
&= E\{x\} \\
&= E\{z(t)u(t-\tau)\} \\
&= R_{uz}(\tau)
\end{aligned} \tag{8.26}
$$

$$R_u(\tau) = E\{u(t)z(t+\tau)\}$$

From Equations (8.24) and (8.25), $E\{x_a(t)\} = R_{uz}(\tau) = \int_0^\infty h(\eta)R_u(t-\eta)d\eta$

Taking Fourier Transforms: $R_{uz}(s) = h(s)R_u(s)$
If the assumption on bandwidth holds: $R_{uz}(s) = kh(s)$, $R_{uz}(\tau) = kh(\tau)$
And if $u(t)$ is white δ_D is a Dirac-delta: $R_u(\tau) = R_u\delta(I)$, $R_u(s) = R_w$
And then, for $R_u = 1$,

$$R_{uz}(\tau) = E\{x_a(t)\} = h(\tau)$$

Complete system identification is subsequently obtained by using N correlators in parallel, such that the quantities
$R_{uz}(\tau_i) = h(\tau_i)$, $i = 1, 2, ...N$ are measured.

Advantages:
   (i) Not critically dependent on normal operating record.
   (ii) By correlating over a sufficiently long period of time, the amplitude of the test signal can be set very low such that the plant is essentially undisturbed by the white noise test signal.
   (iii) No a priori knowledge of the system to be tested is required.

8.5.1 SYSTEM ESTIMATION

Kalman initiated a new formulation of the Wiener (classical) theory expressing the results of the estimation in the time domain rather than in the frequency domain. The modern theory is more fundamental, requires minimum mathematical background, ideal for digital computation, and provides a general estimator, whereas the classical method can only deal with restricted dimension, is rigorous, and has limited applicability to nonlinear systems.

8.5.2 PROBLEM FORMULATION

$$\text{Given} \Theta = \Theta(x) \tag{8.27}$$

where Θ is a vector of m observations and x is a vector whose variables are to be estimated. The estimation problem is continuous if Θ is a continuous function of time; otherwise, it is a discrete estimation problem. *Estimating the past is known as smoothing, estimating the present as filtering, while estimating the future is prediction/forecasting.*

Nomenclature
   (i) An estimate $\hat{x}$ of x is unbiased if $E(\hat{x}) = x$
   (ii) Let $e = \hat{x} - x$ and $C_e = E\left[(\hat{x}-x)(\hat{x}-x)'\right]$

8.5.3 MAXIMUM LIKELIHOOD

Let

$$\Theta = Bx + v \tag{8.28}$$

where v is the noise. The maximum likelihood method takes $\hat{x}$ as the value which maximizes the probability of measurements that actually occurred taking into account known statistical properties of v. The conditional probability density function for Θ given x is the density of v centered around Bx. If v is zero mean Gaussian distributed with covariant matrix C_v, then

$$p(\Theta \mid x) = \frac{1}{(2\pi)^{\frac{1}{2}} |C_v|^{\frac{1}{2}}} e^{\left[\frac{-1}{2}(\Theta - Bx)C_v^{-1}(\Theta - Bx)\right]} \tag{8.29}$$

So that Maximun

$$p(\Theta \mid x) = \max\left[\frac{-1}{2}(\Theta - Bx)C_v^{-1}(\Theta - Bx)\right]$$

Hence,

$$x = \left(B'C_v^{-1}B\right)^{-1} B'C_v^{-1}\Theta \tag{8.30}$$

Least Squares | Weighted Least Squares
The least squares choose $\hat{x}$ as that value which minimizes the sum of squares of the deviations $\theta_i - \hat{\theta}_i$, that is,

$$\text{Minimizes } J = (\Theta - Bx)'(\Theta - Bx) \tag{8.31}$$

$$\text{Thus setting } \frac{\partial J}{\partial x} = 0 \text{ yields } \hat{x} = \left(B'B\right)^{-1} B'\Theta \tag{8.32}$$

For weighted least squares,

$$\text{Minimize } J = (\Theta - Bx)' W^{-1}(\Theta - Bx) \text{ or } J = \|\Theta - Bx\|_{W^{-1}} \tag{8.33}$$

$$\text{Yielding } \hat{x} = \left(B'W^{-1}B\right)^{-1} B'W^{-1}\Theta \tag{8.34}$$

8.5.4 BAYES ESTIMATORS

For Bayes estimators, statistical models for both x and Θ are assumed available. The aposteri conditional density function is $p(x \mid \Theta)$, since it contains all the statistical information of interest.

$$p(x|\Theta) = \frac{p(\Theta|x)p(x)}{p(\Theta)} \quad \left(\text{Bayes'Rule}\right) \tag{8.35}$$

$\hat{x}$ is computed from $p(x \mid \Theta)$.

8.5.5 MINIMUM VARIANCE

$$\text{Minimize } J = \int\limits_{-\infty}^{\infty}\int\limits_{-\infty}^{\infty}\cdots\int\limits_{-\infty}^{\infty}\left(\hat{x}-x\right)'S\left(\hat{x}-x\right)p(x|\Theta)dx_1\ldots dx_n \tag{8.36}$$

where S is an arbitrary, positive semi-definite matrix.

$$\text{Set } \frac{\partial J}{\partial x} = 0 \text{ to yield}$$

$$\hat{x} = \int\limits_{-\infty}^{\infty}\int\limits_{-\infty}^{\infty}\cdots\int\limits_{-\infty}^{\infty}xp(x|\Theta)dx_1\ldots dx_n \tag{8.37}$$

$$\therefore \hat{x} = E[x|\Theta] \tag{8.38}$$

8.5.6 FOR LINEAR MINIMUM VARIANCE UNBIASED (GAUSS-MARKOV)

Let

$$\hat{x} = A\theta,$$

where A is an unknown parameter and
$C_x = E(xx')$, $C_x\theta = E(x\theta')$ and $C\theta = E(\theta\theta')$. Now,

$$C_e = E\left(ee'\right) =$$

$$E\left[\left(\hat{x}-x\right)\left(\hat{x}-x\right)'\right]$$
$$= E\left[\left(A\theta - x\right)\left(A\theta - x\right)'\right]$$
$$= E\left[\left(A\theta\theta'A' - x\theta'A' - A\theta x' + xx'\right)\right]$$
$$= E\left[A\theta\theta'A'\right] - E\left[x\theta'A'\right] - E\left[A\theta x'\right] + E\left[xx'\right]$$
$$= AC_\theta A' - C_{x\theta}A' - AC_\theta x + C_x$$
$$= \left(A - C_{x\theta}C_\theta^{-1}\right)C_\theta\left(A - C_{x\theta}C_\theta^{-1}\right) - C_{x\theta}C_\theta^{-1}C'_{x\theta} + C_x$$

Minimum is obtained when $A - C_{x\theta}C_\theta^{-1} = 0$

$$\text{That is } A = C_{x\theta}C_\theta^{-1} \tag{8.39}$$

$$\text{Therfore, } x = C_{x\theta}C_\theta^{-1}\theta, \quad C_e = C_x - C_{x\theta}C_\theta^{-1}C_{x\theta}' \tag{8.40}$$

(i) If

$$\theta = Bx + v \text{ then} \tag{8.41}$$

$$C_{x\theta} = E\left[x\theta'\right]$$
$$= E\left[x\left(Bx+v\right)'\right]$$
$$= E\left[xx'B'\right] + E\left[xv'\right]$$

$$C_{x\theta} = C_x B' + C_{xv}$$

$$C_\theta = E\left[\theta\theta'\right] = E\left[\left(Bx+v\right)\left(Bx+v\right)'\right]$$

$$= BC_x B' + C_{xv}' B' + BC_{xv} + C_v$$

$$\hat{x} = \left[C_x B' + C_{xv}\right]\left[BC_x B' + \left(BC_{xv}\right)' + BC_{xv} + C_v\right]^{-1}\theta \tag{8.42}$$

$$C_e = C_x - \left(C_x B' + C_{xv}\right)\left[BC_x B' + \left(BC_{xv}\right)' + BC_{xv} + C_v\right]^{-1}\left(C_x B' + C_{xv}\right)' \tag{8.43}$$

(ii) If

$$\theta = Bx + v, C_{xv} = 0 \text{ then} \tag{8.44}$$

$$\hat{x} = \left(C_x B'\right)\left[BC_x B' + C_v\right]^{-1}\theta$$

$$C_e = C_x - \left(C_x B'\right)\left(BC_x B' + C_v\right)^{-1}\left(C_x B'\right)'$$

or

$$\hat{x} = \left(B'C_v^{-1}B\right)^{-1}B'C_v^{-1}\theta$$

$$\hat{x} = \left(C_x^{-1} + B'C_v B\right)^{-1} B'C_v^{-1}\theta \tag{8.45}$$

$$C_e = \left(C_x^{-1} + B'C_v^{-1}B\right)^{-1} \tag{8.46}$$

(iii) If in (ii), $C_x \to \infty$ (no information on state), then

$$\hat{x} = \left(B'C_v^{-1}B\right)^{-1} B'C_v^{-1}\theta \tag{8.47}$$

$$C_e = \left(B'C_v^{-1}B\right)^{-1} \tag{8.48}$$

8.5.7 Partitioned Data Sets

Let θ^r be a set of measurements of dimension r and $\hat{x}^r$ the estimate obtained using θ^r. Let $\theta^r = B^r x + v^r$ such that $\left(C_v^r\right)^{-1}$ exists, $C_{xv}^r = C_{vx}^r = 0;\ C_x^{-1} \to 0$, then

$$\hat{x}^r = \left(B^{r\prime}C_v^r B^r\right)^{-1} B^{r\prime}\left(C_v^r\right)^{-1}\theta^r \tag{8.49}$$

$$C_e = \left(B^{r\prime}\left(C_v^r\right)^{-1}B^r\right)^{-1} \tag{8.50}$$

Suppose now that an additional set of data θ^s is taken

$$\theta^s = B^s x + v^s \tag{8.51}$$

Provided $C_{xv}^s = 0,\ C_{vv}^{rs} = 0$, so that

$$\begin{bmatrix} \theta^r \\ \theta^s \end{bmatrix} = \begin{bmatrix} B^r \\ B^s \end{bmatrix} x + \begin{bmatrix} v^r \\ v^s \end{bmatrix}$$

Let $r + s = m$

So that $\theta^m = (\theta^r, \theta^s)'$, $B^m = (B^r, B^s)'$ and $v^m = (v^r, v^s)'$

$$C_v^m = E\left(v^m, v^{m\prime}\right)$$

$$= E\left[\begin{bmatrix} v^r \\ v^s \end{bmatrix}\begin{bmatrix} v^{r\prime} & v^{s\prime} \end{bmatrix}\right]$$

$$= E\begin{bmatrix} v^r v^{r\prime} & v^r v^{s\prime} \\ v^s v^{r\prime} & v^s v^{s\prime} \end{bmatrix}$$

$$= \begin{bmatrix} C_v^r & C_{vv}^{rs} \\ C_{vv}^{sr} & C_v^s \end{bmatrix}$$

$$= \begin{bmatrix} C_v^r & 0 \\ 0 & C_v^s \end{bmatrix}$$

$$\hat{x}^m = \left(B'^m \left(C_v{}^m \right)^{-1} B^m \right)^{-1} B'^m \left(C_v{}^m \right)^{-1} \theta^m \tag{8.52}$$

$$C_e{}^m = \left(B'^m \left(C_v{}^m \right)^{-1} B^m \right)^{-1} \tag{8.53}$$

$$\hat{x}^m = \left[\begin{pmatrix} B'^r & B^{s'} \end{pmatrix} \begin{bmatrix} C_v^r & 0 \\ 0 & C_v^s \end{bmatrix}^{-1} \begin{pmatrix} B^r \\ B^s \end{pmatrix} \right]^{-1} \begin{pmatrix} B'^r & B^{s'} \end{pmatrix} \begin{bmatrix} C_v^r & 0 \\ 0 & C_v^s \end{bmatrix}^{-1}$$

Thus yielding

$$\hat{x}^m = \left(B'^r \left(C_v^r \right)^{-1} B^r + B^{s'} \left(C_v^s \right)^{-1} B^s \right)^{-1} \left(B'^r \left(C_v^r \right)^{-1} \theta^r + B^{s'} C_v^s \right) \tag{8.54}$$

$$C_e{}^m = \left[\left(C_e^r \right)^{-1} + B^{s'} \left(C_v{}^s \right)^{-1} B^s \right]^{-1} \tag{8.55}$$

8.5.8 KALMAN FORM

Given an estimate $\hat{x}^r$, old error matrix C_e^r, new data $\theta^s = B^s x + v^s$, the new estimate $\hat{x}^m$ based on all the data is found by the sequence:

$$k = C_e^r B^{s'} \left(C_v^s + B^s C_e^r B^{s'} \right)^{-1} \tag{8.56}$$

$$\hat{x}^m = \hat{x}^r + k \left(\theta^s - B^s \hat{x}^r \right) \tag{8.57}$$

$$C_e{}^m = C_e{}^r - k B^s C_e{}^r \tag{8.58}$$

8.5.8.1 Discrete Dynamic Linear System Estimation

$$x_{i+1} = s(i+1,i)x_i + w_i, \quad s(i+1,i) \text{ is transformation matrix}$$
$$E(w_i) = 0, \quad \forall i$$
$$E(w_i w_j^T) = 0, \quad \forall i \ne j$$
$$E(w_i w_j^T) = 0, \quad \forall i \le j \tag{8.59}$$
$$W_i = E(w_i w_j^T)$$

$$\textit{Observation vector } \theta = A_i x_i + q_i, \quad A_i \text{ is transformation matrix}$$
$$E(q_i q_j^T) = 0, \quad \forall i \ne j$$
$$E(q_i w_j^T) = 0, \quad \forall i, j$$
$$E(q_i x_j^T) = 0, \quad \forall i, j \tag{8.60}$$
$$Q_i = E(q_i q_j^T)$$

8.5.9 PREDICTION

$$\hat{X}_p^m = S(p,n)\hat{X}_n^m, \quad p \ge n, \ S(.,.) \text{ is the transformation matrix} \tag{8.61}$$

$$C_p^m = S(p,n)C_n^m S^T(p,n) + \sum_{k=n}^{p-1} S(p,k+1)W_k S^T(p,k+1), \quad p > n \ge m \tag{8.62}$$

8.5.10 FILTERING

$$\hat{X}_{m+1}^m = S(m+1,m)\hat{X}_m^m \tag{8.63}$$

$$C_{m+1}^m = S(m+1,m)C_m^m S^T S(m+1,m) + W_{m+1} \tag{8.64}$$

$$K = C_{m+1}^m \Delta_{m+1}^T \left(Q_{m+1} + \Delta_{m+1}C_{m+1}^m \Delta_{m+1}^T\right)^{-1} \tag{8.65}$$

$$\hat{X}_{m+1}^{m+1} = \hat{X}_m^m + K\left(\theta_{m+1} - \Delta_{m+1}\hat{X}_{m+1}^m\right) \tag{8.66}$$

$$C_{m+1}^{m+1} = C_{m+1}^m - K\Delta_{m+1}C_{m+1}^m \tag{8.67}$$

8.5.11 SMOOTHING

$$\hat{X}_r^m = \hat{X}_r^r + J\left[\hat{X}_{r+1}^m - S(r+1,r)\hat{X}_r^r\right] \tag{8.68}$$

$$C_r^m = C_r^r + J\left[C_{r+1}^m - C_{r+1}^r\right]J^T \tag{8.69}$$

$$J = C_r^r S^T(r+1,r)\left(C_{r+1}^r\right)^{-1} \tag{8.70}$$

$$C_{r+1}^r = S(r+1,r)C_r^r S^T(r+1,r) + W_r \tag{8.71}$$

8.5.11.1 Continuous Dynamic Linear System
Let

$$\dot{x}(t) = Ax(t) + Bu(t) \tag{8.72}$$

Be the system equation and

$$y(t) = Cx(t) + Dv(t) \tag{8.73}$$

be the observation equation. The estimate $\hat{x}(t)$ is restricted as a linear function of $y(\tau)$, $0 \le \tau \le t$, thus

$$\hat{x}(t) = \int_0^t \alpha(\tau)y(\tau)d\tau \tag{8.74}$$

The solution to Equation (8.72) is

$$x(t) = \Phi(t)x(0) + \int_0^t \Phi(t)\Phi^{-1}(s)Bu(s)ds \tag{8.75}$$

where $\Phi(.)$ is the transition matrix. From Equation (8.74)

$$\hat{x}(t+\delta) = \int_0^{t+\delta} \alpha(\tau)y(\tau)d\tau$$

And from Equation (8.4),

$$x(t+\delta) = \Phi(t+\delta)\left\{ x(0) + \int_0^{t+\delta} \Phi^{-1}(\sigma) \times B(\sigma)u(\sigma)d\sigma \right\}$$

$$= \Phi(t+\delta)\Phi^{-1}(t)\Phi(t)\left\{ x(0) + \int_0^t \Phi^{-1}(\sigma)B(\sigma)u(\sigma)d\sigma \right.$$

$$\left. + \int_t^{t+\delta} \Phi^{-1}(\sigma)B(\sigma)u(\sigma)d\sigma \right\}$$

$$x(t+\delta) = \Phi(t+\delta)\Phi^{-1}(t)\left\{ x(t) + \int_t^{t+\delta} \Phi(t)\Phi^{-1}(\sigma)B(\sigma)u(\sigma)d\sigma \right\}$$

Using the orthogonality principle,

$$E\left\{ \left[x(t+\delta) - \hat{x}(t+\delta) y'(\tau) \right] \right\} = 0 \text{ for } 0 \le \tau \le t$$

and recalling that

$$E\left\{ u(\sigma) y'(\tau) \right\} = 0, \sigma > t$$

Hence,

$$E\left\{ \left[\Phi(t+\delta)\Phi^{-1}(t)x(t) - \hat{x}(t+\delta) \right] y'(\tau) \right\} = 0, 0 \le \tau \le t$$

thus

$$\hat{x}(t+\delta) = \Phi(t+\delta)\Phi^{-1}(t)\hat{x}(t),$$

If

$$u(t) \approx N(0, Q(t)),$$
$$\text{and } v(t) \approx N(0, R(t))$$

Given that

$$E\left\{ x(0) \right\} = \hat{x}_0$$

$$E\left\{ \left[x(0) - \hat{x}_0 \right] \left[x(0) - \hat{x}_0 \right]' \right\} = P_0$$

and $R^{-1}(t)$ exists.

The Kalman Filter consists of

$$\text{Estimate}: \dot{\hat{x}}(t) = A\hat{x}(t) + K(t)\left[y(t) - C\hat{x}(t)\right], \hat{x}(0) = \hat{x}_0 \tag{8.76}$$

$$\text{Error Covariance}: \hat{P}(t) = AP(t) + P(t)A' + BQB' - KRK' \tag{8.77}$$

Propagation: $P(0) = P_0$. For steady state $\hat{P}(t) = 0$

$$\text{Kalman Gain Matrix}: K(t) = P(T)C'R^{-1}(t)\text{when } E\left[u(t)'v(\tau)\right] = 0 \tag{8.78}$$

$$\text{and } \dot{K}(t) = \left[P(T)G' + BG\right]R^{-1}(t)\text{when } E\left[u(t)v'(\tau)\right] = G(t)\delta(t-\tau) \tag{8.79}$$

The fixed time smoothing algorithm $\hat{x}_{t|T}$ is as follows:

$$P(t|T) = \left(A + BQB'P^{-1}\right)' - BQB'$$

with $\hat{x}(T|T) = \hat{x}(t = T)$ and $P(T|T) = P(t = T)$ as initial conditions.

8.6 CONTINUOUS NONLINEAR ESTIMATION

The analysis of stochastic dynamic systems often leads to differential equations of the form

$$\dot{x}(t) = f(t,x) + G(t,x)u(t), x_{t_0} = c, 0 \leq t \leq T \leq \infty \tag{8.80}$$

or in integral form

$$x(t) = c + \int_0^t f(s,x)ds + \int_0^t G(s,x)dw(s), 0 \leq t \leq T \leq \infty \tag{8.81}$$

where $\dfrac{dw(t)}{dt} = u(t)$ and $w(t)$ is the Wiener process. $x(t)$ and $f(t, s)$ are n-dimensional, while $G(t, x)$ is (n x m) matrix function and $u(t)$ is m-dimensional.

Itô Rule:

$$\int_0^T G(t,x)dw(t) = \lim_{\Delta \to 0} \sum_{i=0}^N G(t_i, x(t_i))(w(t_{i+1}) - w(t_i)) \tag{8.82}$$

For the partition $t_0 < t_1 < t_i... < t_i < t_{i+1}...t_N = T$ and $\Delta = \max_i(t_{i+1} - t_i)$

Stratonovich Rule:

$$\int_0^T G(t,x)dw(t) = \lim_{\Delta\to 0} \sum_{i=0}^{N-1} G\left(t_i, \frac{x(t_{i+1})-x(t_i)}{2}\right)\left(w(t_{i+1})-w(t_i)\right) \qquad (8.83)$$

Let the observation be of the form
$y(t) = z(t) + v(t)$ where $z(t) = \varphi(x(s),\ s \leq t)$ and $v(t)$ are p-dimensional vectors.
It is further assumed that $E[z(t)z'(t)] < \infty$ and $E[z(t)v'(t)] = 0$ for all t.
Doob obtained the estimator

$$\hat{x}(t|T) = E[x(t)|y(s), t_0 \leq s \leq \tau] \qquad (8.84)$$

where for $t = \tau$ (Filtering), $t < \tau$ (Smoothing), and $t > \tau$ (Prediction).

$$\hat{x}(t) = E[x(t)|y(s), t_0 \leq s \leq \tau]$$
$$= \int_{-\infty}^{\infty} x(t) P_r(x(t)|y(t))dx(t) \qquad (8.85)$$

$$\text{Let } P_t = E\left[\left(x(t)-\hat{x}(t)\right)\left(x(t)-\hat{x}(t)\right)'\right] \qquad (8.86)$$

$$= \int_{-\infty}^{\infty} \left(x(t)-\hat{x}(t)\right)\left(x(t)-\hat{x}(t)\right)' P_r(x(t)|y(t))dx(t)$$

Assume that

$$E\{u(t)\} = E\{v(t)\} = E\{u(t)v'(t)\} = 0$$
$$E\{u(t)u'(t)\} = Q_t, E\left\{v(t)v(t)'\right\} = R_t$$

The Folker Plank stochastic differential equations for the probability density function P_r:

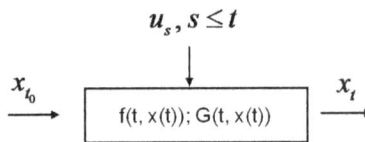

$$u_s, s \leq t$$
$$\downarrow$$

$$x_{t_0} \longrightarrow \boxed{f(t, x(t)); G(t, x(t))} \longrightarrow x_t$$

FIGURE 8.7   Functional configuration.

$$-trace\left(\frac{\partial}{\partial x}\{f(t,x)P_r\}\right)+\frac{1}{2}trace$$

$$\frac{\partial P_r}{\partial t}=\times\left(\frac{\partial}{\partial x}\left[\left[\frac{\partial}{\partial x}\right]'\{G(t,x)QG'(t,x)P_r\}\right]\right)$$
$$+P_r\left(y-\Phi(t,x)\right)R_t^{-1}$$
$$\times\left(\Phi(t,x)-\Phi(t,\hat{x})\right)$$

(8.87)

Using Equation (8.87) in Equations (8.85) and (8.86):

$$d\hat{x}(t)=\hat{f}(t,x)dt+E\left\{(x-\hat{x})\Phi'(t,x)\big|y(t)\right\}R_t^{-1}\left(y(t)-\Phi(t,x)\right)dt \qquad (8.88)$$

$$dP_t+d\hat{x}d\hat{x}=E\left\{f(t,x)(x-\hat{x})\big|y(t)\right\}dt$$
$$+E\left\{(x-\hat{x})f'(t,x)\big|y(t)\right\}dt$$
$$+E\left\{G(t,x)Q_tG'(t,x)\big|y(t)\right\}dt$$
$$+E\left\{(x-\hat{x})(x-\hat{x})'\left[\Phi(t,x)-\Phi(t,\hat{x})\right]R_t^{-1}\left(y(t)-\hat{\Phi}(t,x)\right)\big|y(t)\right\}dt$$

8.7 EXTENDED KALMAN FILTER

The extended Kalman filter results from application of the linear Kalman-Bucy filter to a linearized nonlinear system, where the nonlinear system is relinearlized after each observation.

Let

$$f(t,x)=f(t,\hat{x})-\frac{\partial f}{\partial\hat{x}}(x-\hat{x}) \qquad (8.89)$$

$$\Phi(t,x)=\Phi(t,\hat{x})-\frac{\partial\Phi}{\partial\hat{x}}(x-\hat{x}) \qquad (8.90)$$

and

$$G(t,x)Q_tG'(t,x)=G(t,\hat{x})Q_tG'(t,\hat{x})+(x-\hat{x})\left\{\left(\frac{\partial}{\partial\hat{x}}\right)'G(t,\hat{x})Q_tG'(t,\hat{x})\right\} \qquad (8.91)$$

Substituting the above into the previous equation, we obtain

$$\frac{d\hat{x}}{dt}=f(t,\hat{x})+P_t\frac{\partial\Phi}{\partial\hat{x}}R_t^{-1}\left(y(t)-\hat{\Phi}(t,x)\right) \qquad (8.92)$$

$$\frac{dP_t}{dt} = \frac{\partial f}{\partial \hat{x}} P_t + P_t \frac{\partial f'}{\partial \hat{x}} G(t,\hat{x}) Q_t G'(t,\hat{x}) - P_t \frac{\partial \Phi'}{\partial \hat{x}} R_t^{-1} \frac{\partial \Phi'}{\partial \hat{x}} P_t \qquad (8.93)$$

which can now be solved with appropriate initial conditions

$$\hat{x}\big|_{t_0} = \hat{x}(t_0) \text{ and } P_t\big|_{t_0} = P(t_0)$$

This is the *extended* Kalman Filter for *nonlinear* systems.

8.7.1 PARTITIONAL ESTIMATION

Laintiotis (1971) proposed the partition theorem, which is the general continuous-data Bayes rule for the posterior probability density.

$$\text{Let } P_r(x(\tau)|t,t_0) = \frac{\Lambda(t,t_0|x(\tau))P_r(x(\tau))}{\int \Lambda(t,t_0|x(\tau))P_r(x(\tau))d\tau} \qquad (8.94)$$

where

(i)

$$\Lambda(t,t_0|x(\tau)) = \exp\left\{ \begin{array}{c} \int_{t_0}^{t} \hat{h}'(\sigma|\sigma,t_0;x(\tau)) R_\sigma^{-1} y(\sigma) d\sigma \\ -\frac{1}{2} \int_{t_0}^{t} \left\| \hat{h}(\sigma|\sigma,t_0;x(\tau)) \right\| 2R_\sigma^{-1} d\sigma \end{array} \right\} \qquad (8.95)$$

(ii) $\hat{h}(\sigma|\sigma,t_0;x(\tau)) = E\left\{ h(\sigma,x(\sigma)) \big| y(\sigma); x(\tau) \right\}$
(iii) $P_r(x(\tau))$ is the a priori density of $x(\tau)$

The partitioned algorithm for filtering is given by

$$\hat{x}(t) = \int \hat{x}(\tau) P_r(x(\tau)|t,t_0) dx(\tau) \qquad (8.96)$$

and

$$P_t = \int \left\{ P_\tau + \left[\hat{x}(\tau) - \hat{x}(t) \right]\left[\hat{x}(\tau) - \hat{x}(t) \right]' P_r(x(\tau)|t,t_0) dx(\tau) \right\} \qquad (8.97)$$

where $P_r(x(\tau) \mid t, t_0)$ is given previously and both $\hat{x}(\tau)$ and $P\tau$ are the "anchored" or conditional mean-square error estimate and error-covariance matrices, respectively.

The partitioned algorithm takes its name from the fact that if the observation interval is partitioned into several small subintervals, then repeated use of the filtering equations for each subinterval leads to an effective and computationally efficient algorithm for the general estimation problem.

8.8   INVARIANT IMBEDDING

The invariant imbedding approach provides a sequential estimation scheme, which does not depend on a priori noise statistical assumptions. The concept in invariant imbedding is to find the estimate $\hat{x}(\tau)$ of $x(t)$ such that the cost function

$$J = \frac{1}{2} \int_0^T \left\{ \| y(t) - \Phi(t,\hat{x}(t)) \|_{W_1}^2 + \| \hat{x}(t) - f(t,\hat{x}(t)) \|_{W_2}^2 \right\} dt \qquad (8.98)$$

is minimized where W_1 and W_2 are weighing matrices that afford the opportunity to place more emphasis on the most reliable measurements. The Hamiltonian

$$\begin{aligned} H = \frac{1}{2} W_1 \left(y(t) - \Phi(t,\hat{x}(t)) \right)^2 + \frac{1}{2} W_2 G^2 (t,x(t)) u^2(t) \\ + \lambda^2(t) \left(f(t,\hat{x}(t)) + G(t,\hat{x}(t)) u(t) \right) \end{aligned} \qquad (8.99)$$

The necessary conditions for a minimum are

$$\dot{\hat{x}}(t) = \frac{\partial H}{\partial \lambda}$$

$$\dot{\lambda}(t) = -\frac{\partial H}{\partial \hat{x}} \qquad (8.100)$$

$$\frac{\partial H}{\partial u} = 0$$

which yield the filtering equations

$$\frac{d\hat{x}}{dT} = f(T,\hat{x}(T)) + P \frac{\partial \Phi'}{\partial \hat{x}} \left(y(T) - \Phi(T,\hat{x}(T)) \right) \qquad (8.101)$$

$$\frac{dP}{dT} = \frac{\partial f}{\partial \hat{x}} P + P \frac{\partial f'}{\partial \hat{x}} + P \left(\left[\frac{\partial^2 \Phi'}{\partial \hat{x}^2} \right] \left(y(T) - \Phi(T,\hat{x}(T)) \right) - \frac{\partial \Phi'}{\partial \hat{x}} \frac{\partial \Phi}{\partial \hat{x}} \right) P + \frac{1}{W} \qquad (8.102)$$

8.9   STOCHASTIC APPROXIMATIONS/INNOVATIONS CONCEPT

Stochastic approximation is a scheme for successive approximation of a sought quantity when the observation and the system dynamics involve random errors. It is applicable to the statistical problem of

(i) finding the value of a parameter, which causes an unknown noisy regression function to take on some pre-assigned value.
(ii) finding the value of a parameter, which minimizes an unknown noisy regression function.

Stochastic approximation has wide applications to system modeling, data filtering, and data prediction. It is known that a procedure, which is optimal in decision theoretic sense, can be non-optimal. Sometimes, the algorithm is too complex to implement, for example, in situations where the nonlinear effects cannot be accurately approximated by linearization or the noise process is strictly non-Gaussian. A theoretical solution is obtained by using the concepts of innovations and martingales. Subsequently, a numerically feasible solution is achieved through stochastic approximation. The innovations approach separates the task of obtaining a more tractable expression for the equation

$$\hat{x}(t|T) = E\left\{ x(t) \middle| y(s), 0 \le s \le \tau \right\} \tag{8.103}$$

into two parts:

(i) the data process $\{y(t), 0 \le t \le T\}$ is transformed through a causal and causally invertible filter $v(t) = y(t) - \Phi(\hat{x}(s), s \le t)$ called the innovations process with the same intensity as the observation process.

(ii) the optimal estimator is determined as a functional of the innovations process.

The algorithm given below has been used for several problems:

(i) Pick an α_t^i gain matrix function, such that for each element $\left(\alpha_t^i \right)_{kl}$,

$$\int_0^\infty \left(\alpha_t^i \right)_{kl} dt = \infty, i = 1 \text{ and } \int_0^\infty \left(\alpha_t^i \right)_{kl}^2 dt < \infty \tag{8.104}$$

(ii) Solve the *suboptimal* problem

$$\frac{d\hat{x}}{dt} = f(t, \hat{x}) + \alpha_t^i G(t, \hat{x})\left(y(t) - \Phi(t, \hat{x}) \right) \tag{8.105}$$

where it is assumed without any loss of generality with entries $\left(\alpha_1^i, \alpha_2^i, \dots \alpha_n^i \right)$. The *l*-th component of equation (i) is

$$\frac{d\hat{x}_l}{dt} = f_l(t, \hat{x}) + \sum_{k=1}^m \alpha_i^i g_{lk}(t, \hat{x})\left(y_k(t) - \Phi_k(t, \hat{x}) \right)$$

(iii) Compute the innovations process

$$v^i(t) = y(t) - \Phi(t, \hat{x}^i) \tag{8.106}$$

and check for its whiteness (within a prescribed tolerance level) by computing the auto-correlation function as well as the power spectrum.
(iv) If the result of the test conducted in step (iii) is positive, STOP.
ELSE, iterate on α_t^i thus

$$\alpha^{i+1}(t) = \alpha^i(t) + \gamma^i(t)\Psi\left(v^i(t)\right) \qquad (8.107)$$

where

$$\alpha^1(t) = \gamma^i(t)$$
$$= \left\{-\frac{a}{t} \text{ or } -\frac{a}{b+t} \text{ or } -\frac{a+t}{b+t^2}\right\} \qquad (8.108)$$

and

$$\Psi\left(v^i(t)\right) = v^i(t) - E\left\{v^i(t)\right\} \qquad (8.109)$$

(v) Go to step (ii)

The optimal trajectories constitute a martingale process, and the convergence of the
approximate algorithm depends on the assumption that the innovations of the observations is a martingale process, thus the martingale convergence theorem states that:
    If $\{x^n\}_k$ is a submartingale, and if least upper bond(l.u.b.) $E\{x^n\}_k < \infty$, then there is
a random discretized system.

8.10   MODEL CONTROL – MODEL REDUCTION, MODEL ANALYSIS

8.10.1   INTRODUCTION

One of the challenges in Systems Modeling and subsystem estimation is the need
to reduce large-scale systems to lower dimensions in order to effectively control the
models.
    Consider an *nth*-order linear system S_1 defined by

$$S_1 : \dot{x} = Ax + Bu \qquad (8.110)$$

where x is the n-dimensional state vector, A is an (nXn)system matrix, and u is a
p-dimensional input vector. Let z be an m-vector $(m < n)$ related to x by

$$z = Cx. \qquad (8.111)$$

In model reduction, it is desirable to find an *m*th-order system S_2described by

$$S_2 : \dot{z} = Fz + Gu. \qquad (8.112)$$

The $(m \times n)$ matrix C in Equation (8.111) is the aggregation matrix and S_2 is the aggregated system or the reduced model. It is easy to show that $G = CD$ and that F must satisfy the matrix equation

$$FC = CA. \tag{8.113}$$

Equation (8.113) defines an over-specified system of equation for the unknown matrix F, and hence, F must be approximated. A multivariate linear regression scheme is used to yield a "best" approximation for F in the form

$$\hat{F} = CAC^T \left(CC^T \right)^{-1}, \tag{8.114}$$

where T and -1 denote matrix transpose and matrix inverse, respectively. The rank of C is assumed to be m. The result given by Equation (8.114) is interpreted as a linear, unbiased, minimum-variance estimate of F and its form agrees with that given by Aoki (1968), following an ad hoc procedure.

In addition, the covariance of $\hat{F}$ is found to be

$$con\left(vec\, \hat{F} \right) = \sigma^2 \left[\left(CC^T \right)^{-1} \otimes I_m \right] \tag{8.115}$$

and it is shown that this covariance matrix can be used for model reduction error assessment. In Equation (8.115), the Kronecker product $\otimes$ and the "vec" operator are defined as

$$P \otimes Q = \left[P_{ij} \quad Q \right] \tag{8.116}$$

$$vec(P) = \left[P_1, \; P_2, \; \ldots \right]^T, \tag{8.117}$$

where P and Q are matrices of arbitrary dimensions and P_k is the kth column of matrix P.

From the computational point of view, it is desirable to circumvent the use of matrix inverses in Equations (8.114) and (8.115), particularly for systems with large aggregation matrices. In what follows, this is accomplished through the use of matrix singular value decomposition (SVD), which has found useful application in several linear least squares problems.

The SVD concept gives the Moore-Penrose pseudo-inverse of C as

$$C^+ = VAU^T \tag{8.118}$$

where U and V are unitary matrices whose columns are the eigenvectors of matrices DD^T, and D^TD, respectively, and

$$A = \begin{bmatrix} \sigma_1^{-1} & & & 0 & & \\ & \sigma_2^{-1} & & & & \\ & & \ddots & & & \\ 0 & & & \sigma_m^{-1} & & \\ & & & & 0 & \\ & & & & & \ddots \\ & & & & & & 0 \end{bmatrix}_{n \times n} \qquad (8.119)$$

where $\sigma_1 \geq \sigma_2 \geq \cdots \geq \sigma_m \geq 0$, called singular values, are the nonnegative square roots of the eigenvalues of $D^T D$. A discussion of this decomposition and its properties can be found in Stewart (1993).

Now, Equations (8.113) gives

$$\hat{F} = CAC^+ \qquad (8.120)$$

And, using SVD, we can write

$$\hat{F} = CA \left(VAU^T \right). \qquad (8.121)$$

The matrix C can also be written in the form

$$D = U \Sigma V^T \qquad (8.122)$$

where

$$\Sigma = \begin{bmatrix} \sigma_1 & & & & & \\ & \sigma_2 & & 0 & & \\ & & \ddots & & & \\ & 0 & & \sigma_m & & \\ & & & & 0 & \\ & & & & & \ddots \\ & & & & & & 0 \end{bmatrix}_{n \times n} \qquad (8.123)$$

and we have

$$\hat{F} = U \Sigma V^T A V A U^T \qquad (8.124)$$

Compared with Equation (8.114), either Equation (8.121) or Equation (8.124) provides a more efficient method of computation for $\hat{F}$ due to elimination of the matrix inverse.

Similarly, advantages are realized in the calculation of $con\left(vec \hat{F} \right)$. Following the SVD scheme,

$$
\begin{aligned}
\left(DD^T \right)^{-1} = \left(D^T \right)^{+} D^{+} \\
= \left(UAV^T \right) \left(VAU^T \right) \\
= \left(UA^2U^T \right)
\end{aligned}
\tag{8.125}
$$

Equation (8.6) now takes the form

$$
con\left(vec\ \hat{F} \right) = \sigma^2 \left[UA^2U^T \otimes I_m \right],
\tag{8.126}
$$

which is clearly of a simpler structure compared to Equation (8.115).

8.11 MODAL APPROACH FOR ESTIMATION IN DISTRIBUTED PARAMETER SYSTEMS

We can expand this concept to those describable by distributed parameter systems: based on a scheme, on eigenmode expansion, and for the estimation of system responses in distributed parameter systems with nonlinear sensors. The associated joint conditional probability distribution is realized by using the partition theorem and Wiener functional expansions. The system state parameters are subsequently computed by obtaining the innovation processes from the observations, thus leading to structural estimation.

A technique for optimal estimation of system parameters in a stochastic environment with nonlinear sensors which are hereby presented. It is based on the eigenmode representation of generalized displacements and orthogonal expansion of the conditional joint probability distribution. The algorithm proposed in this chapter for realization of the innovations and system state is the starting point in the search of nonlinear distributed-parameter estimation techniques. A linear dynamical system is used to illustrate the correspondence between the innovations process, the observations process, and the system response.

The proposed algorithm may be applied to several civil engineering and space structures, best described by the distributed parameter equation

$$
m\left(x \right) u_{ii}\left(x,t \right) + Bu_i\left(x,t \right) + Cu\left(x,t \right) = F_0\left(x,t \right)
\tag{8.127}
$$

with the initial conditions

$$
u\left(x,t_0 \right) = u^0 \quad \text{and} \quad u_i\left(x,t_0 \right) = u_i{}^0
$$

where $m(x)$ is the mass per unit length (assumed unity); $u(x, t)$ is the vector of generalized displacements; B and C are in general spatial differential operators representing the damping and restoring forces, respectively; and $F_0(x, t)$ is the external impressed forces (wind loads, earthquake excitations, etc.).

8.12 THE MODAL CANONICAL REPRESENTATION

Let the displacement be expressed in terms of structural dominant mode shapes $\{\phi_i(x)\}_1^N$ with associated frequencies $\{w_i\}_1^N$

$$u(x,t) = \sum_1^N u_i(t)\phi_i(x); \quad F_0(x,t) = \sum_1^N F_1(t)\phi_i(t)$$

so that
  and

$$
\left.
\begin{aligned}
u(x,t) &= U^T(t)\Phi(x) \\[2mm]
F_0(x,t) &= F_0{}^T(t)\Phi(x)
\end{aligned}
\right\}
\tag{8.128}
$$

where

$$U = (u_1, u_2 \ldots, u_N)^T, \quad F_0 = (F_1^0, F_2^0, \ldots, F_N^0)^T$$

and

$$\Phi = (\phi_1, \phi_2, \ldots, \phi_N)^T$$

Substituting Equation (8.128) into Equation (8.127) gives

$$U^T(t)\Phi(x) + BU^T(t)\Phi(x) + CU^T(t)\Phi(x) = F_0{}^T(t)\Phi(x)$$

thus obtaining the ith mode amplitude equation

$$\ddot{u}_i(t) + 2\xi_i\omega_i\dot{u}_i(t) + \omega_i^2 u_i(t) = F_i^0(t) \tag{8.129}$$

where $C\phi_i = \omega_i^2\phi_i$ and $B\phi_i = 2\xi_i\omega_i\phi_i$ with appropriate initial conditions.
  Equation (8.129) can now be put in the state space form

$$\dot{q}(t) = Aq(t) + F \tag{8.130}$$

such that

$$q = (q_1, q_2)^T$$

with

$$q_1 = u_i, \quad q_2 = \dot{u}_i$$

$$A = \begin{pmatrix} 0 & 1 \\ -\omega_i^2 & -2\xi_i\omega_i \end{pmatrix}$$

and

$$F = \left(0, F_i^0\right)^T$$

The sensors are assumed to have the form

$$y(t) = z(t) + v(t); \quad z(t) = h\left[q(s); \ s \le t\right] \tag{8.131}$$

where

$$E\{v(t)\} = 0;$$
$$E\{v(t)v^T(s)\} = R(t)\delta(t-s);$$
$$E\{z(t)\} = 0;$$
$$\left[z(t)\right] \le M < \infty \text{ for all } t;$$

$$\int_0^T E\{z(t)z^T(t)\} \ dt < \infty; \quad E\{v(t)F^T\} = 0 \quad \text{and } h \text{ is a functional of } q(s).$$

It is further assumed that the forcing function *F* is additively (preferably Gaussian) random.

The least squares estimate of $q(t)$ is

$$\hat{q}(t|\tau) = E\{q(\tau)|y(s), \ 0 \le s \le \tau\} \tag{8.132}$$

Assuming that the property of causal equivalence holds, the sensor equations may be transformed into a white Gaussian process $v(t)$, called the strict sense innovations process with a simpler structure so that Equation (8.132) can now be put in the form

$$\hat{q}(t|\tau) = \int_0^\tau E\{q(t)v^T(s)|v(\sigma), \ 0 \le \sigma < s\}v(s)ds \tag{8.133}$$

where $v(t) = y(t) - \hat{z}(t \mid t)$, $0 \le t < \tau$ and $\int$ is the $It\hat{o}$ (1951) integral. A differential structure for Equation (8.7) is obtained as

$$\dot{\hat{q}}(t \mid t) + A\hat{q}(t \mid t) + K(t)v(t) \qquad (8.134)$$

where

$$K(t) = E\left\{\hat{q}(t)v^T(t) \mid v(\sigma), \; 0 \le \sigma < t\right\} \qquad (8.135)$$

Conditional joint probability distribution functions are required to obtain explicit solution to the stochastic differential Equation (8.134).

A method for the evaluation of the probability densities is obtained in the form of a partition theorem (Laintiotis 1971)

$$p\left(q \mid f^t\right) = A\left(t \mid q\right)p\left(q_0\right)\Big| \int P\left(q_0\right)A\left(t \mid q\right)dq \qquad (8.136)$$

with $f^t = \{y(\tau)\}_0^t$ and

$$A\left(t \mid q\right) = \exp\left\{\int_0^t \hat{h}(s,q)R^{-1}(s)y(s)ds - \frac{1}{2}\int_0^t \left\|\hat{h}(s,q)\right\|^2_{R^{-1}(s)} ds\right\} \qquad (8.137)$$

where

$$\hat{h}(s,q) = E\left\{h(s,q) \mid f^s\right\}$$

and $p(q_0)$ is the priori density function.

Equation (8.136) can be approximated by expanding the distributions $A(t \mid q)$ in a Volterra (Barrett and Lampard 1955, Biglieri 1973) power series with functional terms

$$A\left(t \mid q\right) = \sum_{i=0}^{\infty} A_i\left(t \mid q\right) \qquad (8.138)$$

where

$$A_i\left(t \mid q\right) = \frac{1}{i!}\int_0^{\infty} d\tau_1 \dots \int_0^{\infty} d\tau_i g_i\left(\tau_1, \tau_2, \dots, \tau_i\right)\prod_i^{k-1} v\left(t - \tau_k\right) \qquad (8.139)$$

such that $g_i(...)$, $i = 1, ..., n$ are the integral kernels describing the system and $v(\cdot)$ is the innovations process. It has been assumed that in order to satisfy the requirements of physical realizability, the kernels are zero for any argument less than zero. The kernels can be identified in the following manner.

(i) Let $v(t) = A\delta_0(t)$ be an impulse of strength A so that

$$\Lambda(t|q) = Ag_1(t) + A^2 g_2(t,t) + A^3 g_3(t.t,t) + ...$$ (8.140)

and

$$g_1(t_1) = \frac{\delta\Lambda(t_1|q)}{\delta A}\bigg|_{A=0}$$ (8.141)

(ii) Second-order kernel: $g_2(t, t)$

Let $P(t) = A\delta_0(t) + B\delta_0(t + \tau)$ be two impulses at t and $t + \tau$ of strength A and B, respectively, where A and B are specified constants as above ; then

$$A(t|\theta) = Ag_1(t) + Bg_1(t+\tau) + A^2 g_2(t,t) + 2ABg_2(t,t+\tau) + B^2 g_2(t+\tau,t+\tau) + ...$$

and the second-order kernel

$$g_2(t,\ t+\tau) = \frac{1}{2}\frac{\delta^2\Lambda(t_1|q)}{\delta A\delta B}\bigg|_{\substack{A=0\\B=0}}$$ (8.142)

Higher-order kernels can be computed in similar manners.
The system state estimates can now be obtained as

$$\hat{q}(t|t) = J(t|t_0)\hat{q}(t|t_0) + \int_{t_o}^{t} J(t|\tau)k(\tau)v(\tau)d\tau$$ (8.143)

where

$$J(t|t_0) = \exp\{A(t-t_0)\}$$ (8.144)

is the transition/fundamental matrix; $k(\tau)$ having been approximated using Equations (8.136) and (8.138).

As an example take a hypothetical linear dynamical system that is solved below, using the Kalman filtering equations, to illustrate the close correlation (in a non-statistical sense) between the observations process, the innovations process, and the systems state. It has been assumed that the distributed parameter system is reducible to a linear ordinary differential equation system as follows:

System dynamics:

$$\dot{x}(t) = 0.5x(t) + u(t), \quad x(0) = 0$$

Observation process:

$$y(t) = x(t) + v(t)$$

Figure 8.8 represents the observation process, which is simply taken here as the system response plus a white Gaussian noise. Figure 8.9 is the innovations

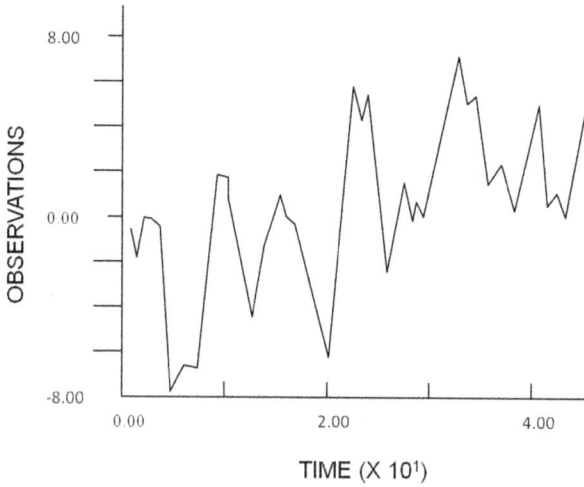

FIGURE 8.8   Observation process.

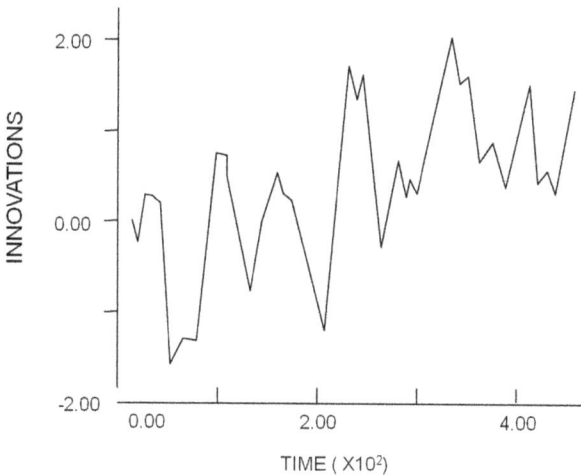

FIGURE 8.9   Innovations process.

representation. Figure 8.10 is the innovations auto-correlation, which indicates that the process is a white noise. Finally, Figure 8.11 shows the system state $x(t)$.

An approximation technique becomes imperative in the nonlinear case since the innovations sequence generated is not a true innovations process, and hence, a scheme for adaptive improvement has to be employed.

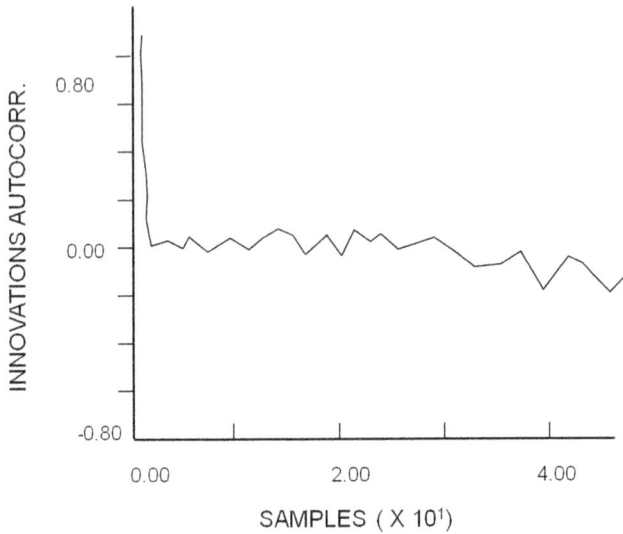

FIGURE 8.10   Autocorrelation for innovations process.

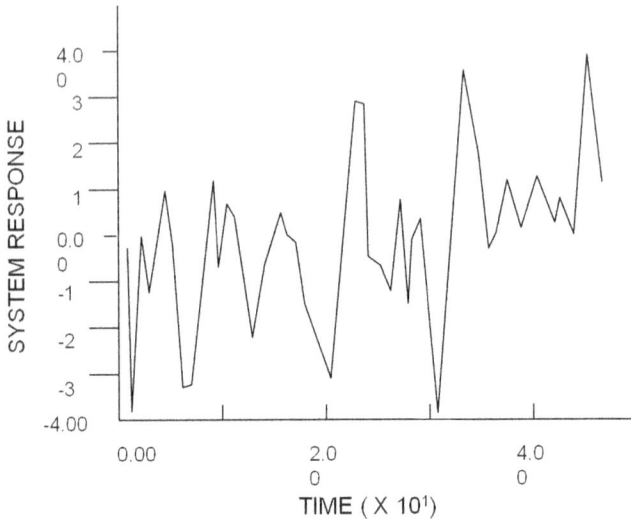

FIGURE 8.11   System response.

The modal approach for the decomposition of response estimates and conditional probability distributions provides an innovative method for the optimal estimation in some nonlinear distributed parameter systems. The generation and determination of the innovations for the general (non-Gaussian) case is obtained through a stochastic approximation technique as is illustrated by an example given in this paper. The method is expected to provide a viable alternative to similar Markov diffusion problems since it is easier to understand and computationally more efficient. Further details on the contents of this chapter can be found in Albert and Gardner 1967; Bekey, 1970; Clark, 1969; Liebelt, 1967; Gelb, 1974; Sage and Melsa 1971; Golub and Reinsch, 1970; Golub and Van Loan 1979; Itô, 1951; Lee, 1964; Rashwan and Ahmed 1982; Soong (1977).

REFERENCES

Albert, A. E. and L. A. Gardner Jr (1967). *Stochastic Approximation and Nonlinear Regression.* John Wiley, New York, NY.

Aokl, M. (1968). "Control of Large Scale Dynamic Systems B Aggregation". *IEEE Transactions on Automatic Control* AC-13: 246–258.

Badiru, A. B., Oye Ibidapo-Obe and B. J. Ayeni (2012). *Industrial Control Systems: Mathematical and Statistical Models and Techniques.* Taylor & Francis Group/CRC Press, Boca Raton, FL.

Barrett, J. F. and D. G. Lampard (1955). "Statistical Distributions". *IRE Transactions on Information Theory* 1: 10, 27–35.

Bekey, G. A. (1970). "System Identification – An Introduction and a Survey". *Simulation* 45(9): 35–43.

Biglieri, E.(1973). "Power Series with Functional Terms". *Proceedings of the Institution of Electrical Engineers,* 61: 251.

Clark, J. M. C. (1969). *Computational Statistics for Control, Technical Report.* Department of Computing and Control, Imperial College, London, UK.

Liebelt, P. B. (1967). *An introduction to Optimal Estimation.* Addison-Wiley, New York, NY.

Gelb, A. (1974). *Applied Optimal Estimation.* M.I.T Press, Boston, MA.

Sage A. P. and Melsa J. L. (1971). *System Identification.* Academic Press, New York, NY.

Golub, G. H. and Reinsch, C. (1970). "Singular Value Decomposition and Least Squares Solutions". *Numerische Mathematik* 14: 403–420.

Golub G. H. and Van Loan C. (1979). "Total least squares," in *Smoothing Techniques for Curve Estimation* (T. Gasser and M. Rosenblatt, Eds.), pp. 62–76. Springer Verlag, New York.

Itô, K. (1951). "On Stochastic Differential Equations". *Memoirs of the American Mathematical Society* 4 .

Laintiotis, D. G. (1971). "Statistics of Simulation Analysis". *International Journal of Control* 14: 1137.

Lee, R. C. K. (1964). *Optimal Estimation, Identification and Control.* M.I.T. Press.

Rashwan, A. H. and A. S. Ahmed (1982). Simulation Analysis Conference, *Proc. I. F. A. C.,* India.

Soong T. T. (1977). "On Model Reduction of Large Scale Systems". *Journal of Mathematical Analysis and Applications* 60: 477–482.

Stewart G. W. (1993). *Introduction to Matrix Computations.* Academic Press, New York, 1978.

9 Mathematical Utility Modeling for AI Application

9.1 INTRODUCTION TO UTILITY MODELING

We, by default, believe that every innovation has utility to the same desirable extent. Unfortunately, it is not always the case. Every innovation has utility on differing scales and to differing extents. In fact, for social reasons, some innovations may not be desirable at all. For example, in warfare history, the invention of the Gatling Gun was hailed by some, while abhorred by others. So, innovation can mean different things to different groups. For this reason, the technique of utility modeling is applicable for the assessment of innovation. The advantage of using utility modeling for innovation is that assessment metrics are imposed on the expected performance of innovation. How do we know if and when innovation has occurred and to what extent? It is easy to claim to be innovative or to proclaim commitment to innovation, but it is a different thing to be able to present a quantifiable proof of innovation. See Badiru (2008, 2012) and Saaty (1980).

9.2 INNOVATION INVESTMENT CHALLENGE

Innovation investment selection is an important aspect of investment planning. The right investment must be undertaken at the right time to satisfy the constraints of time and resources. A combination of criteria can be used to help in investment selection, including technical merit, management desire, schedule efficiency, cost-benefit ratio, resource availability, criticality of need, availability of sponsors, and user acceptance.

Many aspects of investment selection cannot be expressed in quantitative terms. For this reason, investment analysis and selection must be addressed by techniques that permit the incorporation of both quantitative and qualitative factors. Some techniques for investment analysis and selection are presented in the sections that follow. These techniques facilitate the coupling of quantitative and qualitative considerations in the investment decision process. Techniques such as net present value, profit ratio, and equity break-even point, which have been presented in the preceding sections, are also useful for investment selection strategies.

9.3 UTILITY MODELS

The term "utility" refers to the rational behavior of a decision-maker faced with making a choice in an uncertain situation. The overall utility of an investment can be measured in terms of both quantitative and qualitative factors. This section presents an approach to investment assessment based on utility models that have been developed within an extensive body of literature. The approach fits an empirical utility function to each factor that is to be included in a multi-attribute selection model. The specific utility values (weights) that are obtained from the utility functions are used as the basis for selecting an investment see Badiru (2008), Badiru (2012), and Saaty (1980).

Utility theory is a branch of decision analysis that involves the building of mathematical models to describe the behavior of a decision-maker faced with making a choice among alternatives in the presence of risk. Several utility models are available in the management science literature. The utility of a composite set of outcomes of n decision factors is expressed in the general form below:

$$U(x) = U(x_1, x_2, x_n),$$

where x_i = specific outcome of attribute X_i, $i = 1,2,...,n$ and $U(x)$ is the utility of the set of outcomes to the decision-maker. The basic assumption of utility theory is that people make decisions with the objective of maximizing those decisions' *expected utility*. Drawing on an example presented by Park and Sharp-Bette (1990), we may consider a decision-maker whose utility function with respect to investment selection is represented by the following expression:

$$u(x) = 1 - e^{-0.0001x},$$

where x represents a measure of the benefit derived from an investment. Benefit, in this sense, may be a combination of several factors (e.g., quality improvement, cost reduction, or productivity improvement) that can be represented in dollar terms. Suppose this decision-maker is faced with a choice between two investment alternatives, each of which has benefits as specified below:

Investment 1: Probabilistic levels of investment benefits.

Benefits (x): –$10,000, $0, $10,000, $20,000, $30,000
Probabilities (P(x)): 0.2, 0.2, 0.2, 0.2, 0.2

Investment 2: A definite benefit of $5000.

Assuming an initial benefit of zero and identical levels of required investment, the decision-maker must choose between the two investments. For Investment 1, the expected utility is computed as shown below:

$$E[u(x)] = \Sigma u(x)\{P(x)\}.$$

For Investment 1, using the utility function calculations and the probability values provided, we have the following series of calculations for the expected utility:

$$-\$10,000(-1.7183)(0.2) = -0.3437$$

$$\$0(0)(0.2) = 0$$

$$\$10,000(0.6321)(0.2) = 0.1264$$

$$\$20,000(0.8647)(0.2) = 0.1729$$

$$\$30,000(0.9502)(0.2) = 0.1900$$

Thus, $E[u(x)_1] = 0.1456$.

For Investment 2, we have $u(x)_2 = u(\$5000) = 0.3935$. Consequently, the investment providing a certain amount of $5000 is preferred to the riskier Investment 1, even though Investment 1 has a higher expected benefit of $\Sigma xP(x) = \$10,000$. A plot of the utility function used in the above example is presented in Figure 9.1.

If the expected utility of 0.1456 is set equal to the decision-maker's utility function, we obtain the following:

$$0.1456 = 1 - e^{-0.0001x*},$$

which yields $x* = \$1574$, referred to as the *certainty equivalent* (*CE*) of Investment 1 ($CE_1 = 1574$). The certainty equivalent of an alternative with variable outcomes is a *certain amount* (*CA*), which a decision-maker will consider to be desirable to the same degree as the variable outcomes of the alternative. In general, if *CA* represents the certain amount of benefit that can be obtained from Investment 2, then the criteria for making a choice between the two investments can be summarized as follows:

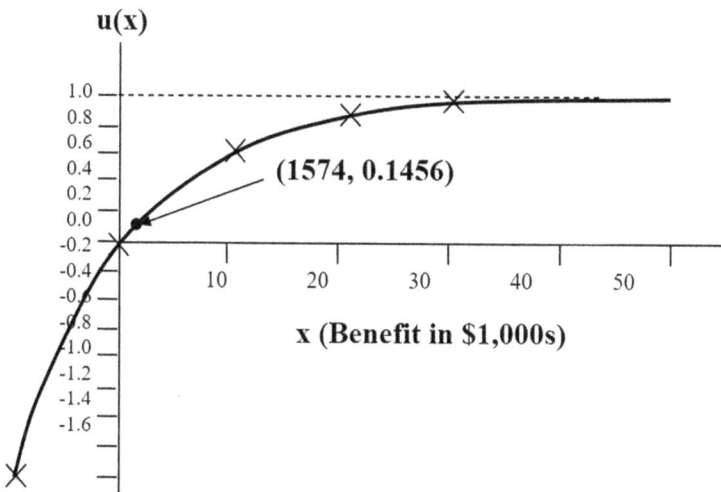

FIGURE 9.1 Utility function and certainty equivalent.

If $CA < \$1574$, select Investment 1.
If $CA = \$1574$, select either investment.
If $CA > \$1574$, select Investment 2.

The key in using utility theory for investment selection is choosing the proper utility model. The sections that follow describe two simple but widely used utility models: the *additive utility model* and the *multiplicative utility model*.

9.3.1 ADDITIVE UTILITY MODEL

The additive utility of a combination of outcomes of n factors $(X_1, X_2, \ldots, X_n)$ is expressed as follows:

$$U(x) = \sum_{i=1}^{n} U\left(x_i, \bar{x}_i^0\right)$$

$$= \sum_{i=1}^{n} k_i U\left(x_i\right),$$

where

x_i = measured or observed outcome of attribute i
n = number of factors to be compared
x = combination of the outcomes of n factors
$U(x_i)$ = utility of the outcome for attribute i, x_i
$U(x)$ = combined utility of the set of outcomes, x
k_i = weight or scaling factor for attribute $i (0 < k_i < 1)$
X_i = variable notation for attribute i
x_i^0 = worst outcome of attribute i
x_i^* = best outcome of attribute i
$\bar{x}_i^0$ = set of worst outcomes for the complement of x_i
$U\left(x_i, \bar{x}_i^0\right)$ = utility of the outcome of attribute I and the set of worst outcomes
    for the complement of attribute i
$k_i = U\left(x_i^*, \bar{x}_i^0\right)$

$\sum_{i=1}^{n} k_i = 1.0$ (required for the additive model).

Example 9.1

Let **A** be a collection of four investment attributes defined as **A** = {Profit, Flexibility, Quality, Productivity}. Now define **X** = {Profit, Flexibility} as a subset of A. Then, $\bar{X}$ is the complement of **X** defined as $\bar{X}$ = {Quality, Productivity}. An example of the comparison of two investments under the additive utility model is summarized in Table 9.1 and yields the following results:

$$U(x)_A = \sum_{i=1}^{n} k_i U_i(x_i) = .4(.95) + .2(.45) + .3(.35) + .1(.75) = 0.650$$

$$U(x)_B = \sum_{i=1}^{n} k_i U_i(x_i) = .4(.90) + .2(.98) + .3(.20) + .1(.10) = 0.626$$

since $U(x)_A > U(x)_B$, Investment A is selected.

9.3.2 MULTIPLICATIVE UTILITY MODEL

Under the multiplicative utility model, the utility of a combination of outcomes of n factors $(X_1, X_2, \dots, X_{n1})$ is expressed as

$$U(x) = \frac{1}{C}\left[\prod_{i=1}^{n}\left(Ck_i U_i(x_i) + 1\right) - 1\right],$$

where C and k_i are scaling constants satisfying the following conditions:

$$\prod_{i=1}^{n}\left(1 + Ck_i\right) - C = 1.0$$

$$-1.0 < C < 0.0$$

$$0 < k_i < 1.$$

The other variables are as defined previously for the additive model. Using the multiplicative model for the data in Table 9.1 yields $U(x)_A = 0.682$ and $U(x)_B = 0.676$. Thus, Investment A is the best option.

9.3.3 FITTING A UTILITY FUNCTION

An approach presented in this section for multi-attribute investment selection is to fit an empirical utility function to each factor to be considered in the selection process. The specific utility values (weights) that are obtained from the utility functions

TABLE 9.1
Example of Additive Utility Model

Attribute (i)	Weight (k_i)	Investment A $U_i(x_i)$	Investment B $U_i(x_i)$
Profitability	0.4	0.95	0.90
Flexibility	0.2	0.45	0.98
Quality	0.3	0.35	0.20
Throughput	0.1	0.75	0.10
	1.00		

may then be used in any of the standard investment justification methodologies. One way to develop empirical utility function for an investment attribute is to plot the "best" and "worst" outcomes expected from the attribute and then to fit a reasonable approximation of the utility function using concave, convex, linear, S-shaped, or any other logical functional form.

Alternately, if an appropriate probability density function can be assumed for the outcomes of the attribute, then the associated cumulative distribution function may yield a reasonable approximation of the utility values between 0 and 1 for corresponding outcomes of the attribute. In that case, the cumulative distribution function gives an estimate of the cumulative utility associated with increasing levels of attribute outcome. Simulation experiments, histogram plotting, and goodness-of-fit tests may be used to determine the most appropriate density function for the outcomes of a given attribute. For example, the following five attributes are used to illustrate how utility values may be developed for a set of investment attributes. The attributes are return on investment (ROI), productivity improvement, quality improvement, idle time reduction, and safety improvement.

Example 9.2

Suppose we have historical data on the return on investment (ROI) for investing in a particular investment. Assume that the recorded ROI values range from 0% to 40%. Thus, the worst outcome is 0% and the best outcome is 40%. A frequency distribution of the observed ROI values is developed and an appropriate probability density function (*pdf*) is fitted to the data. For our example, suppose the ROI is found to be exceptionally distributed with a mean of 12.1%. That is:

$$f(x) = \begin{cases} \dfrac{1}{\beta} e^{-x/\beta}, & \text{if } x \geq 0 \\ 0, & \text{otherwise} \end{cases}$$

$$F(x) = \begin{cases} 1 - e^{-x/\beta}, & \text{if } x \geq 0 \\ 0, & \text{otherwise} \end{cases}$$

$$\approx U(x),$$

where $\beta = 12.1$. $F(x)$ approximates $U(x)$. The probability density function and cumulative distribution function are shown graphically in Figure 9.2. The utility of any observed ROI within the applicable range may be read directly from the cumulative distribution function.

For the productivity improvement attribute, suppose it is found (based on historical data analysis) that the level of improvement is normally distributed with a mean of 10% and a standard deviation of 5%. That is,

$$f(x) = \dfrac{1}{\sqrt{2\pi}\sigma} e^{-\frac{1}{2}\left(\frac{x-\mu}{\sigma}\right)^2}, \quad -\infty < x < \infty$$

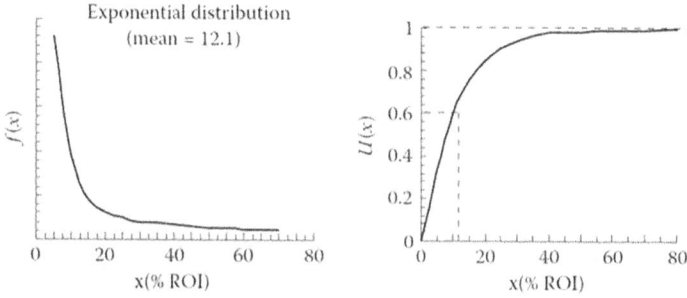

FIGURE 9.2   Estimated utility function for investment ROI.

where $\pi = 10$ and $\sigma = 5$. Since the normal distribution does not have a closed-form expression for $F(x)$, $U(x)$ is estimated by plotting representative values based on the standard normal table. Figure 9.3 shows $f(x)$ and the estimated utility function for productivity improvement. The utility of productivity improvement may also be evaluated on the basis of cost reduction.

Suppose quality improvement is subjectively assumed to follow a beta distribution with shape parameters $\alpha = 1.0$ and $\beta = 2.9$. That is,

$$f(x) = \frac{\Gamma(\alpha+\beta)}{\Gamma(\alpha)\Gamma(\beta)} \cdot \frac{1}{(b-a)^{\alpha+\beta-1}} \cdot (x-a)^{\alpha-1}(b-x)^{\beta-1},$$

$$\text{for } a \le x \le b \text{ and } \alpha > 0, \ \beta > 0.$$

where

a = lower limit for the distribution
b = upper limit for the distribution
α, β are the shape parameters for the distribution.

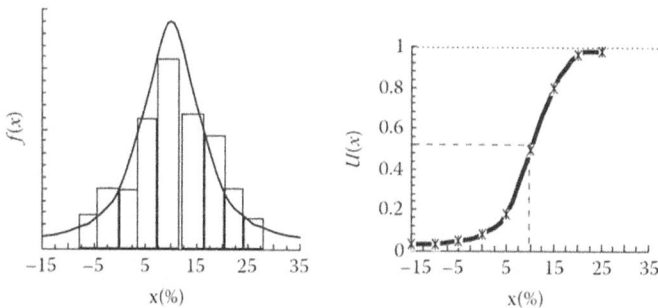

FIGURE 9.3   Utility function for productivity improvement.

As with the normal distribution, there is no closed-form expression for $F(x)$ for the beta distribution. However, if either of the shape parameters is a positive integer, then a binomial expansion can be used to obtain $F(x)$. Figure 9.4 shows a plot of $f(x)$ and the estimated $U(x)$ for quality improvement due to the proposed investment.

Based on work analysis observations, suppose idle time reduction is found to be best described by a log normal distribution with a mean of 10% and standard deviation of 5%. This is represented as shown below:

$$f(x) = \frac{1}{x\sqrt{2\pi\sigma^2}} e^{\left[\frac{-(\ln x - \mu)^2}{2\sigma^2}\right]}, \qquad x > 0.$$

There is no closed-form expression for $F(x)$. Figure 9.5 shows $f(x)$ and the estimated $U(x)$ for idle-time reduction due to the investment.

For example, suppose safety improvement is assumed to have a previously known utility function, defined as follows:

$$U_p(x) = 30 - \sqrt{400 - x^2},$$

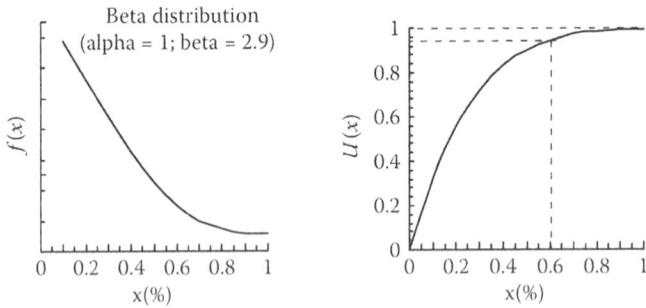

FIGURE 9.4 Utility function for quality improvement.

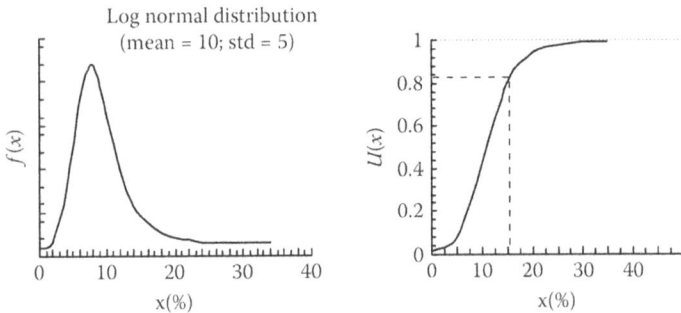

FIGURE 9.5 Utility function for idle time reduction.

where x represents percent improvement in safety. For the expression, the un-scaled utility values range from 10 (for 0% improvement) to 30 (for 20% improvement). To express any particular outcome of an attribute i, x_i, on a scale of 0.0 to 1.0, it is expressed as a proportion of the range of best to worst outcomes shown below:

$$X = \frac{x_i - x_i^0}{x_i^* - x_i^0},$$

where X = outcome expressed on a scale of 0.0 to 1.0

x_i = measured or observed raw outcome of attribute i
x_i^0 = worst raw outcome of attribute i
x_i^* = best raw outcome of attribute i.

The utility of the outcome may then be represented as $U(X)$ and read off of the empirical utility curve. Using the above approach, the utility function for safety improvement is scaled from 0.0 to 1.0. This is shown in Figure 9.6. The numbers within parentheses represent the scaled values.

The respective utility values for the five attributes may be viewed as relative weights for comparing investment alternatives. The utility obtained from the modeled functions can be used in the additive and multiplicative utility models discussed earlier. For example, Table 9.2 shows a composite utility profile for a proposed investment.

Using the additive utility model, the *composite utility* (CU) of the investment, based on the five attributes, is given by:

$$U(X) = \sum_{i=1}^{n} k_i U_i(x_i)$$
$$= .30(.61) + .20(.49) + .25(.93) + .15(.86) + .10(.40) = 0.6825.$$

This composite utility value may then be compared with the utilities of other investments. On the other hand, a single investment may be evaluated independently

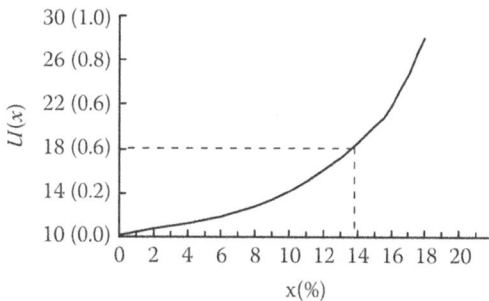

FIGURE 9.6   Utility function for safety improvement.

TABLE 9.2
Composite Utility for a Proposed Investment

Attribute (*i*)	k_i	Value (%)	$U_i(x_i)$
Return on investment	0.30	12.1	0.61
Productivity improvement	0.20	10.0	0.49
Quality improvement	0.25	60.0	0.93
Idle-time reduction	0.15	15.0	0.86
Safety improvement	0.10	15.0	0.40
	1.00		

on the basis of some minimum acceptable level of utility (MALU) desired by the decision-maker. The criteria for evaluating an investment based on MALU may be expressed by the following rule:

Investment j is acceptable if its composite utility, $U(X)_j$, is greater than MALU;
Investment j is not acceptable if its composite utility, $U(X)_j$, is less than MALU.

The utility of an investment may be evaluated on the basis of its economic, operational, or strategic importance to an organization. Utility functions can be incorporated into existing justification methodologies. For example, in the analytic hierarchy process, utility functions can be used to generate values that are, in turn, used to evaluate the relative preference levels of attributes and alternatives. Utility functions can be used to derive component weights when the overall effectiveness of investments is being compared. Utility functions can generate descriptive levels of investment performance as well as indicate the limits of investment effectiveness, as shown by the S-curve in Figure 9.7.

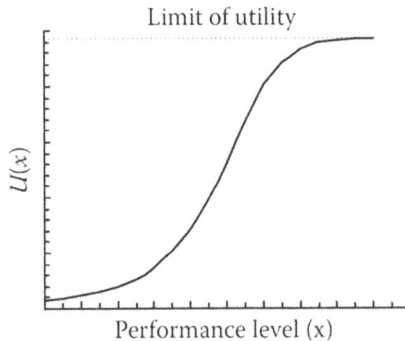

FIGURE 9.7   S-curve model for investment utility.

9.3.4 INVESTMENT VALUE MODEL

A technique that is related to the utility modeling is the investment value model (PVM), which is an adaptation of the manufacturing system value (MSV) model presented by Troxler and Blank (1989). The model is suitable for the incorporation of utility values, and an example is presented below. The model provides a heuristic decision aid for comparing investment alternatives. "Value" is represented as a determined vector function that indicates the value of tangible and intangible "attributes" that characterize an alternative. Value can be expressed as follows:

$$V = f\left(A_1, A_2, \ldots, A_p\right)$$

where V = value, $A = (A_1, \ldots, A_n)$ = vector of quantitative measures or attributes, and p = the number of attributes that characterize the investment. Examples of investment attributes include quality, throughput, capability, productivity, and cost performance. Attributes are considered to be a combined function of "factors," x_i, expressed as

$$A_k\left(x_1, x_2, \ldots, x_{m_k}\right) = \sum_{i=1}^{m_k} f_i\left(x_i\right)$$

where $\{x_i\}$ = set of m factors associated with attribute A_k $(k = 1,2,\ldots,p)$ and f_i = contribution function of factor x_i to attribute A_k. Examples of factors are market share, reliability, flexibility, user acceptance, capacity utilization, safety, and design functionality. Factors are themselves considered to be composed of "indicators," v_i, expressed as

$$x_i\left(v_1, v_2, \ldots, v_n\right) = \sum_{j=1}^{n} z_i\left(v_i\right)$$

where $\{v_j\}$ = set of n indicators associated with factor x_i $(i = 1,2,\ldots,m)$ and z_j = scaling function for each indicator variable v_j. Examples of indicators are debt ratio, investment responsiveness, lead time, learning curve, and scrap volume. By combining the above definitions, a composite measure of the value of an investment is given by the following:

$$PV = f\left(A_1, A_2, \ldots, A_p\right)$$

$$= f\left\{ \left[\sum_{i=1}^{m_1} f_i\left(\sum_{j=1}^{n} z_j\left(v_j\right)\right)\right]_1, \left[\sum_{i=1}^{m_2} f_i\left(\sum_{j=1}^{n} z_j\left(v_j\right)\right)\right]_2, \ldots, \left[\sum_{i=1}^{m_k} f_i\left(\sum_{j=1}^{n} z_j\left(v_j\right)\right)\right]_p \right\}$$

where m and n may assume different values for each attribute. A subjective measure to indicate the utility of the decision-maker may be included in the model by using an attribute weighting factor, w_i, to obtain the following:

TABLE 9.3
Comparison of Technology Values

Alternatives	Suitability $(k = 1)$	Capability $(k = 2)$	Performance $(k = 3)$	Productivity $(k = 4)$
Investment A	0.12	0.38	0.18	0.02
Investment B	0.30	0.40	0.28	1.00
Investment C	0.53	0.33	0.52	1.10

$$\text{PV} = f\left(w_1 A_1, w_2 A_2, \ldots, w_p A_p\right)$$

where

$$\sum_{k=1}^{p} w_k = 1, \quad \left(0 \le w_k \le 1\right).$$

As an example, an analysis using the above model to compare three investments on the basis of four attributes is presented in Table 9.3. The four attributes, *capability*, *suitability*, *performance*, and *productivity*, require careful interpretation before relative weights for the alternatives can be developed.

9.4   CAPABILITY

The term "capability" refers to the ability of equipment to produce certain features. For example, a certain piece of equipment may only produce horizontal or vertical slots, flat finishes, and so on. But a multi-axis machine can produce spiral grooves, internal metal removal from prismatic or rotational parts, thus increasing the variety of parts that can be made. In Table 9.3, the levels of increase in part variety from the three competing investments are 38%, 40%, and 33% respectively.

9.5   SUITABILITY

"Suitability" refers to the appropriateness of the investment to company operations. For example, chemical milling is more suitable for making holes in thin, flat metal sheets than in drills. Drills need special fixtures to hold the thin metal down and protect it from wrinkling and buckling. The parts that the three investments are suitable for are, respectively, 12%, 30%, and 53% of the current part mix.

9.6   PERFORMANCE

"Performance" in this context refers to the ability of the investment to produce high-quality outputs, or the ability to meet extra-tight performance requirements. In our example, the three investments can, respectively, meet tightened standards on 18%, 28%, and 52% of the normal set of jobs.

9.7 PRODUCTIVITY

"Productivity" can be measured by a simulation of the performance of the current system with the proposed technology at its current production rate, quality level, and part application. For the example in Table 9.3, the three investments, respectively, show increases of 0.02, -1.0, and -1.1 on a uniform scale of productivity measurement.

A plot of the histograms of the respective "values" of the three investments is shown in Figure 9.8. Investment C is the best alternative in terms of suitability and performance. Investment B shows the best capability measure, but its productivity is too low to justify the needed investment. Investment A, then, offers the best productivity, but its suitability measure is low.

The relative weights used in many justification methodologies are based on subjective propositions of the decision-maker(s). Some of those subjective weights can be enhanced by the incorporation of utility models. For example, the weights shown in Table 9.3 could be obtained from utility functions.

9.8 POLAR PLOTS

Polar plots provide a means of visually comparing investment alternatives (Badiru, 1991). In a conventional polar plot, as shown in Figure 9.9, the vectors drawn from the center of the circle are on individual scales based on the outcome ranges for each attribute. For example, the vector for NPV (Net Present Value) is on a scale of $0 to $500,000, while the scale for Quality is from 0 to 10. It should be noted that the overall priority weights for the alternatives are not proportional to the areas of their respective polyhedrons.

A modification of the basic polar plot is presented in this section. The modification involves a procedure that normalizes the areas of the polyhedrons with respect to the total area of the base circle. With this modification, the normalized areas of the polyhedrons are proportional to the respective priority weights of the alternatives, so, the alternatives can be ranked on the basis of the areas of the polyhedrons. The steps involved in the modified approach are presented below:

1. Let n be the number of attributes involved in the comparison of alternatives, such that $n \geq 4$. The number of attributes is in a preferred order $(1,2,3,\dots, n)$.

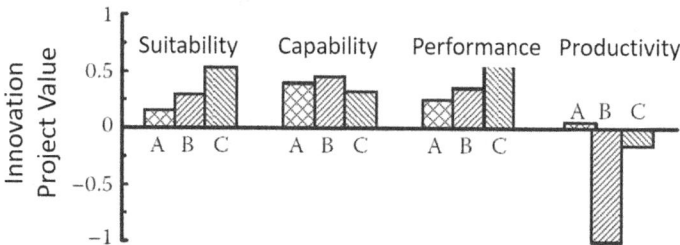

FIGURE 9.8 Relative system value weights of three alternatives.

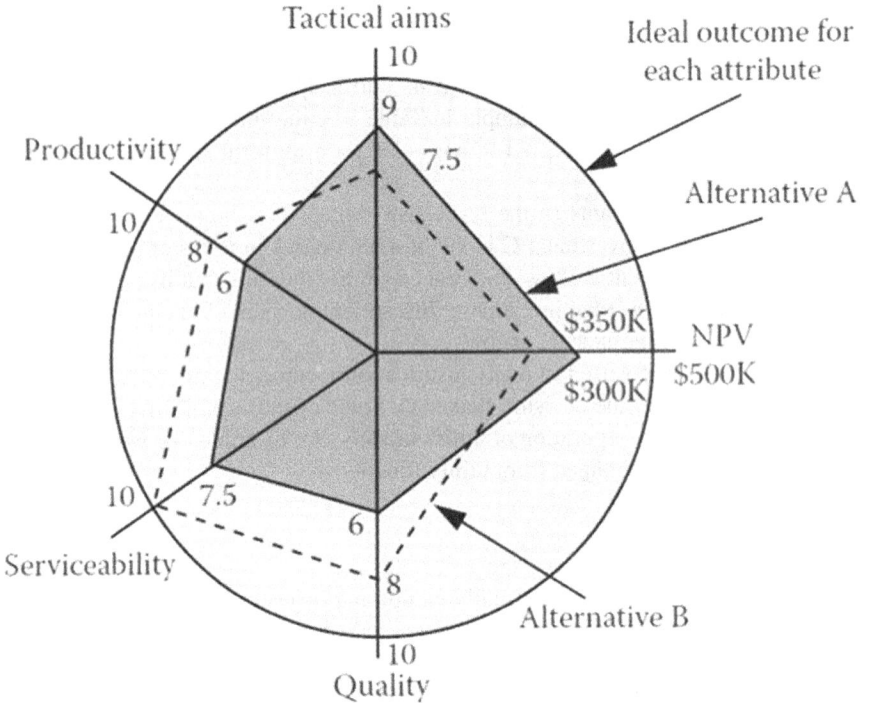

FIGURE 9.9 Basic polar plot.

2. If the attributes are considered to be equally important (i.e., equally weighted), compute the sector angle associated with each attribute as

$$\theta = \frac{360^\circ}{n}.$$

3. Draw a circle with a large enough radius. A radius of 2 inches is usually adequate.

4. Convert the outcome range for each attribute to a standardized scale of 0 to 10 using appropriate transformation relationships.

5. For Attribute 1, draw a vertical vector up from the center of the circle to the edge of the circle.

6. Measure θ clockwise and draw a vector for Attribute 2. Repeat this step for all attributes in the numbered order.

7. For each alternative, mark its standardized relative outcome with respect to each attribute along the attribute's vector. If a 2-inch radius is used for the base circle, then we have the following linear transformation relationship:
   0.0 inches = rating score of 0.0
   2.0 inches = rating score of 10.0

8. Connect the points marked for each alternative to form a polyhedron. Repeat this step for all alternatives.
9. Compute the area of the base circle as follows:

$$\Omega = \pi r^2$$
$$= 4\pi \quad \text{squared inches}$$
$$= 100\pi \quad \text{squared rating units}$$

10. Compute the area of the polyhedron corresponding to each alternative. This can be done by partitioning each polyhedron into a set of triangles and then calculating the areas of the triangles. To calculate the area of each triangle, note that we know the lengths of two sides of the triangle and the angle subtended by the two sides. With these three known values, the area of each triangle can be calculated through basic trigonometric formulas.

For example, the area of each polyhedron may be represented as λ_I ($I = 1,2,\ldots,m$), where m is the number of alternatives. The area of each triangle in the polyhedron for a given alternative is then calculated as

$$\Delta_t = \frac{1}{2}(L_j)(L_{j+1})(\sin\theta),$$

where
L_j = standardized rating with respect to attribute j
L_{j+1} = standardized rating with respect to attribute $j+1$
L_j and L_{j+1} are the two sides that subtend θ.

Since $n \geq 4$, θ will be between 0 and 90 degrees, and $\sin(\theta)$ will be strictly increasing over that interval.

The area of the polyhedron for alternative i is then calculated as

$$\lambda_i = \sum_{t(i)=1}^{n} \Delta_{t(i)}.$$

Note that θ is constant for a given number of attributes and the area of the polyhedron will be a function of the adjacent ratings (L_j and L_{j+1}) only.

11. Compute the standardized area corresponding to each alternative as

$$w_i = \frac{\lambda_i}{\Omega}(100\%).$$

12. Rank the alternatives in a decreasing order of λ_i. Select the highest ranked alternative as the preferred alternative.

Example 9.3

The problem presented here is used to illustrate how modified polar plots can be used to compare investment alternatives. Table 9.4 presents the ranges of possible evaluation ratings within which an alternative can be rated with respect to each of five attributes. The evaluation rating of an alternative with respect to attribute j must be between the given range a_j to b_j. Table 9.5 presents the data for raw evaluation ratings of three alternatives with respect to the five attributes specified in Table 9.4.

The attributes of Quality (I), Profit (II), and Productivity (III) are quantitative measures that can be objectively determined. The attributes of Flexibility (IV) and Customer Satisfaction (V) are subjective measures that can be intuitively rated by an experienced investment analyst. The steps in the solution are presented below.

Step 1:

It is given that $n = 5$. The attributes are numbered in the following preferred order:
Quality: Attribute I
Profit: Attribute II
Productivity: Attribute III
Flexibility: Attribute IV
Satisfaction: Attribute V.

TABLE 9.4
Ranges of Raw Evaluation Ratings for Polar Plots

Attribute (j)	Description	Rank k_j	Evaluation Range Lower Limit a_j	Upper Limit b_j
I	Quality	1	0.5	9
II	Profit (× $1000)	2	0	100
III	Productivity	3	1	10
IV	Flexibility	4	0	12
V	Satisfaction	5	0	10

TABLE 9.5
Raw Evaluation Ratings for Modified Polar Plots

Alternatives	I (j = 1)	II (j = 2)	III (j = 3)	IV (j = 4)	V (j = 5)
A (i = 1)	5	50	3	6	10
B (i = 2)	1	20	1.5	9	2
C (i = 3)	8	75	4	11	1

Step 2:

$$\theta = 360° / n$$
$$= 72°.$$

The sector angle is computed as

Step 3: This step is shown in Figure 9.10.

Step 4: Let Y_{ij} be the raw evaluation rating of alternative i with respect to attribute j (see Table 9.5).

Let Z_{ij} be the standardized evaluation rating.

The standardized evaluation ratings (between 0.0 and 10.0) shown in Table 9.6 were obtained by using the following linear transformation relationship:

$$Z_{ij} = 10\left[\frac{(Y_{ij} - a_j)}{b_j - a_j}\right].$$

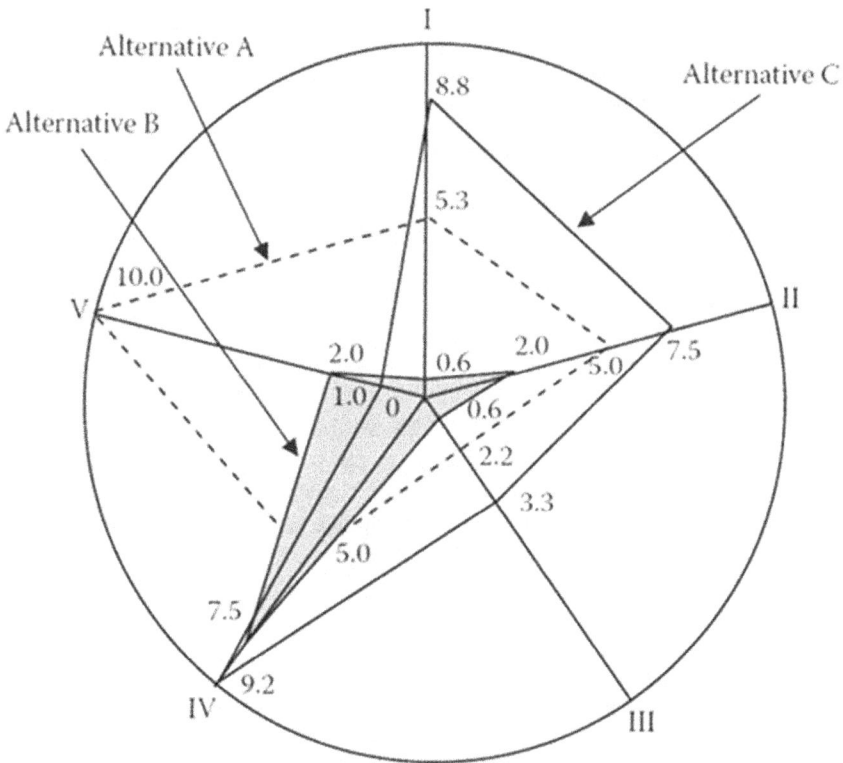

FIGURE 9.10   Modified polar plot.

TABLE 9.6
Standardized Evaluation Ratings for Modified Polar Plots

	Attributes				
Alternatives	I ($j = 1$)	II ($j = 2$)	III ($j = 3$)	IV ($j = 4$)	V ($j = 5$)
A ($i = 1$)	5.3	5.0	2.2	5	10.0
B ($i = 2$)	0.6	2.0	0.6	7.5	2.0
C ($i = 3$)	8.8	7.5	3.3	9.2	1.0

Steps 5, 6, 7, 8: These are shown in Figure 9.10.

Step 9: The area of the base circle is $\Omega = 100\pi$ squared rating units.

Note that it is computationally more efficient to calculate the areas in terms of rating units rather than inches.

Step 10: Using the expressions presented in Step 10, the areas of the triangles making up each of the polyhedrons are computed and summed up. The respective areas are

$\lambda_A = 72.04$ squared units
$\lambda_B = 10.98$ squared units
$\lambda_C = 66.14$ squared units.

Step 11: The standardized areas for the three alternatives are as follows:

$w_A = 22.93\%$
$w_B = 3.50\%$
$w_C = 21.05\%$.

Step 12: On the basis of the standardized areas in Step 11, Alternative A is found to be the best choice.

As an extension to the modification presented above, the sector angle may be a variable indicating relative attribute weights, while the radius represents the evaluation rating of the alternatives with respect to the weighted attribute. That is, if the attributes are not equally weighted, the sector angles will not all be equal. In that case, the sector angle for each attribute is computed as

$$\theta_j = p_j\left(360°\right),$$

where
p_j = relative numeric weight of each of n attributes

$$\sum_{j=1}^{n} p_j = 1.0.$$

TABLE 9.7

Relative Weighting of Attributes for Polar Plots

Attribute (i)	Weight p_i	Angle θ_j
I	0.333	119.88
II	0.267	96.12
III	0.200	72.00
IV	0.133	47.88
V	0.067	24.12
	1.000	360.00

Suppose the attributes in the preceding example are considered to have unequal weights, as shown in Table 9.7.

The resulting polar plots for weighted sector angles are shown in Figure 9.11. The respective weighted areas for the alternatives are

$\lambda_A = 51.56$ squared units
$\lambda_B = 9.07$ squared units
$\lambda_C = 60.56$ squared units.

The standardized areas for the alternatives are as follows:

$w_A = 16.41\%$
$w_B = 2.89\%$
$w_C = 19.28\%$.

Thus, if the given attributes are weighted as shown in Table 9.7, then Alternative C will turn out to be the best choice. However, it should be noted that the relative weights of the attributes are too skewed, resulting in some sector angles being greater than 90 degrees. It is preferable to have the attribute weights assigned in such a way that all sector angles are less than 90 degrees. This leads to more consistent evaluation since $\sin(\theta)$ is strictly increasing between 0 and 90 degrees.

It should also be noted that the weighted areas for the alternatives are sensitive to the order in which the attributes are drawn in the polar plot. Thus, a preferred order of the attributes must be defined prior to starting the analysis. The preferred order may be based on the desired sequence in which alternatives must satisfy management goals. For example, it may be desirable to attend to product quality issues before addressing throughput issues. The surface area of the base circle may be interpreted as a measure of the global organizational goal with respect to performance indicators such as available capital, market share, capacity utilization, and so on. Thus, the weighted area of the polyhedron associated with an alternative may be viewed as the degree to which that alternative satisfies organizational goals.

Some of the attributes involved in a selection problem might constitute a combination of quantitative and/or qualitative factors or a combination of objective and/or subjective considerations. Prioritizing the factors and considerations is typically

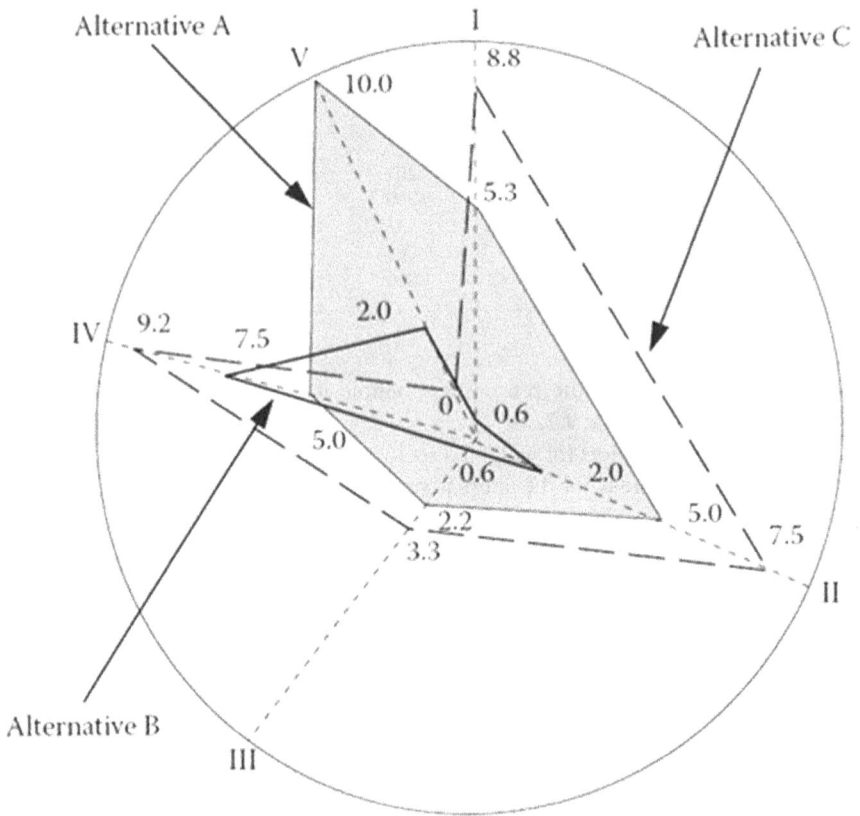

FIGURE 9.11　Polar plot with weighted sector angles.

based on the experience, intuition, and subjective preferences of the decision-maker. Goal programming is another technique that can be used to evaluate multiple objectives or criteria in decision problems.

9.9　TECHNICAL INNOVATION BENCHMARKING

The techniques presented in the preceding sections can be used for benchmarking quantitative investments. For example, to develop a baseline schedule, evidence of successful practices from other investments may be needed. Metrics based on an organization's most critical investment implementation issues should be developed. *Benchmarking* is a process whereby target performance standards are established based on the best examples available. The objective is to equal or surpass the best example. In its simplest term, benchmarking means learning from and emulating a superior example. The premise of benchmarking is that if an organization replicates the best quality examples, it will become one of the best in the industry. A major approach of benchmarking is to identify performance gaps between investments.

Figure 9.12 shows an example of such a gap. Benchmarking requires that an attempt be made to close the gap by improving the performance of the subject investment.

Benchmarking requires frequent comparison with the target investment. Updates must be obtained from investments already benchmarked, and new investments to be benchmarked must be selected on a periodic basis. Measurement, analysis, feedback, and modification should be incorporated into the performance improvement program. The Benchmark-Feedback model presented in Figure 9.13 is useful for establishing a continuous drive toward performance benchmarks.

The figure shows the block diagram representation of input-output relationships of the components in a benchmarking environment. In the model, $I(t)$ represents the set of benchmark inputs to the subject investment. The inputs may be in terms of data, information, raw material, technical skill, or other basic resources. The index t denotes a time reference. $A(t)$ represents the feedback loop actuator. The actuator facilitates the flow of inputs to the various segments of the investment. $G(t)$ represents the forward transfer function, which coordinates input information and resources to produce the desired output, $O(t)$. $H(t)$ represents the management control process that monitors the status of improvement and generates the appropriate feedback information, $B(t)$, which is routed to the input transfer junction. The feedback information is necessary to determine what control actions should be taken at the next improvement phase. The primary responsibility of an economic analyst is to ensure proper forward and backward flow of information concerning the performance of an investment on the basis of the benchmarked inputs.

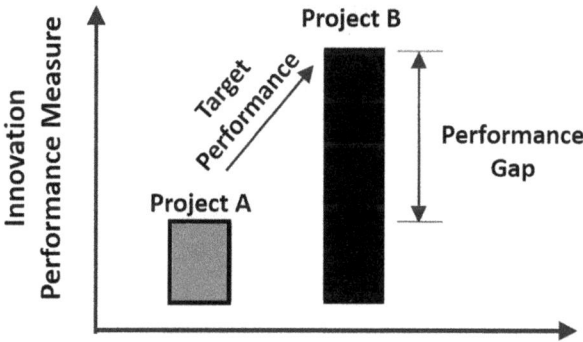

FIGURE 9.12 Identification of benchmark gaps.

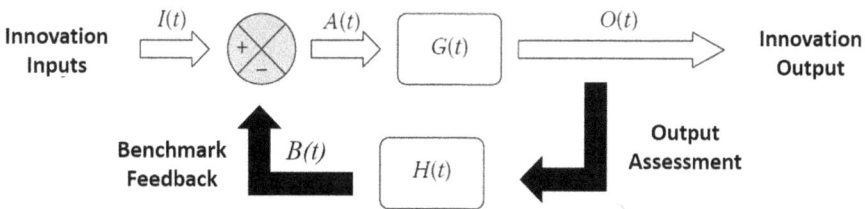

FIGURE 9.13 Innovation benchmark-feedback model.

REFERENCES

Badiru, Adedeji B. (1991). *Project Management Tools for Engineering and Management Professionals*. Industrial Engineering & Management Press, Norcross, GA.

Badiru, Adedeji B. (2008) *Triple C Model of Project Management: Communication, Cooperation, and Coordination*. CRC Press/Taylor & Francis Group, Boca Raton, FL.

Badiru, A. B. (2012). *Project Management: Systems, Principles, and Applications*. Taylor & Francis Group/CRC Press, Boca Raton, FL.

Park, Chan S. and Gunter P. Sharp-Bette (1990). *Advanced Engineering Economics*. Wiley & Sons, New York.

Saaty, Thomas L. (1980). *The Analytic Hierarchy Process*. McGraw-Hill, New York.

Troxler, Joel W. and Leland Blank (1989). "A Comprehensive Methodology for Manufacturing System Evaluation and Comparison". *Journal of Manufacturing Systems* 8(3): 176–183.

10 Artificial Intelligence and Human Factors Integration in Additive Manufacturing

10.1 INTRODUCTION

This chapter is based on the work of Doran et al. (2022), which specifically considers the integration of human factors into artificial intelligence (AI) for additive manufacturing applications. Much has been stated and written about the current and future potential of additive manufacturing. However, one area that has not been adequately addressed is Human Factors and Ergonomics in additive manufacturing concerning promoting better system performance. As additive manufacturing transitions from a technology of manufacturing prototypes to rapid manufacturing, more human factor considerations must be assessed and integrated for improved work design. This paper provides an overview of human-machine integration for human factors, cognitive ergonomics, and AI to improve the performance output in the additive manufacturing process.

10.2 BACKGROUND OF ADDITIVE MANUFACTURING

Additive manufacturing (AM) is a fabrication process that deposits, cures, or consolidates material layer upon layer to create a product (Hom and Harrison, 2012). The product formed is based on a three-dimensional (3D) model made from computer-aided design (CAD) software like SolidWorks. When AM first originated in the 1980s, its primary use was for rapid prototyping products due to reducing both time and cost compared to other traditional manufacturing methods such as milling and drilling (Gardan, 2016). Today, AM's primary use remains for rapid prototyping. Although AM has not become more common in the manufacturing environment outside of prototyping due to the technology remaining expensive and unable to mass-produce (Bailey and Bosworth, 2014), the industry is slowly growing (Bikas et al., 2016). General Electric is one company that has led the growth of metal AM with its recent use and development of 3D printing for components used in its aircraft engines (Sand, 2019). The success of General Electric's use of AM could lead to lowered cost and shortened lead times for them as a company due to the consolidation of manufacturing their components (Sand, 2019), but this could be due to the small batch size of their engines compared to other industries. Although more research is necessary, there are known environmental, ecological, and design benefits with a potential for even more depending on the development of AM technologies.

Some of the studied ecological benefits of using AM are less raw materials needed for production, which leads to a reduction in the weight of transported products and wasteful manufacturing processes, along with the declining need for centralised locations in part manufacturing (Peng et al., 2018). Other potential ecological benefits exist for AM such as limited material waste, a recyclable material, and higher energy efficiency, but these benefits are dependent on the design of the product, the material of the product, the amount of excess material used for supporting the part during printing, and the number of products being produced (Rejeski et al., 2018). In terms of economic benefits, the foremost economic interests coincide with AM's primary function of rapid prototyping with reduced development costs and material waste during prototyping compared to traditional subtractive manufacturing processes (Bailey and Bosworth, 2014). The other prospective cost savings that come with AM processes have similar variability to environmental benefits with factors such as build failure rates or incorrect fabrication of the product (Baumers et al., 2017). The major advantage of using AM is the ease of design and production of models in short product cycles (Hom and Harrison, 2012).

With the direct use of a three-dimensional CAD model to manufacture a product, AM can create a single component that has intricacies that other manufacturing processes cannot replicate. The ability to compose a complex component allows the designer to create a model with sophisticated structures instead of designing for manufacturability and creating a series of pieces requiring final assembly (Mantyjarvi et al., 2018). This freedom of design provided with AM allows engineers to design products with higher performance for numerous applications, such as modular weaponry for military use (Schrand, 2016) or aeroplane blades made of high-performance alloys (Hom and Harrison, 2012). Although AM has clear and significant advantages for manufacturers beyond rapid prototyping, there are disadvantages to AM that have prevented AM from growing into a more prominent role in the manufacturing environment.

The significant reasons why AM has not become commonplace in the manufacturing environment outside of rapid prototyping is due to its inability to match the throughput, quality, quantity, and product consistency of traditional manufacturing processes that comes at a manageable cost for mass production (Hom and Harrison, 2012). AM has grown with the development of multiple fabrication processes in laser-based processes, extrusion-based processes, material jetting processes, adhesive-based processes, and electron beam processes as well as expanded availability of metals and plastics for producing parts. However, machining is still needed for post-processing of the 3D printing for removing the excess material for the part. Each of these process types has favourable applications compared to the others, but all create parts layer by layer using a CAD model (Bikas et al., 2016). But until all these technologies for AM become more reliable and less expensive, it will remain in its current role in rapid prototyping. However, as the AM technologies advance to produce at rapid manufacturing rates, the human-in-the-loop must be considered and the use of AI. This chapter will review and leverage published literature regarding human factors and cognitive assessment methods to evaluate operators' performance and the human-system design in AM tasks and workstation functions.

10.3 COGNITIVE ERGONOMICS

In general, ergonomics is the study of human work that seeks to improve the safety, comfort, and production of humans (Venda et al., 2000). Since its origination, ergonomics has been a topic for numerous pieces of research to enhance the work of humans in a variety of work environments. Research in ergonomics was solely on physical ergonomics during its initial stages of development. This research led to the application of physical ergonomic standards in most working environments, thanks to the creation of the Occupational Safety and Health Administration (OSHA). With the growth of technology came a new and more relevant field of ergonomics for today's work environments: cognitive ergonomics.

Cognitive ergonomics analyses human work from a mental work perspective and, like physical ergonomics, seeks to improve humans' work and productivity (Venda et al., 2000). The mental work perspective is an essential aspect of designing and evaluating occupational tasks because the interaction between an operator and an assigned task is critical. Mental work measures provide awareness of where increased task demands could negatively impact human performance (Bommer and Fendley, 2018).

Two significant aspects of cognitive ergonomics are people perceive processed information through sight, sound, or feel and how people make decisions based on this information (MacLeod, 2004). The use of cognitive ergonomics is seen every day from how a light switch is designed and mounted vertically on a wall (MacLeod, 2004) or how a door unlocks when the door handle is pushed down on instead of up. Subtle changes such as a door handle and light switch are designed to be easily used and follow perception and decision-making that humans are accustomed to. Cognitive ergonomics are also applied to high-stress environments such as an airplane's cockpit, its endless number of buttons and switches being designed and placed for a typical pilot's decisionmaking (MacLeod, 2004). Cognitive ergonomics is used for developing a product comfortably for humans and increasing the performance of a product and its user and the safety of the person producing the product.

As stated previously, one of the central focuses of ergonomics is to increase the safety of humans (Venda et al., 2000), which is vital in hazardous environments such as manufacturing or construction. An example of designing for safety is the construction of safety signs (Chen et al., 2018). In a study done on the construction safety signs, results showed that if the sign was the colour red, the signs were easier to identify with shorter response time compared to other coloured signs (Chen et al., 2018). Other safety precautions follow this knowledge of red being an ideal colour, such as how a computer numerical control (CNC) has a red emergency button that can completely shut down a machine. A CNC can also produce a noise like a siren to increase the attention of an operator to alert that operator that something is incorrect with the setup or operation of the machine and not to precede with the operation (Wei et al., 2018). Lastly, cognitive ergonomics is used to increase human production (Venda et al., 2000) such as presenting information in appropriate detail (MacLeod, 2004) when working on a computer interface to avoid mentally overloading the human operating the computer interface. Cognitive ergonomics' central focuses revolve around the idea of cognitive load and avoiding mental overload and understimulation for humans.

Cognitive load theory (CLT) defines human limitations in processing information in working memory from three distinct loads: intrinsic, extraneous, and germane (Galy and Melan, 2015). First, intrinsic cognitive load is the quantity of material required for processing and the difficulty of the material being processed (Galy and Melan, 2015). Second, extraneous cognitive load is the element of one's learning environment that can negatively or positively affect one's workload (Galy and Melan, 2015). Lastly, the germane mental workload is the load that is generated by the applications of problemsolving techniques (Galy and Melan, 2015). All three of these loads are factored into both types of memory: working memory and long-term memory (Kalyuga, 2011). Working memory is temporary memory that one contemplates based on the information they perceive (Wickens and Lee, 2004), while long-term memory is information that can be retrieved from the past, whether it be minutes later to years later (Wickens and Lee, 2004). Understanding these different types of workload and memory provides insight into the various cognitive abilities required for an operator. Furthermore, knowledge of these workloads as well as the multiple resources that factor into one's mental workload allows one to measure the cognitive workload of an operator and acquire a more holistic understanding of their workload and whether one's performance is a result of their task demands (Bommer and Fendley, 2018).

The Multiple Resource Theory (MRT) assesses task demand by considering the resources that factor into the workload of humans and a human's ability to perform in high-workload environments (Wickens, 2002). The five resources of the human mind that factor into a human's performance are visual processing, auditory processing, cognitive processing, speech, and motor (Bommer and Fendley, 2018). With these five resources in mind, the MRT analyses three specific components within a compilation of the five resources: task demands, availability of resources, and resource overlap (Bommer and Fendley, 2018). By analysing these three components, the evaluation of an operator will be holistic by completely breaking down all the elements into one's task performance. This analysis will also allow one to predict an operator's performance and their ability to complete their given task or tasks (Bommer and Fendley, 2018). The CLT provides insight into the distinct three mental workloads. The MRT breaks down a human's performance by analysing the five resources that factor into mental workload. These two theories can be applied by selecting an appropriate resource or resources to measure a human's performance and the distinct mental workload needed to complete the operator's given tasks.

10.4   COMPUTATIONAL METHODS

For measuring mental or cognitive workload, there are three types of measures: physiological, subjective, and performance. Physiological measures are measures of the body's physical responses that range from eye fixation rate to the pitch of one's voice (Badiru and Bommer, 2017). These physiological measures include measuring the resources outlined in the MRT by using bodily activities of the following: cardiac, brain, respiratory, speech, and eye (Badiru and Bommer, 2017).

The cardiac measures for mental workload include heart rate (HR), heart rate variability (HRV), and absolute interbeat intervals (Grassmann et al., 2017). Some of the physiological measures of the brain can be measured using an electroencephalogram (EEG), electrocardiogram (ECG), electrooculogram (EOG), or an electromyogram (EMG) (Zhang et al., 2017). Respiratory measures for mental workload include the respiratory rate, airflow, and volume, along with respiratory gas analysis (Charles and Nixon, 2019). Speech measures used for measuring mental workload include the pitch, rate, and loudness of speech (Badiru and Bommer, 2017). Lastly, the eye measures used in calculating mental workload include pupil dilation, blink rate, fixation rate, saccadic rate, and pupil diameter (Yan et al., 2019).

Generally, physiological measures have been used in measuring mental workload due to the measurements providing accurate and objective data (Zhao et al., 2018). Physiological measures are also known for having better performances in measurement facets such as sensitivity, diagnostic ability, and non-intrusiveness when comparing it to subjective measures (Zhao et al., 2018). Subjective measures provide a different dynamic to measuring one's cognitive workload that physiological measures do not offer with the measure being based on the operator's perception (Badiru and Bommer, 2017).

Subjective measures are easier to administer and analyse compared to physiological measures but are administered after the task or tasks have been completed, which can affect the reliability of the results if the task or tasks are extensive (Longo, 2015). Subjective mental workload measures include the Workload Profile (WP), Cooper Harper scales, Bedford scale, subjective workload assessment technique (SWAT), subjective workload dominance technique (SWORD), and the National Aeronautics and Space Administration Task Load (NASA-TLX) (Badiru and Bommer, 2017).

Workload Profile collects data on demands of the task, which includes perceptual processing, response selection execution, spatial processing, verbal processing, visual processing, auditory processing, manual output, and speech output using a rating system imposed on all of their tasks with these dimensions being analysed (Badiru and Bommer, 2017). Like the Workload Profile, Cooper-Harper scales use a rating system that is filled out by operators after the task is completed but can use a decision tree in conjunction with the rating system to analyse the workload of an operator (Moorhouse, 1990). The Bedford workload scale uses a 10-point rating system assigned based on the mental workload of the operator (Wang et al., 2013). For example, ratings 1 to 3 indicate a low workload and the operator has spare mental capacity doing these tasks, while a task with a rating of 10 indicates that there is no spare capacity and no available attention that can be permitted to be used on other tasks (Wang et al., 2013).

The SWAT is a subjective questionnaire-type analysis that looks at three dimensions: time load, mental effort load, and psychological stress load (Rubio et al., 2004). These three dimensions are then rated on a three-level scale: low (1), medium (2), and high (3) that determines the mental workload of an operator (Rubio et al., 2004). The next step of the SWAT analysis is having an operator rank all 27 possible combinations of the three dimensions and three levels in their perception of increasing workload (Rubio et al., 2004). Following the operator ranking all possible combinations, conjoint scaling procedures are followed to create a single rating scale with interval properties (Rubio et al., 2004). Lastly, actual ratings of workload for given tasks are completed and converted into a numeric score between 0 and 100 using the interval scale (Rubio et al., 2004). The next subjective measure for mental workload that will be covered is the SWORD method.

The SWORD method measures workload using a series of relative judgements that compare the workload of various task conditions (Vidulich et al., 1991). SWORD follows three distinct steps: collecting raw judgement data, constructing the judgement matrices, and calculating SWORD ratings (Vidulich et al., 1991). In collecting raw judgement data, the rater is given the rating sheet that lists all possible paired comparison of tasks after completing all tasks (Vidulich et al., 1991). This rating sheet has one task on the left side of the evaluation and the other on the right side of the evaluation, and the rater has to compare the levels of workload for both of the tasks and determine which task has a higher workload or workload dominant and how dominant it is (Vidulich et al., 1991). The rater does this for all of his tasks before constructing the judgement matrices to compare task difficulty (Vidulich et al., 1991). The rater marks the judgement matrices based off of which task was the dominant (Vidulich et al., 1991). If the left-side task was more dominant, the rater would use a 2 to 9 scale depending on how dominant the leftside task is compared to the right-side task, with two being the least dominant and nine being the most dominant (Vidulich et al., 1991). If the right-side task was more dominant, the rater would use the 1/2 to 1/9 with 1/2 being the least dominant and 1/9 being the most dominant (Vidulich et al., 1991). The SWORD rating for each task is then calculated by using the geometric mean for each row of the matrix and normalising the means (Vidulich et al., 1991). The final subjective measurement method for mental workload covered will be NASA-TLX.

Lastly, NASA-TLX has operators rate the mental demand, physical demand, temporal demand, overall performance, effort, and frustration of their tasks in a questionnaire format following the completion of all tasks (Alaimo et al., 2018). The first step to obtaining the rating for each dimension is assigning a score on 20-step bipolar scales for each dimension (Rubio et al., 2004). A score from 0 to 100 is received on each scale, and a paired comparison between all six dimensions is completed to consolidate the six individual ratings into a single score (Rubio et al., 2004). The pairwise comparison requires an operator to pick the dimension that is most relevant to workload out of all the pairs of the six dimensions (Rubio et al., 2004). The weighting of a dimension scale for each task is determined by the number of times the dimension is picked as the most relevant (Rubio et al., 2004). Finally, the workload score is computed by multiplying the weight by the individual dimension scale score, summing across all scales, and dividing by 15, which is the total number of paired comparisons (Rubio et al., 2004). Performance measures are discussed in the next section.

The first thing to note with performance measures for mental workload is that it assumes that increased task difficulty leads to an increase in task demand (Rubio et al., 2004). To meet this assumption, measuring mental workload based on performance requires the use of two types of measures: primary task and secondary task (Longo, 2015). Primary task measures are used to indicate performance, while secondary task measures are used to calculate one's spare attentional capacity with short periods of workload (Longo, 2015).

A performance measure for mental workload is centred around human error probabilities (HEPs), which are calculated by taking the number of observed errors and dividing it by the number of possibilities for an error (Badiru and Bommer, 2017a). There are plenty of analyses used for forecasting HEP, such as the Technique for Human Error Rate Prediction (THERP) and Cognitive Reliability and Error Analysis Method (CREAM) (Liang et al., 2010).

For THERP, the overall HEP is calculated by completing a task analysis that is then converted into a tree diagram that displays all the human errors possible for each task (Shirley et al., 2015). HEPs are then subjectively estimated for each possible error, and the overall HEP is computed using all the HEPs from each task (Shirley et al., 2015). The CREAM method is a human error classification method system that divides human errors into eight error modes and three headings of influence factors that gives a framework for evaluating human error (Liao et al., 2016). Like THERP, CREAM forecasts HEPs subjectively through the opinions of experts (Liao et al., 2016). The use of performance measures does allow for a detailed analysis of the resources competing between one another in tasks (Longo, 2015) but lacks the objectivity of physiological measures and input of operators that subjective measures include.

Cognitive ergonomics have made significant progress from its origination till now with several theories being developed and applied to a multitude of studies and experiments. The development of the theories such as the CLT and MRT has helped in distinguishing the various mental workloads that humans deal with, the resources that go into receiving and retaining information, and how to measure mental workload overall. Subjective, physiological, and performance measures have been developed and used in a variety of work environments to analyse human performance. The results of these measurements have been analysed and tested to produce modifications to tasks, workstations, and/or product designs to adapt to human perception (MacLeod, 2004) and potentially increasing productivity and/or safety of the human (Chen et al., 2018).

10.5 HUMAN CONSIDERATIONS IN ADDITIVE MANUFACTURING

Human considerations in AM stem from the three goals of human factors: increasing safety, enhancing performance, and increasing user satisfaction (Wickens and Lee, 2004). Of these three factors, currently, the most prominent issue with AM is the safety hazards that could damage a human's health and, eventually, the environment (Bours et al., 2017). The three main safety hazards for humans are the exposure of toxic chemicals throughout the manufacturing process, the harmful emissions produced in some of the processes, and potential static causing hazardous materials to react (Bours et al., 2017). For enhancing the performance of AM, there is definite room for growth in the performance of AM systems in their throughput, quality, and product consistency of the 3D printer (Hom and Harrison, 2012) as well as the energy consumed in AM's unit processes (Kellens et al., 2017).

In terms of the human performance in these systems, training and gaining experience with 3D modelling and the process of printing 3D models on a 3D printer safely and properly are the two primary enhancement methods for a human's performance in AM (Shelton, 2019). Improvements in the user's interface for the computer used to create the 3D models, and the interface for the 3D printers along with how the AM process is managed and analysed are refinements to the technology that can increase human performance. The idea of incorporating more AI in the process should also increase human performance in AM. User satisfaction can be increased mostly through making the user interfaces on three-dimensional software like SolidWorks or Autodesk Inventor and the 3D printer more user-friendly by providing the right amount of feedback and simple design (Lee et al., 2010). The advantages and disadvantages slightly differ on the AM process used whether that is selective laser melting, selective laser sintering, fused deposition modelling, or stereolithography (Bikas et al., 2016), but the safety potential benefits and concerns for each of these processes are similar. Some of the potential safety benefits for AM include there is less material requirement, less waste, and the process is more self-contained compared to the conventional manufacturing processes (Rejeski et al., 2018). Unlike popular manufacturing methods, AM methods present similar safety concerns from the hazardous materials used to managing the harmful emissions produced during the manufacturing processes (Rejeski, et al., 2018).

A major difference between AM and modern-day manufacturing processes is the materials used, which present different health hazards such as higher toxicity in materials used, an overabundance of static produced in the process that could potentially cause a reaction between materials used in the process (Shelton, 2019), and the different emissions that are released (Rejeski et al., 2018). The toxicity hazard is minimised by handling materials with no direct contact, such as setting and cleaning up the 3D printer through gloves that are attached to the glass window (Shelton, 2019). Oxygen is consistently monitored in the room to assure that there are no harmful toxins that can be inhaled and in other processes such as the filtration process where excess powder filters out any toxic materials so that the excess powder can be reused (Shelton, 2019). If the percentage of oxygen in the room goes below a certain threshold, anyone in the room is alarmed to evacuate immediately (Shelton, 2019). If the percentage of oxygen goes below an operator set level during the filtration process, the machine automatically shuts off to prevent any exposure to toxic particles (Shelton, 2019).

For reducing the chance of a reaction, static is reduced with the use of full-body suits that are fire retardant and electrostatic discharge (ESD) shoes during the transfer of the metal powder along with an ESD floor (Shelton, 2019). Another form of personal protection equipment (PPE) used during the transfer of the metal powder is a positive pressure helmet that breaks down toxic particles and pushes them out of the helmet to prevent the particles from being inhaled (Shelton, 2019). Another safety hazard that AM has is the harmful emissions that are produced in the 3D printing process. Research has been done regarding reducing emissions in the AM process, and one tactic that has been found to reduce the emissions is to use a high-efficiency filter in an enclosed, well-ventilated indoor environment that could help alleviate this safety concern (Kwon et al., 2017).

The other safety and health concerns with AM reside in the 3D modelling portion of AM.

The safety concerns that come with 3D modelling are like that of a typical office worker such as the potential for computer vision syndrome (CVS) due to the extended use of a computer (Teo et al., 2019) or carpal tunnel syndrome (CTS) due to the use of keyboard or mouse that lack in ergonomic design (Liu et al., 2016). The health concerns with excess computer use have been studied extensively with well-known mitigation strategies in place for these. For computer vision syndrome, reduction methods include optimising lighting within the room, the position of your computer, taking breaks throughout the day, using lubricating eye drops, and computer glasses that reduce exposure to blue light (Teo et al., 2019). The use of a pad for both the mouse and keyboard has also proven to provide support for the wrists to improve their posture, thus helping prevent carpal tunnel syndrome (Liu et al., 2016). There are some methods for enhancing the performance of humans in AM outside of more training and experience like improving the interface of the 3D printer and increasing the use of AI in AM and the AI's effectiveness.

Since most of the work for humans in AM comes from the time needed to make a part three-dimensional, proper training and experience with 3D modelling are the primary ways to enhance the performance of a human. The 3D modelling software's interface, as well as the interface of the AM machine or 3D printer, is another way the performance of a human can be enhanced, which can also lead to higher user satisfaction with the software. Using cognitive ergonomic techniques and guidelines to design both user interfaces can lead to potential rises in performance and satisfaction, such as consistent dialogue box designs, maximising screen space on the interface, and customisability with the system's interface (Lee et al., 2010). The failure analysis of AM could also use improvement by utilising more effective AI in AM systems to effectively improve human performance.

AI is being utilised in three tasks in AM: the computer is able to select the point of contact for the part on the 3D printer's platform, the amount of supportive material needed, and the supportive material's placement on the part to print properly and providing images of each layer of the part after the part has been printed (Shelton, 2019). These three uses of AI in AM have not been particularly effective with operators choosing a better point of contact for the part and operators choosing the amount of support material needed and its placement, which is typically less wasteful than the computer's support material suggested along with saving time by reducing the time needed to remove the excess material (Shelton, 2019). The images provided for each layer of the part do allow failure analysis post-process but do not monitor the part during the printing process (Shelton, 2019).

Failure analysis is also a tedious process with a vast number of images that are produced and organised in a chronological order of layer printed, which needs to be analysed when a failure occurs (Shelton, 2019). To reduce the number of failures and reprinting of parts, a potential new application of AI in AM could be used to analyse if a part can be successfully printed instead of having a human operator take a significant amount of time to assess the build and revise it if needed (Shelton, 2019). Instituting this technology would remove a stressful task for the operator and reduce the mental workload for the operator while potentially increasing the performance of the operator if the AI is effective in determining if a build is printable or not (Shelton, 2019). The current AI technologies in AM need to be improved to meet the performance of a human to reduce the workload on the human in strenuous tasks such as 3D build assessment.

10.6   ARTIFICIAL INTELLIGENCE CASE STUDIES

Artificial intelligence was first thought of by the philosopher Thomas Hobbes in the 1650s with his philosophy on thinking consisting of symbolic operations that can be deduced mathematically (Badiru and Cheung, 2002). The development of the computer and its capabilities led to growing of machine intelligence and methods for how this intelligence can be evaluated with tests such as the Turing test (Badiru and Cheung, 2002). The Turing test tasks a human interrogator with asking questions to two parties (a male and female for the first portion of the test and a computer and human for the second portion of the test) through a computer with both parties responding through a separate computer in a different location (Badiru and Cheung, 2002). For both portions of the test, the human interrogator must interpret the responses of the two parties to distinguish which responses are coming from which party with one party responding truthfully to the human interrogator's questions, while the other party is attempting to fool the human interrogator (Badiru and Cheung, 2002). The results of the Turing test should indicate whether the computer shows signs of intelligence. If the human interrogator's success rate in distinguishing between the two parties is lower for the computer and the human portion compared to the male and female portion, the computer is considered to be intelligent (Badiru and Cheung, 2002). The intelligence of computers and its programming only continued to grow, which eventually lead to the birth of the term 'artificial intelligence' by John McCarthy in 1956 (Badiru and Cheung, 2002).

After its origination, AI began to rapidly grow with the invention of the general problem solver (GPS), LISP (list processing), the computer programming language with great memory organisation and control structure, neural networks, and modern-day expert systems (Badiru and Cheung, 2002).

Herbert Shaw, Allen Newell, and Cliff Shaw introduced GPS, a system that uses means-end analysis to complete tasks such as solve theorems or play chess (Badiru and Cheung, 2002). Around the same time, John McCarthy developed LISP, a new computer programming language that organised its memory with interconnections between memory groups and then controlled its program by starting with a goal and determining the requirements for achieving that goal (Badiru and Cheung, 2002). A neural network is a network where all neurons in a layer are connected to all the neurons in the following layer (Bazrafkan and Corcoran, 2018). Data travels through these layers sequentially to eventually provide an interpretation of one feature of the input data structure (Bazrafkan and Corcoran, 2018). Lastly, expert systems can mimic a human expert's knowledge and reasoning as well as draw interfaces that normal computer programs cannot (Pandit, 1994). These technologies have continued to develop since their inceptions and become a more integrated society.

Today, AI is changing humans' lives by increasing automation in both our personal and work environments (Calp, 2019). AI has affected numerous work environments from hospitals to manufacturing plants with its ability to assist problems involving issues such as classification or optimisation (Calp, 2019). The algorithms and mathematical models that are the basis for AI have led it solving highly advanced optimisation problems whether the optimisation is continuous or static (Calp, 2019). Mechanisms such as random walk, swarm intelligence, algorithmic

TABLE 10.1

Artificial Intelligent Technologies in Additive Manufacturing

AI Technology	Task Utilisation	Citation
Augmented reality (AR)	Design assessment, real-time monitoring	Malik et al. (2019)
Bayesian networks (BN)	Failure diagnosis	Bacha et al. (2019)
Convolutional neural network (CNN)	Data acquisition, data processing, generating patches/labels, image segmentation	Minnema et al. (2018)
Ensemble-based multi-gene roughness genetic programming (EN-MGGP) and waviness	Prediction of surface	Garg et al. (2018)
Evolutionary algorithms (EAs)	Solving a variety of optimisation and problems (e.g., machine set up and Martinsen process planning)	Simon (2013) and Leirmo and Martinsen (2019)
Graphics processing unit (GPU)	Manipulation of computer graphics, image processing, processing big data	Ongsulee (2017)
Machine learning (ML) Predictive analytics	Image processing, text classification, speech recognition	Gardner et al. (2019) and Singh et al. (2016)
	Predictive modelling, data mining, pattern recognition	Ongsulee (2017)

flow, and heuristic and metaheuristic can be used to solve these optimisation problems (Calp, 2019).

The use of AI has been started in AM for a variety of purposes that have been well-documented and researched. The use of cognitive ergonomics can assist decision-making on where to deploy AI technologies to effectively support the human operator. Table 10.1 outlines different forms of AI that are being researched or used in AM.

10.7 COGNITIVE ERGONOMICS IN ADDITIVE MANUFACTURING

With technological advancements continually being made and implemented in manufacturing, more focus is needed on cognitive ergonomics that is constantly evaluated like physical ergonomics (Bommer, 2017). One of the most significant technological advancements in manufacturing is AM, with its ability to create a product based on a three-dimensional model made using software such as SolidWorks. AM, like many technology-driven processes, requires more analysis on the cognitive ergonomics compared to traditional manufacturing processes due to its constant use of a computer interface. Some research has been done on the cognitive workload that comes with AM from the workload needed for the spatial design in the CAD (Dadi and Goodrum, 2014) to data organisation and analysis in lifecycle management (Muller et al., 2017), but there is a need for more specified research on the cognitive ergonomics in AM. This section will cover some of the current studies completed on cognitive ergonomics in AM and other studies conducted on cognitive ergonomics applicable to AM processes.

The current trend and changes in manufacturing have led to an increased cognitive workload for various process operators due to an emphasis on problem-solving and reasoning skills in manufacturing tasks and processes (Bommer, 2017). This is especially true in AM, which does not feature many traditional manufacturing processes such as milling and drilling. Instead, most of the work processes in AM are done through a human-computer interface, whether on a computer doing the 3D modelling of the product or setting up the 3D printer to print the 3D model part. Of those two human-computer interfaces, the most time-consuming element for the human is the 3D modelling, so understanding the cognitive workload of 3D modelling using CAD software is critical. Therefore, cognitive ergonomics can be applied to increase productivity for the human operator and ease the mental workload for CAD activities.

One study helped frame the workload required in 3D modelling by analysing the cognitive workload needed for interpreting 2D drawings, a 3D CAD interface, and a 3D printed model by having a person looking at all three models of the product in one of the six potential sequences (Dadi and Goodrum, 2014). In this study, mental workload was measured using the subjective measurement method, NASA-TLX (Dadi and Goodrum, 2014). Of the three models, the 3D CAD interface had the highest mean for composite workload, mental workload, effort, and frustration, along with the lowest performance compared to the other two. The study also found that more training and experience in CAD helped reduce those workloads and increase performance (Dadi and Goodrum, 2014). Ultimately, this study outlines the need for using cognitive ergonomics when designing CAD software and the need for training to approve the efficiency of humans in AM (Dadi and Goodrum, 2014), while other research has provided insight on the cognitive workload that goes into the of lifecycle management in AM (Muller et al., 2017).

Additive manufacturing's ability to quickly produce parts is a major benefit, but the short lifecycle that features multiple datasets with different data formats that have been gathered in manufacturing process makes it difficult to interpret the multitude of data points and draw correlations between product behaviour, product design, and process settings based on these data points (Muller et al., 2017). In a study done on the lifecycle design and management of AM, a cognitive workload issue was identified and tested in the management of AM technology with the organisation of millions of data points and the consolidation of these points into a single format to analyse the 3D printed prototype (Muller et al., 2017). Although this was believed to be an issue for the demonstrator of the AM process, the demonstrator was able to collect a large amount of data and organise it into a singular and easily accessible database (Muller et al., 2017). Even with the demonstrator's ability to organise and consolidate the multitude of data points into a single format, there are questions on whether this is able to translate into an actual manufacturing environment and the security and safety of the data (Muller et al., 2017).

The CAD software systems use the core principles that other human-computer interactions (HCIs) use such as easy to learn, easy to remember, and pleasant to use to make humans more proficient in using the software (Muller et al., 2017). There are other factors that affect a human's cognitive ability in an HCI, such as emotions (Liu et al., 2014). A study was conducted on the emotions of a human working on a CAD system to discover if there is a correlation between the human's emotions and associated CAD tasks (Muller et al., 2017). This research provides more insight into the mental workload involved in CAD-based work, such as AM and the 3D modelling done in the process (Liu et al., 2014). The study provided a framework of psycho-physiological analysis in engineering task analysis in CAD operation with the results of the study providing insight on how CAD software could be improved for better performance such as a better interface for completing designs on CAD and separate dialogue boxes for requirements and available configurations (Liu et al., 2014). Along with these suggestions for improving CAD software's interface and management of the AM lifecycle, the interface for the 3D printer can also be approved to reduce mental workload by using findings from other research on computer-based procedure systems (Lee et al., 2005).

As previously stated, most of the work done in 3D printing of a part involves HCI, whether it is when modelling the 3D part or setting up the 3D printer to print the 3D modelled part. The previous recommendations for the CAD interface can also be applied to the 3D printer interface (Liu et al., 2014) along with other suggestions such as navigation aid and embedded controls (Lee et al., 2005). In one study done on reducing cognitive workload for a computer-based system, it is found that embedded controls produce a better performance time for humans and are easier to use compared to separated controls, but they come at the cost of limited information for operators and are harder to develop and maintain (Lee et al., 2005). For designing the 3D printer's interface, using embedded controls when an abundance of information is not needed for an operation and keeping the computer interface simplistic should reduce the setup time and improve the machine and human performance.

10.8 HUMAN–MACHINE INTEGRATION

This section highlights the Design-Evaluate-Justify-Integrate (DEJI) model. The DEJI model (Figure 10.1) is a systems engineering model of work design, evaluation, justification, and integration (DEJI™) (Badiru and Bommer, 2017) that encourages the practice of building relevance into a product in the beginning to increase the success of the integration of the system later (Badiru and Racz, 2018). Originally developed for product development, the DEJI model can be applied to numerous types of programs due to every program going through the four stages of the DEJI model: design, evaluation, justification, and integration (Badiru, 2012). Some of the applications that the DEJI model has been applied to include a product acquisitions life cycle (Badiru, 2012), distance learning (DL) in graduate programs (Badiru and Jones, 2012), and a design for quality engineering (Badiru, 2014). One of the factors that make the DEJI model an effective systems engineering model is its use of both qualitative and quantitative assessment techniques throughout the process such

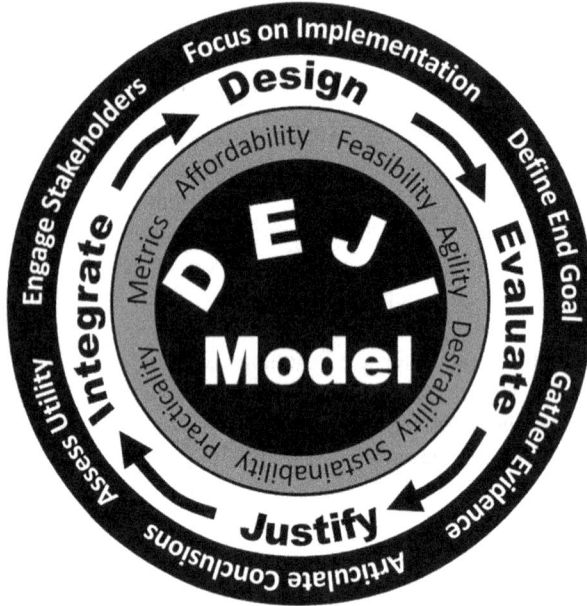

FIGURE 10.1   Design-evaluate-justify-integrate (DEJI) model.

as Pareto analysis and process mapping used in the design stage (Badiru and Jones, 2012). Another component that separates the DEJI model from other systems engineering models such as the V Model or Waterfall Model is its emphasis on integration as the final step in the structure of the system (Badiru and Jones, 2012).

The first stage of the DEJI model is the work design phase. In the design phase, activities such as planning, organising, and coordination of work elements take place to guide work designers into strategic thinking about work elements with a futuristic mindset instead of a mindset solely focused on the present needs of the system (Badiru and Bommer, 2017). The design stage uses two different product states to track the pointto-point transformations of the design: product state and produce state-space (Badiru, 2014). The product state is a set of conditions that describe the product at a specified point in time, while the product state-space is the set of all possible states of the product lifecycle (Badiru, 2014). Product state can be measured by analysing the input and expected output of the system (Badiru, 2014). Product state-space can be measured by using mathematical models that use potential product state variables such as cost or operational efficiency (Badiru, 2014). The quantitative metric that can be used for the state-space is shown in Equation (10.1) (Badiru, 2014):

$$Z = f(z,x); Y = g(z,x) \qquad (10.1)$$

Z = intermediate vector relating x to y
f = vector-valued function

z = output
x = input
g = vector-valued function
Y = output vector

For a product that transitions from one state to another, a driving function that creates a transitional relationship is shown in Equation (10.2) (Badiru, 2014):

$$Ss = f\left(x\,I\,Sp\right) + e \tag{10.2}$$

Ss = subsequent state
f = given action(s) applied to product
x = state variable
Sp = the preceding state
e = error component

The first mathematical model for state-space can analyse which actions are needed to achieve the next desired product state, while the second mathematical model allows one to see the effect of state variables on the product from going from one state to next, which can be expanded into greater detail if needed (Badiru, 2014). The next mathematical model is used if a product (P) is described by state variables (s;) where the composite state of the product can be represented at any time by the vector (S) containing P elements shown in Equation (10.3) (Badiru, 2014):

$$S = \left\{s1, s2, \ldots, sP\right\} \tag{10.3}$$

Lastly, the DEJI model includes a mathematical model that can be used to monitor stateby-state transformations using the mathematical model shown in Equation (10.4) (Badiru, 2014):

$$Sn = Tn\left(Sn - 1\right) \tag{10.4}$$

Sn = final state
Tn = transformation

All these quantitative models are further investigated in the next stage of the DEJI model, which is the evaluation stage.

Work evaluation assesses the intended purpose of the work and its various work elements being done in the organisation (Badiru and Bommer, 2017). Like product state variables, the evaluation of a product can be done based on the cost, quality, schedule, and/or meeting requirements (Badiru, 2014). Another identified technique for product evaluation is learning curve productivity due to the measurement being based on the concept of growth and decay, which factors into the half-life properties of the learning curve (Badiru, 2012). After the evaluation(s) has been concluded, justification of the program and its work element is needed (Badiru, 2014).

The third stage of the DEJI model requires rigorous justification of the program and its work elements (Badiru and Bommer, 2017). The work justification stage is necessary to assure that errant and non-value added work elements are not added into the organisational pursuits (Badiru and Bommer, 2017). An important note for work elements is that a value-added work element does not just include work elements that generate physical products, but also adds value to the worker's well-being (Badiru and Bommer, 2017). The value of these work elements can be shown as a deterministic vector function to designate the value of both tangible and intangible attributes that characterise the project seen in Equation (10.5) (Badiru, 2014):

$$V = f\left(Al, A2,...,Ap\right) \qquad (10.5)$$

V = assessed value
A = quantitative measures or attributes

The basis of the justification stage is that all the work elements are necessary and not hampering the organisation's pursuits (Badiru and Bommer, 2017). The last stage of the DEJI model is the integration phase, and without the integration phase, a system will be isolated and potentially worthless (Badiru, 2012).

The integration stage attempts to incorporate all the work elements that have been justified in the product system (Badiru and Bommer, 2017). All justified work elements of a system then must be properly integrated to align with the system's functional goals to remain sustainable for the organisation (Badiru, 2014). The DEJI model assures that sustainable work elements are the ones that fit within the flow of the organisation's operations (Badiru and Bommer, 2017). Without the integration of a new work element within the flow of the organisation's operations, the new work element will have short longevity (Badiru and Bommer, 2017). With the DEJI model, the model's structure makes it essential for the work elements to be associated with the organisation's end goals.

10.8.1 IMPLEMENTATION STRATEGY FOR DEJI MODEL IN HUMAN FACTORS FOR ADDITIVE MANUFACTURING

The strategy for implementing the DEJI model in human factors for AM can follow similar applications of the DEJI model covered in the past. Like previous applications of the DEJI model, the start will be with the design phase by defining the product's state at specified points in time, and in this case, one would start with looking at the state of the operator during various parts in the 3D printing process (Badiru, 2014). The specified points in time for the human operator in the additive process would include the conceptualisation of the 3D model, the review, revision, and finalisation of the 3D model, the setup for printing of the 3D model, monitoring the product during the printing stage, cleaning the 3D printer and organising excess powder after the product has completed, filtering of the excess powder, and the post-process analysis of the product. With each of these specified points, the next step would be analysing the input and the expected output on each of these specified points using Equation (10.1) to understand the state of the product with these state variables: product cost, final product due date, output quality, throughput, resource utilisation, and operational efficiency (Badiru, 2012).

Using Equation (10.2) to measure the transitional relationship from one product state to another (Badiru and Racz, 2018), the state inputs include planning, defining, designing, revising, finalising, preparing, printing, and filtering, cleaning, machining, and analysing. The analysis of these state inputs will provide needed actions to advance to the next product state (Badiru, 2012). Next, Equation (10.3) can describe the potential states of a 3D printed part from a 3D modelled part, and metal powder/plastic to a 3D printed part if more in-depth product analysis is needed (Badiru, 2014).

Following this potential product analysis, the last mathematical model used would be Equation (10.4) to monitor state-by-state transformation using the state inputs previously listed and these outputs: product specifications, problem statement, 3D part layout, revised 3D part layout, final 3D part layout, machine setup, fabrication, reusable material, prepared machine, finalised part, and product complete (Badiru, 2014). After these design stage equations have been set up and calculated, the evaluation stage follows.

For human factors in AM, the evaluation stage should include evaluating the operator's efficiency in successfully printing a quality part, and the time needed to complete all tasks within the 3D printing process. An evaluation on these two metrics would allow for an analysis of the performance and mental workload of the human operator to see how the AM process currently affects an operator and identify human factor issues in AM and what tasks are having causing these issues to arise. An analysis of these will allow for AM to explore solutions to these problems and how tasks can be adjusted to alleviate these issues, such as using more effective AI. Following this evaluation, these tasks require justification to assure that these are value-added activities.

The justification dimension of the DEJI model starts with justifying each task in the 3D printing process and making sure each task adds value, either tangible or intangible (Badiru and Racz, 2018). Equation (10.5) assesses the value of these tasks by using quantitative measures such as operator efficiency, reliability, and part quality (Badiru, 2014). The other justification needed for human factors in AM is the various precautions used in the 3D printing process, such as the PPE used to reduce static and oxygen monitoring for the room and filtration process (Shelton, 2019). These two justifications are needed to validate the implementation of DEJI model for human factors in AM and continue onto the last stage of the DEJI model, which is the integration stage.

The integration of the DEJI model for human factors in AM should be seamless with all the safety precautions already used in AM. The system of 3D printing parts also revolves around the operator's ability to effectively design a 3D model that can successfully print and properly set up the 3D printer to make a fully functional replica of the 3D model, so maximising the human's performance is of the utmost importance. Some of the previous quantitative metrics used to perform an effective assessment of the product state are cost and resource utilisation (Badiru, 2012). Following this strategy for implementing the DEJI model for human factors for AM will help assure that the AM process is safe, user-friendly, and effective for human operators.

10.9 CONCLUSION

An overview of the literature to advance human–machine integration for improving the performance output in the AM process was presented in this chapter. This chapter begins with introducing the topics of AM, cognitive ergonomics, and AI. Then, it discusses human considerations in the AM process and how cognitive ergonomics support AI techniques. Finally, case studies for the integration of human factors, cognitive ergonomics, and AI are provided. Also, the DEJI systems engineering model for human–machine integration is examined. It is anticipated that the contents of this review paper will pave the way for further research into the integration of human factors and cognitive aspects in the future wave of AI in AM.

REFERENCES

Alaimo, A., A. Esposito and C. Orlando (2018). Cockpit Pilot Warning System: A Preliminary Study, *2018 IEEE 4th International Forum on Research and Technology for Society and Industry (RTSI)*, Palermo, Italy, pp. 1–4. DOI: 10.1109/RTSI.2018.8548518.

Bacha, A., A. H. Sabry and J. Benhra (2019). "Fault Diagnosis in the Field of Additive Manufacturing (3D Printing) using Bayesian Networks". *International Journal of Online and Biomedical Engineering* 15(03): 110–123.

Badiru, A. B. (2012). "Application of the DEJI Model for Aerospace Product Integration". *Journal of Aviation and Aerospace Perspectives* 2(2): 20–34.

Badiru, A. B. (2014). "Quality Insights: The DEJI Model for Quality Design, Evaluation, Justification, and Integration". *Quality Engineering and Technology* 4(4): 369–377.

Badiru, A. B. and Bommer, S. C. (2017a). Application of Cognitive Systems Engineering for More Effective Work Design, Pittsburgh, *Proceedings of the IISE Conference*, Pittsburgh, PA.

Badiru, A. B. and S. C. Bommer (2017b). *Work Design: A Systematic Approach*, CRC Press, Boca Raton, FL.

Badiru, A. B. and J. Y. Cheung (2002). *Fuzzy Engineering Expert Systems with Neutral Network Applications*. John Wiley & Sons, New York.

Badiru, A. B. and R. R. Jones (2012). "A Systems Framework for Distance Learning in Engineering Graduate Programs". *Systems Engineering* 15(2): 191–201.

Badiru, A. B. and L. Racz (2018). Integrating Systems Thinking in Interdisciplinary Education Programs: A Systems Integration Approach. *Proceedings of ASEE Annual Conference*, Salt Lake City, UT.

Bailey, M. N. and B. Bosworth (2014). "US Manufacturing: Understanding Its Past and Its Potential Future". *Journal of Economic Perspectives* 28(1): 3–25.

Baumers, M., L. Beltrametti, A. Gasparre and R. Hague (2017). "Informing Additive Manufacturing Technology Adoption: Total Cost and the Impact of Capacity Utilisation". *International Journal of Production Research* 55(23): 6957–6970.

Bazrafkan, S. and P. Corcoran (2018). "Pushing the AI Envelope: Merging Deep Networks to Accelerate Edge Artificial Intelligence in Consumer Electronics Devices and Systems". *IEEE Consumer Electronics Magazine*, March, pp. 55–61.

Bikas, H., P. Stavropoulos and G. Chryssolouris (2016). "Additive Manufacturing Methods and Modeling Approaches: A Critical Review". *International Journal of Advanced Manufacturing Technology* 83(1–4): 389–405.

Bommer, S. C. (2017) Going cognitive with manufacturing ergonomics, *SE Magazine*, January, pp. 41–43.

Bommer, S. C. and M. Fendley (2018). "A Theoretical Framework for Evaluating Mental Workload Resources in Human Systems Design for Manufacturing Operations". *International Journal of Industrial Ergonomics* 63: 7–17.

Bours, J., B. Adzima, S. Gladwin, J. Cabral and S. Mau (2017). "Addressing Hazardous Implications of Additive Manufacturing: Complementing Life Cycle Assessment with a Framework for Evaluating Direct Human Health and Environmental Impacts". *Journal of Industrial Ecology* 21(Sl): S25–S36.

Calp, M. H. (2019). "Evaluation of Multidisciplinary Effects of Artificial Intelligence with Optimization Perspective". *BRAIN: Broad Research in Artificial Intelligence and Neuroscience* 10(1): 20–29.

Charles, R. L. and J. Nixon (2019). "Measuring Mental Workload using Physiological Measures: A Systematic Review". *Applied Ergonomics* 74: 221–232.

Chen, J., R. Q. Wang, Z. Lin and X. Guo (2018). "Measuring the Cognitive Loads of Construction Safety Sign Designs during Selective and Sustained Attention". *Safety Science* 105: 9–21.

Dadi, G. B. and P. M. Goodrum (2014). "Cognitive Workload Demands using 2D and 3D Spatial Engineering Information Formats". *Journal of Construction Engineering Management* 140(5): 04014001-1-8.

Doran, E., S. Bommer and A. Badiru (2022). "Integration of Human Factors, Cognitive Ergonomics, and Artificial Intelligence in the Human-Machine Interface for Additive Manufacturing". *International Journal of Quality Engineering and Technology, International Journal of Mechatronics and Manufacturing Systems* 15(4): 310–330.

Galy, E. and C. Melan (2015). "Effects of Cognitive Appraisal and Mental Workload Factors on Performance in an Arithmetic Task". *Applied Psychophysiology and Biofeedback* 40(4): 313–325.

Gardan, J. (2016). "Additive Manufacturing Technologies: State of the Art and Trends". *International Journal of Production Research* 54(10): 3118–3132.

Gardner, J. M., K. A. Hunt, A. B. Ebel, E. S. Rose, S. C. Zylich, B. D. Jensen, K. E. Wise, E. J. Siochi and G. Sauti (2019). "Machines as Craftsmen: Localized Parameter Setting Optimization for Fused Filament Fabrication 3D Printing". *Advanced Materials Technologies* 4(3): 1800653.

Garg, A., J. S. Lam and M. M. Savalani (2018). "Laser Power Based Surface Characteristics Models for 3-D Printing Process". *Journal of Intelligent Manufacturing* 29(6): 1191–1202.

Grassmann, M., E. Vlemincx, A. von Leupoldt and Van den Bergh (2017). "Individual Differences in Cardiorespiratory Measures of Mental Workload: An Investigation of Negative Affectivity and Cognitive Avoidant Coping in Pilot Candidates". *Applied Ergonomics* 59:. 274–282.

Hom, T. J. and O. L. A. Harrison (2012). "Overview of Current Additive Manufacturing Technologies and Selected Applications". *Science Progress* 95(3): 255–282.

Kalyuga, S. (2011). "Cognitive Load Theory: How Many Types of Load Does It Really Need?". *Educational Psychology Review* 23(1): l–19.

Kellens, K., R. Mertens, D. Paraskevas, W. Dewulf and J. R. Duflou (2017). "Environmental Impact of Additive Manufacturing Processes: Does AM Contribute to a More Sustainable Way of Part Manufacturing?" *Procedia Cirp* 61: 582–587.

Kwon, O., C. Yoon, S. Ham, J. Park, J. Lee, D. Yoo and Y. Kim (2017). "Characterization and Control of Nanoparticle Emission During 3D Printing". *Environmental Science & Technology* 51(18): 10357–10368.

Lee, G., C. M. Eastman, T. Taunk and C. -H. Ho (2010). "Usability Principles and Best Practices for the User Interface Design of Complex 3D Architectural Design and Engineering Tools". *International Journal of Human-Computer Studies* 68(1): 90–104.

Lee, Y. -L., S. -L. Hwang and E. Min-Yang Wang (2005). "Reducing Cognitive Workload of a Computer-Based Procedure System". *International Journal of Human-Computer Studies* 63(6): 587–606.

Leirmo, T. S. and K. Martinsen (2019). "Evolutionary Algorithms in Additive Manufacturing Systems: Discussion of Future Prospects". *Procedia CIRP* 81: 671–676.

Liang, G. -F., J. -T. Lin, S. -L. Hwang and E. M. -Y. P. P. Wang (2010). "Preventing Human Errors in Aviation Maintenance using an On-line Maintenance Assistance Platform". *International Journal of Industrial Ergonomics* 40(3): 356–367.

Liao, P. -C., X. Luo, T. Wang and Y. Su (2016). "The Mechanism of How Design Failures Cause Unsafe Behavior: The Cognitive Reliability and Error Analysis Method (CREAM)". *Procedia Engineering* 145: 715–722.

Liu, B. -S., K. -N. Huang, H. -J. Chen and K. -C. Yang (2016). Ergonomic Evaluation of New Wrist Rest on Using Computer Mouse, *2016 International Conference on Advanced Materials for Science and Engineering (ICAMSE)*, Tainan, Taiwan.

Liu, Y., J. M. Ritchie, T. Lim, Z. Kosmadoudi, A. Sivanathan and R. C. Sung (2014). "A Fuzzy Psycho-Physiological Approach to Enable the Understanding of an Engineer's Affect Status During CAD Activities". *Computer-Aided Design,* CRC Press, Boca Raton, FL 54: 19–38.

Longo, L. (2015). "A Defeasible Reasoning Framework for Human Mental Workload Representation and Assessment". *Behavior and Information Technology* 34(8): 758–786.

MacLeod, D. (2004). "Cognitive Ergonomics". *Industrial Engineer*, March, pp. 26–30.

Malik, A., H. Lhachemi, J. Ploennigs, A. Ba and R. Shorten (2019). "An Application of 3D Model Reconstruction and Augmented Reality for Real-Time Monitoring of Additive Manufacturing". *Procedia Cirp* 81: 346–351.

Mantyjarvi, K., T. Iso-Junno, H. Nemi and J. Makikangas (2018). "Design for Additive Manufacturing in Extended DFMA Process". *Key Engineering Materials* 786: 342–347.

Minnema, J., M. van Eijnatten, W. Kouw, F. Diblen, A. Mendrik and J. Wolff (2018). "CT Image Segmentation of Bone for Medical Additive Manufacturing Using a Convolutional Neural Network". *Computers in Biology and Medicine* 103: 130–139.

Moorhouse, D. J. (1990). "On the Level 2 Ratings of the Cooper-Harper Scale". *Journal ofGuidancce, Control, and Dynamics* 13(1): 189–191.

Muller, J. R., M. Panarotto, J. Malmqvist and O. Isaksson (2017). *Lifecycle Design and Management of Additive Manufacturing Technologies*, Taylor & Francis Group, Boca Raton, FL.

Ongsulee, P. (2017). *Artificial Intelligence, Machine Learning and Deep Learning*, Prentice Hall, Upper Sadle River, NJ.

Pandit, V. (1994). *Artificial Intelligence and Expert Systems: A Technology Update*, CRC Press, Boca Raton, FL.

Peng, T., K. Kellens, R. Tang, C. Chen and G. Chen (2018). "Sustainability of Additive Manufacturing: An Overview on its Energy Demand". *Additive Manufacturing* 21: 694–704.

Rejeski, D., F. Zhao and Y. Huang (2018). "Research Needs and Recommendations on Environmental Implications of Additive Manufacturing". *Additive Manufacturing* 19: 21–28.

Rubio, S., E. Diaz, J. Martin and J. M. Puente (2004). "Evaluation of Subjecctive Mental Workload: A Comparison of SWAT, NASA-TLX, and Workload Profile Methods". *Applied Psychology: An International Review* 53(I): 61–86.

Sand, M. (2019). "The Environmental Benefits of Additive Manufacturing". *ISE: Industrial and Systems Engineering at Work* 51(2): 46–49.

Schrand, A. M. (2016). "Additive Manufacturing: From Form to Function". *Strategic Studies Quarterly* 10(3): 74–90.

Shelton, T. (2019) *Human Factors in Additive Manufacturing* [Interview] (18 July, 2019).

Shirley, R. B., C. Smidts, M. Li and A. Gupta (2015). "Validating THERP: Assessing the Scope of a Full-Scale Validation of the Technique for Human Error Rate Prediction". *Annals of Nuclear Energy* 77: 194–211.

Simon, D. (2013). *Evolutionary Optimization Algorithms: Biologically-Inspired and PopulationBased Approaches to Computer Intelligence*, John Wiley & Sons Inc., Hoboken, NJ.

Singh, A., N. Thakur and A. Sharma (2016). *A Review of Supervised Machine Learning Algorithms*, McGraw-Hill, London, UK.

Teo, C., P. Giffard, V. Johnston and J. Treleaven (2019). "Computer Vision Symptoms in People With and Without Neck Pain". *Applied Ergonomics* 80: 50–56.

Venda, V. F., R. J. Trybus and N. I. Venda (2000). "Cognitive Ergonomics: Theory, Laws, and Graphic Models". *International Journal of Cognitive Ergonomics* 4(4): 331–349.

Vidulich, M. A., G. F. Ward and J. Schueren (1991). "Using the Subjective Workload Sominance (SWORD) Technique for Projective Workload Assessment". *Human Factors* 33(6): 677–691.

Wang, Y., M. White, I. H. S. Owen and G. Barakos (2013). "Effects of Visual and Motion Cues in Flight Simulation of Ship-Borne Helicopter Operation". *CEAS Aeronautical Journal* 4(4): 385–396.

Wei, L., J. Xu, X. Jia, X. Zhang and H. Li (2018). "Effects of Safety Facilities on Driver Distance Perception in Expressway Tunnels". *Journal of Advanced Transportation* 20(7): 1–10.

Wickens, C. (2002). "Multiple Resources and Performance Prediction". *Theoretical Issues in Ergonomics Science* 3(2): 159–177.

Wickens, C. D., J. D. Lee, Y. Liu and S. E. Gordon Becker (2004). *An Introduction to Human Factors Engineering*, Pearson Education, Inc., Upper Saddle River, NJ.

Yan, S., Y. Wei and C. C. Tran (2019). "Evaluation and Prediction Mental Workload in User Interface of Maritime Operations Using Eye Response". *International Journal of Industrial Ergonomics* 71: 117–127.

Zhang, J., Z. Yin and R. Wang (2017). "Nonlinear Dynamic Classification of Momentary Mental Workload using Physiological Features and NARX-Model-Based Least-Square Support Vector Machines". *IEEE Transactions on Human-Machine Systems* 47(4): 536–549.

Zhao, G., Y. -J. Liu and Y. Shi (2018). "Real-Time Assessment of the Cross-Task Mental Workload Using Physiological Measures During Anomaly Detection". *IEEE Transactions on HumanMachine Systems* 48(2): 149–160.

11 AI Systems Optimization Techniques

11.1 INTRODUCTION

Optimization encompasses areas such as the theory of ordinary maxima and minima, calculus of variations, linear and nonlinear programming, dynamic programming, maximum principles, discrete and continuous games, and differential games of varying degree of complexity.

In this chapter (Badiru et al., 2012), the methodical development of the optimization problem is examined in relation to applications in Applied Sciences from classical techniques through stochastic approaches to contemporary and recent methods of Intelligent MetaHeuristic. Indeed, computational techniques have grown from classical techniques, which afford closed form solutions to approximation and search techniques and more recently intelligent search techniques. Such intelligent heuristics include Tabu Search, Simulated Annealing, Fuzzy systems, Neural Networks and Genetic Algorithms, among myriad of evolving modern computational intelligent techniques. Furthermore, important applications of these heuristics to the evolution of self-organizing adaptive systems such as Modern Economic Models, Transportation, and Mobile Robots do exist.

Central to all problems in optimization theory are the concepts of payoff, controllers, or players, system and information sets. In order to define a solution to an optimization problem, the concept of payoff must be defined, and the controllers must be identified. If there is only one person on whose decision the outcome of some particular process depends and the outcome can be described by a single quantity, then the meaning of payoff (and hence solution to the optimization problem) and controller or player is clear.

The simplest, of course, is the problem of parameter optimization, which includes the classical theory of maxima and minima, linear and nonlinear programming. In parameter optimization, there is one (deterministic or probabilistic) criterion, one controller, one complete information set, and the system state described by static equations and/or inequalities in the form of linear or nonlinear algebraic or difference equations.

On the next rung of complexity are optimization problems of dynamic systems where the state is defined by ordinary or partial differential equations. These can be thought of as limiting cases of multistage (static) parameter optimization problems where the time increment between steps tends to zero. In this class, developed extensively as **optimal control**, we encounter the classical calculus of variation problems and their extension through various maximum and optimality principles, that is, Pontryagin's Minimum Principle, and the dynamic programming principle. We are still concerned with one criterion, controller and information set but have added dimension in that the problem is dynamic and might be deterministic or stochastic.

DOI: 10.1201/9781003247395-11

The next level would introduce two controllers (players) with a single conflicting criterion. Here, we encounter elementary or finite matrix game theory where the controls and payoff are continuous functions. Each player at this level has complete information regarding the payoff for each strategy but may or may not have knowledge of his opponent's strategy. Such games are known as zero-sum games since the sum of the payoffs to each player for each move is zero, what one player gains, the other player loses. If the order in which the players act does not matter, that is, the minimum and maximum of the payoff are equal (this minimax is called the Value of the game and is unique), the optimal strategies of the players are unaffected by knowledge or lack of knowledge of each other's strategy. A solution to this game involves the Value and at least one optimal strategy for each player.

Next, we can consider extensions to dynamical systems where the state is governed by differential equations; we have one conflicting criterion and two players, that is, zero-sum, two player differential games. The information available to each player might be complete or incomplete. In cases with complete information, the finite game concept of a solution is directly applicable. For the incomplete information case, it is reasonable to expect mixed strategies to form the solution, but not much is known of solution methods or whether a solution always exists in the finite game theoretic sense. See Cadenas and Jimenez 1994; Darwin, 1964; Davis, 1991; Gill, et al., 1983; Glover and Macmillan,1986; Gray et al., 2022; Haykin, 1993; Ingber, 1993; Luenberger, 1984; Marti, 1996; Powell, 1964; Rayward-Smith, 1995; Siarry, et al. 1997; Wasan, 1969.

In the last group or uppermost rung of the hierarchy, we identify a class of optimization problems where the concept of a solution is far from clear. To this class belong multiple criteria, n-person games with complete or incomplete information, nonzero sum (either or both players may lose or gain).

The goal of an optimization problem can be formulated as follows: find the combination of parameters (independent variables) which optimize a given quantity, possibly subject to some restrictions on the allowed parameter ranges. The quantity to be optimized (maximized or minimized) is termed the *objective function*, the parameters which may be changed in the quest for the optimum are called *control or decision variables*, the restrictions on allowed parameter values are known as *constraints*. A maximum of a function f is a minimum of –f. The general optimization problem may be stated mathematically as:

$$\text{Minimize } f(X), X = (x_1, x_2, \ldots x_n)^T \tag{11.1}$$

Subject to $C_i(X) = 0, i = 1, 2, \ldots m'$

$$C_i(X) > 0, i = m'+1, m'+2, \ldots m$$

where $f(X)$ is the objective function, X is the column vector of the n independent variables, and $C_i(X)$ is the set of constraints. Constraint equations of the form $C_i(X) = 0$ are termed equality constraints and those of the form $C_i(X) > 0$ are inequality constraints. Taken together, $f(X)$ and $C_i(X)$ are known as the problem functions. See Tables 11.1 and 11.2.

TABLE 11.1
Optimization Problem Classifications

Characteristics	Property	Classification
No. of decision variables	One	Univariate
	More than one	Multivariate
Types of decision variables	Continuous real numbers	Continuous
	Integers	Discrete
	Both continuous real numbers and integers	Mixed integer
	Integers in permutation	Combinatorial
Objective functions	Linear functions of decision variables	Linear
	Quadratic functions of decision variables	Quadratic
	Other nonlinear functions of decision variables	Nonlinear
Problem formulation	Subject to constraints	Constrained
	Not subject to constraints	Unconstrained
Decision variable realization within the optimization model	Exact	Deterministic
	Subject to random variation	Stochastic
	Subject to fuzzy uncertainty	Fuzzy
	Subject to both random variation and fuzzy uncertainty	Fuzzy-stochastic

TABLE 11.2
Typical Applications

Field	Problem	Classification
Nuclear Engineering	In-core nuclear fuel management	Nonlinear Constrained Multivariate Combinatorial
Computational Chemistry	Energy minimization for 3D structure prediction	Nonlinear Unconstrained Multivariate Continuous
Computational Chemistry and Biology	Distance geometry	Nonlinear Constrained Multivariate Continuous

The strict definition of the global optimum X^* of $f(X)$ is that

$$f\left(X^*\right) < f\left(Y\right) \forall Y \in V\left(X\right), Y \neq X^* \tag{11.2}$$

where $V(X)$ is the set of feasible values of the control variables X. Obviously, for an unconstrained problem, $V(X)$ is infinitely large.

A point Y^* is a strong local minimum of $f(X)$ if

$$f\left(Y^*\right) < f\left(Y\right) \forall Y \in N\left(Y^*, \eta\right) Y \neq Y^* \tag{11.3}$$

where (Y^*, η) is defined as the set of feasible points contained in the neighborhood of Y, that is, within some arbitrary small distance of Y. For Y^* to be a weak local minimum, only an equality needs to be satisfied:

$$f\left(Y^*\right) \leq f\left(Y\right) \forall Y \in N\left(Y^*, \eta\right) Y \neq Y^* \tag{11.4}$$

If $f(X)$ is a smooth function with continuous first and second derivatives for all feasible X, then, a point X^* is a stationary point of $f(X)$ if

$$g\left(X^*\right) = 0 \tag{11.5}$$

where $g(X)$ is the gradient of $\hat{f}\left(X\right)$. This first derivative vector $\hat{f}\left(X\right)$ has components given by

$$g_i\left(X\right) = \frac{\partial f\left(X\right)}{\partial x_i} \tag{11.6}$$

The point X is also a strong local minimum of $f(X)$ if the Hessian matrix $H(X)$, the symmetric matrix of second derivatives with components

$$H_{ij}\left(X\right) = \frac{\partial^2 f\left(X\right)}{\partial x_i \partial x_j} \tag{11.7}$$

is positive-definite at X^*, that is, if for a vector $\boldsymbol{u}$;

$$\boldsymbol{u}^T H\left(X^*\right) u > 0, \forall u \neq 0 \tag{11.8}$$

This condition is a generalization of convexity, or positive curvature to higher dimensions. Figure 11.1 illustrates the different types of stationary points for unconstrained univariate functions. See also Figures 11.2–11.4.

As shown in Figure 11.4, the situation is slightly more complex for constrained optimization problems. The presence of a constraint boundary, in Figure 11.4, in the form of a simple bound on the permitted values of the control variable, can cause the global minimum to be an extreme value, an extremum (i.e., an end point), rather than a true stationary point. Some methods of treating constraints transform the optimization problem into an equivalent unconstrained one, with a different objective function.

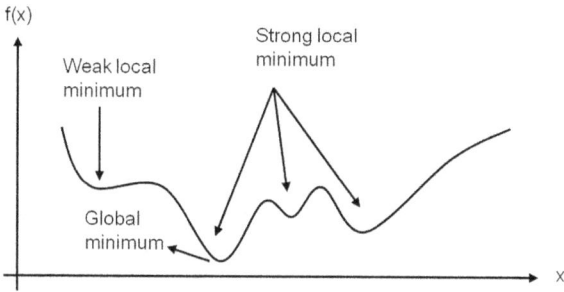

FIGURE 11.1 Types of minima for unconstrained optimization problems.

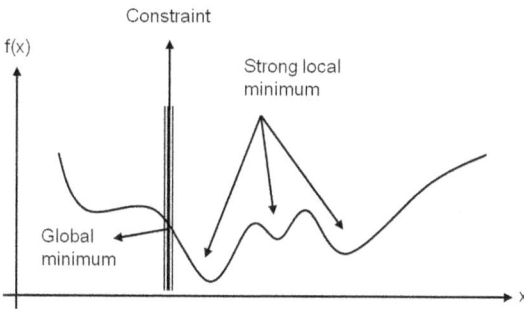

FIGURE 11.2 Types of minima for constrained optimization problems.

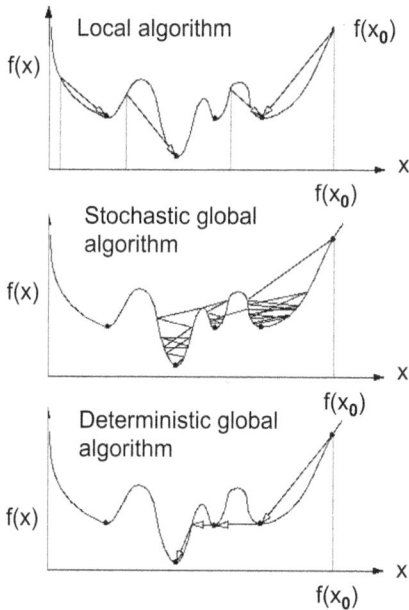

FIGURE 11.3 Types of structure of local and global minimization algorithms.

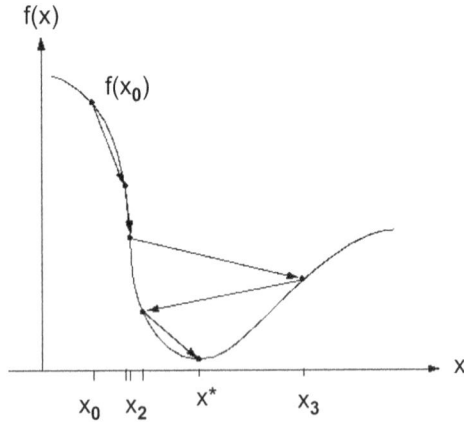

FIGURE 11.4　The descent structure of local minimization algorithms.

11.2　BASIC STRUCTURE OF LOCAL METHODS

A starting point is chosen; a direction of movement is prescribed according to some algorithm, and a line search or trust region is performed to determine an appropriate next step. The process is repeated at the new point and the algorithm continues until a local minimum is found. Schematically, a model local minimizer method can be described as follows:

Algorithm 3.1

Basic Local Optimizer
　Supply an initial guess x_0
　For $k = 0, 1, 2, \ldots$ until convergence,

　　　1. Test x_k for convergence
　　　2. Calculate a search direction p_k
　　　3. Determine an approximate step length λ_k (or modified step s_k)
　　　4. Set x_{k+1} to $x_k + \lambda_k p_k$ (or $x_k + s_k$)

11.3　DESCENT DIRECTIONS

It is reasonable to choose a search vector p that will be a descent direction, that is, a direction leading to function value reduction. A descent direction P is defined as one along which the directional derivative is negative:

$$g(X)^T p < 0 \qquad\qquad (11.9)$$

when we write the approximation

$$f(X + \lambda p) \approx f(X) + \lambda g(X)^T p \qquad\qquad (11.10)$$

we see that the negativity of the right-hand side guarantees that a lower function value can be found along p for sufficiently small λ.

11.4 STEEPEST DESCENT (SD)

SD is one of the oldest and simplest methods. At each iteration of SD, the search direction is taken as $-g_k$, the negative gradient of the objective function at x_k. Recall that a descent direction p_k satisfies $g_k^T p_k < 0$. The simplest way to guarantee the negativity of this inner product is to choose $p_k = -g_k$. This choice also minimizes the inner product $-g_k^T p$ for unit-length vectors, and thus gives rise to the name *Steepest Descent*. SD is simple to implement and requires modest storage, $O(n)$. However, progress toward a minimum may be very slow, especially near a solution.

11.5 CONJUGATE GRADIENT (CG)

The first iteration in CG is the same as in SD, but successive directions are constructed so that they form a set of mutually conjugate vectors with respect to the (positive-definite) Hessian $\mathbf{A}$ of a general convex quadratic function $qA(X)$.

Algorithm 11.2.2

CG method to solve $AX = -b$

1. Set $r_0 = -(Ax_0 + b)$, $d_0 = r_0$
2. For $k = 0, 1, 2, \ldots$ until r is sufficiently small, compute:

$$\lambda_k = r_k^T r_k / d_k^T A d_k$$
$$x_{k+1} = x_k + \lambda_k d_k$$
$$r_{k+1} = r_k - \lambda_k d_k$$
$$\beta_k = r_{k+1}^T r_{k+1} / r_k^T r_k$$
$$d_{k+1} = r_{k+1} + \beta_k d_k$$

11.6 NEWTON METHODS

All Newton methods are based on approximating the objective function locally by a quadratic model and then minimizing that function approximately. The quadratic model of the objective function f at x_k along p is given by the expansion

$$f\left(x_k + p\right) \approx f\left(x_k\right) + g_k^T p + \frac{1}{2} p^T H_k p \tag{11.11}$$

The minimization of the right-hand side is achieved when p is the minimum of the quadratic function:

$$qH_k\left(p\right) = g_k^T p + \frac{1}{2} p^T H_k p \tag{11.12}$$

Alternatively, such a *Newton direction* P satisfies the linear system of n simultaneous equations, known as the *Newton equation*:

$$H_k p = -g_k \qquad (11.13)$$

In the "classic" Newton method, the Newton direction is used to update each previous iterate by the formula $x_{k+1} = x_k + p_k$, until convergence. For the one-dimensional version of Newton's method for solving a nonlinear equation $f(X) = 0$:

$$x_{k+1} = x_k - f(x_k) / f'(x_k) \qquad (11.14)$$

The analogous iteration process for minimizing $f(X)$ is:

$$x_{k+1} = x_k - f'(x_k) / f''(x_k) \qquad (11.15)$$

Newton variants are constructed by combining various strategies for the individual components above.

11.6.1 ALGORITHM: MODIFIED NEWTON

For $k = 0, 1, 2, \ldots$, until convergence, given x_0:

1. Test x_k for convergence
2. Compute a descent direction p_k so that $\|H_k p_k + g_k\| \leq \eta_k \|g_k\|$,
     where η_k controls the accuracy of the solution and some symmetric matrix $\bar{H}_k$ may represent H_k.
3. Compute a step length λ so that for $x_{k+1} = x_k + \lambda p_k$,

$$f(x_{k+1}) \leq f(x_k) + \alpha \lambda g_k^T p$$
$$\left| g_{k+1}^T p_k \right| \leq \beta \left| g_k^T p_k \right|,$$

    with $0 < \alpha < \beta < 1$
4. Set $x_{k+1} = x_k + \lambda p_k$

Newton variants are constructed by combining various strategies for the individual components.

11.7 THE STOCHASTIC CENTRAL PROBLEMS

An object that changes randomly both in time and space is said to be **stochastic**. A basic characteristic of applied optimization problems treated in engineering is the fact that the data for these problems, for example, parameters of the material (yield stress, allowable stresses, moment capacities, specific gravity, etc.), external loadings, manufacturing errors, cost factors, etc., are not known at the planning stage but have to be considered to be random variables with a certain probability distribution; in addition, there is always some uncertainty in the mathematical modeling of practical problems. Typical problems of this type are:

- The limit (collapse) load analysis and the elastic or plastic design of mechanical structures represented mathematically by means of
  1. The equilibrium equation
  2. Hooke's law
  3. The member displacement equation
- Optimal trajectory planning of robots by offline programming such that the control strategy, based on the optimal open loop control, causes only low online correction expenses. Here, the underlying mechanical system is described by the kinematics and the dynamic equation, and the optimal velocity profile and configuration variables are determined for fixed model parameters by a certain variation problem.

  Since the (online) correction of a decision, for example, the decision on the design of a mechanical structure or on the selection of a velocity profile, after the realization/observation of the random data might be highly expensive and time-consuming, the already known prior and statistical information about the underlying probability mechanism generating the random data should be taken into account already in the planning phase. Hence, taking into account stochastic parameter variations already in the planning phase, for offline programming, that is, applying stochastic programming instead of ordinary mathematical programming methods, the original optimization problem with random data is replaced using appropriate decision criteria by a deterministic substitute problem, for example,
- Using a chance-constrained programming approach, the objective function is replaced by its mean value, and the random constraints are replaced by chance constraints.
- Evaluating the violation of the random constraints by means of penalty functions, a weighted sum of the expectation of the primary objective function and the total expected penalty costs are minimized subject to the remaining deterministic constraints, for example, box constraints or the mean value of the objective function is minimized subject to constraints involving upper bounds for the expected penalty cost arising from violations of the original constraints.

A main problem in the solution of these problems is the numerical computation and differentiation of risk functions.

11.7.1 STOCHASTIC APPROXIMATION

While some non-classical optimization techniques are able to optimize on discontinuous objective functions, they are unable to do so when complexity of the data becomes very large. In this case, the complexity of the system requires that the objective function be estimated. Furthermore, the models that are used to estimate the objective function may be stochastic due to the dynamic and random nature of the system and processes.

The basic idea behind the stochastic approximation method is the gradient descent method. Here, the decision variable is varied in small increments, and the impact of this variation (measured by the gradient) is used to determine the direction of the next step. The magnitude of the step is controlled to have larger steps when the perturbations in the system are small and vice versa. Stochastic approximation algorithms based on various techniques have been developed recently. They have been applied to both continuous and discrete objective functions.

11.7.2   GENERAL STOCHASTIC CONTROL PROBLEM

The control of a random, dynamic system in some optimal fashion using imperfect measurement data is the general problem. It also constitutes a problem about which it is very difficult to obtain any meaningful insights. Although feedback is used in order to compensate for unmodeled errors and inputs, most controllers are designed and analyzed in a deterministic context. The control inputs for the system generally must be based on imperfect observations of some of the variables, which describe the system. The control policy that is utilized must be based on a priori knowledge of the system characteristics, on the time history of the input variables.

The mathematical model of the system is described by a nonlinear difference equation

$$x_{k+1} = f\left(x_k,\ H_k\right) + w_k, \quad k = 0,1,\ldots N \tag{11.16}$$

The noise w has been assumed to be additive primarily for reasons of convenience. The state x is n-dimensional and the input u is p-dimensional. In general, a probabilistic model for the initial state xo and for the plant w_k is assumed to be known except for some unknown parameters. With rare exception, these variables are regarded as having a Gaussian distribution such that

$$E\left[x_0\right] = U_0, \quad E\left[w_k\right] = 0, \quad \forall k \tag{11.17}$$

$$E\left[\left(x_0 - H_0\right)\left(x_0 - U_0\right)\right] = M_0, \quad E\left[x_0 w_k^T\right] = 0, \quad \forall k \tag{11.18}$$

$$E\left[w_k w_k^T\right] = O_k \delta_k \tag{11.19}$$

Thus, the plant noise sequence is white and independent of the initial state.

The measurement system is described by a nonlinear algebraic relation to the state. The m-dimensional measurement vector is given by

$$z_k = h_k\left(x_k\right) + v_k, \quad k = 0,1,\ldots N \tag{11.20}$$

The noise v is considered to be additive for reasons of convenience. It is assumed to be a zero mean, white Gaussian sequence, which is independent of the initial state and the plant noise sequence.

$$E\left[v_k\right] = 0, \quad \forall k \tag{11.21}$$

$$E\left[v_k v_i^{T}\right] = R_k \delta_{ki}, \quad \forall k \tag{11.22}$$

$$E\left[v_k x_0^{T}\right] = 0, \quad \forall k \tag{11.23}$$

$$E\left[v_k w_j^{T}\right] = 0, \quad \forall k, j \tag{11.24}$$

The equations above provide the mathematical description of the system. It is this part of the complete system that represents the physical system that must be controlled. The structure of the controller, of course, depends on the exact form of the system model equations $f(\bullet, \bullet)$ and $h(\bullet)$.

The behavior of the system is controlled through the input signal u_k, which are introduced at each sampling time t_k. The manner in which the controls are generated can be accomplished in a limitless number of ways. Certainly, the controls are constrained by the objectives that are defined for the control action and by the restrictions on the control and state variables themselves. Generally, there will be more than one control policy that satisfies the system constraints and achieves the prescribed objectives. Then, it is reasonable to attempt to select the control policy among all these admissible policies that is "best" according to some well-defined performance measure. Optimal stochastic control theory is concerned with the determination of the best admissible control policy for the given system.

The following performance index is assumed:

$$J_0 = E \sum_{i=0}^{N-1} w_i\left(x_{i+1}, u_i\right) \tag{11.25}$$

Notice that the summation $E \sum_{i=0}^{N-1} w_i\left(x_{i+1}, u_i\right)$ is a random variable. Consequently, it is appropriate to consider its minimization; instead, it is mapped into a deterministic quantity by considering its expected value.

11.8   THE INTELLIGENT HEURISTIC MODELS

There had been continuing vast advances in optimal solution techniques for intelligent systems in the last two decades. The heuristic methods offer a very viable approach; however, the design and implementation of problem-specific heuristic can be a long and expensive process and the result is often domain-dependent and not flexible enough to deal with changes, which may occur over time. Hence, considerable interest is focused on general heuristic techniques that can be applied to a variety of different combinatorial problems. This has yielded some new generation of intelligent heuristic techniques such as Tabu Search, Simulated Annealing, Evolutionary Algorithms such as genetic algorithms, neural networks, etc.

11.8.1 HEURISTICS

Heuristics are the knowledge used to make good judgments, or strategies, tricks or "rules of thumb" used to simplify the solution of problems. They include "trial and error" (experience based) knowledge and intelligent guesses/procedures for domain-specific problem solving. They are particularly suitable for ill-defined or poorly posed problems, and poor models when there is incomplete data. Heuristics play an important role in such strategies because of the exponential nature of most problems. They help to reduce the number of alternatives from an exponential number to a polynomial number, and thereby obtain a solution in tolerable amount of time.

11.8.2 INTELLIGENT SYSTEMS

Intelligence is the ability to acquire, understand, and apply knowledge or the ability to exercise thought or reasons. It also embodies both conscious and unconscious knowledge and feats, which animate beings have acquired through study and experience. (Artificial) intelligent systems are thus machines and coded programs aimed at mimicking such feat and knowledge. Systems have been designed to perform many types of intelligent tasks. These can be physical systems like robots or mathematical computational systems such as scheduling systems, which solve diverse tasks, systems used in planning complex strategies for military and for business, in medical diseases diagnosis and control, and so on.

11.8.3 THE GENERAL SEARCH PARADIGM

The General Search Algorithm (11.5) is of the form:

11.8.3.1 General Search

Objective is to maximize $f(x), x \in U$
   X, Y, Z: multiset of solutions $\subset U$
   Initialize (X);
While not finish (X) do

```
Begin
Y: = select (X)
Z: = create (Y)
X: = merge (X, Y, Z)
End
```

where

   X = the initial pool of one or more potential solutions to the problem. Since X may contain multiple copies of some solutions, it is more appropriately called a multiset.
   Y is a selection from X
   Z is created from Y

When a new solution is **created** either initially or by using the Operator "**create**," the function value, $f(x)$ is applied to determine the value of the solution. X is reconstructed from the penultimate pool of X, Y, and Z by the Operator "**merge**." The process is repeated until the pool, X, is deemed satisfactory.

11.8.4 INTEGRATED HEURISTICS

Modern approaches to local search have incorporated varying degrees of intelligibility. The contribution of intelligent search techniques should not be solely viewed in terms of improved performance alone as the traditional systems engineers or analysts expect (even though very much desirable). The trust of the contribution of intelligent search techniques should be in terms of **improved intelligibility, flexibility, and transparency of these emerging computational techniques**. A synergy between intelligibility and performance is normally of utmost importance in assessing the efficiency of an intelligent heuristic.

11.8.5 TABU SEARCH

Tabu Search is one successful variant of the Neighborhood search paradigm designed to avoid the problem of becoming trapped in a local optimum.

The Tabu Search Paradigm 11.5 is shown as below:

11.8.5.1 Tabu Search

Objective is to maximize $f(x)$, $x \in U$

```
X, Z: multiset of solutions ⊂U
Tabu set of rules of type U → {true, false}
Initialize (X);
Initialize (Tabu); {very often to Φ}
While not finish (X) do
Begin
Z: = create (X, Tabu)
Tabu = update(Tabu)
X: = merge(X, Z)
End
```

The difficulty in Tabu Search is in constructing the set of rules. Considerable expertise and experimentation is required to construct the rules and to ensure its dynamic nature is correctly controlled. If the expertise is available, the resulting search can be efficient. Aspiration criteria are often included to help the Tabu Search in not being too restrictive. These criteria are rules, which say that certain moves are to be preferred over others. Some form of expert rules may also serve as Tabu Search rules. Each rule may have an associated weight, negative if Tabu and positive if an aspiration. The combined set of rules thus associates a weight to each neighbor. A large positive weight suggests it is a desirable move, while a large negative weight suggests it can be discounted. Tabu Search has found applications to real-world problems such as Packing and Scheduling problems (flow shop problems, employee scheduling problem, machine scheduling, etc.); Traveling Salesman; Vehicle Routing and Telecommunications.

11.8.6 SIMULATED ANNEALING (SA)

The Simulated Annealing (SA) exploits an analogy between the way in which a metal cools and freezes into a minimum energy crystalline structure (the annealing process) and the search for a minimum in a more general system.

The SA is essentially a local search technique in which a move to an inferior solution is allowed with a probability that decreases as the process progresses, according to some Bolzmann-type distribution. The inspiration for SA approach is the law of thermodynamics, which states that at temperature t, the probability of an increase in energy of magnitude, δE, is given by:

$$P(\delta E) = \exp(-\delta E / kt) \qquad (11.26)$$

where k is the physical constant known as the Bolzmann constant. The equation is applicable to a system that is cooling until it converges to a "frozen" state. The system is perturbed from current state and the resulting energy change δE is calculated. If the energy has decreased, the system moves to the new state; otherwise, the new state is only accepted with the probability given above. The cycle can be repeated for a number of iterations at each temperature and subsequently reduced and the number of cycles repeated for the new lower temperature. This whole process is repeated until the system freezes to its steady state.

We can associate the potential solutions of an optimization problem with the system states; the cost of a solution corresponds to the concept of energy and moving to any neighbor corresponds to a change of state. A simple version of the SA Paradigm 11.6 is of the form:

11.8.6.1 Simulated Annealing

```
X = x₀where x₀ is some initial solution
temp = temp₀ where temp₀ is some initial temperature
      while not finish f(X) do
          For i = 1 to n
             Begin
             Randomly select x′, a neighbor of x
             Improvement = f(x′) - f(x)
                 If improvement > 0 then x = x′
                 Else
                     Begin
                     Generate x ∈ Q[0, 1];
                     If x < exp(-improvement/temp) then x
                        = x′
                 End {else}
             End {for}
             t = reduced (t)
         End {while}
```

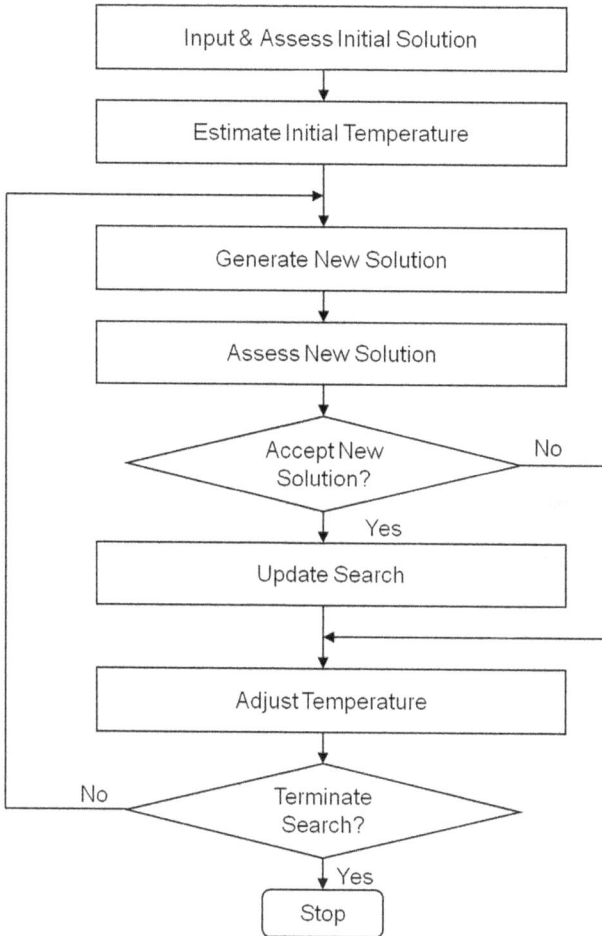

FIGURE 11.5 The structure of the Simulated Annealing algorithm.

See Figure 11.5.

Myriad of practical applications of SA include Graph Coloring, Packing and Scheduling Problems, Traveling Salesman, Vehicle Routing, and Quadratic Assignment problems.

11.8.7 GENETIC ALGORITHMS (GA)

Genetic algorithms are search techniques, based on an abstracted model of Darwinian evolution. Fixed length strings represent solutions over some alphabet. Each such string is thought of as a "chromosome." The value of the solution then represents "fitness" of the chromosome. The concept of "survival of the fittest" is then used to allow better solutions to combine to produce offspring. The Genetic Algorithm paradigm follows closely the same search concept exploited in Tabu Search and SA. The paradigm is:

11.8.7.1   Genetic Algorithm (11.7)

Objective is to maximize $f(x)$, $x \in U$

```
X, Y, Z: multiset of solutions ⊂U
Initialize (X);
While not finish (X) do
Begin
Y: = select (X)
Z: = create (Y)
X: = merge(X, Y, Z)
End
```

Here, the Operators **select, create**, and **merge** correspond to the Operators select, reproduction, crossover. The iteration loop of a basic GA hence looks like.

11.8.7.2   Procedure GA

```
Begin
    Generate initial population, P(0); t = 0;
    Evaluate Chromosomes in P(0);
    Repeat
        t = t + 1'
        Select P(t) from P(t - 1);
        Recombine chromosomes in P(t) using genetic operators;
        Evaluate Chromosomes in P(t);
    Until termination condition is satisfied;
End
```

In natural evolution, each species searches for beneficial adaptations in an ever-changing environment. As species evolve, these new attributes are encoded in the chromosomes of individual members. This information does change by random mutation, but the real driving force behind evolutionary development is the combination and exchange of chromosomal material during breeding.

GAs differ from traditional optimization algorithms in four important respects:

- They work using an encoding of the control variables rather than the variables themselves.
- They search from one population to another rather than from individual to individual.
- They use only objective function information, not derivatives.
- They use probabilistic, not deterministic transition rules.

11.8.8   GENETIC ALGORITHM OPERATORS

They are rather known as evolutionary algorithms than genetic algorithms.

Mutation: Bit-wise change in strings at randomly selected points. See Figure 11.6.

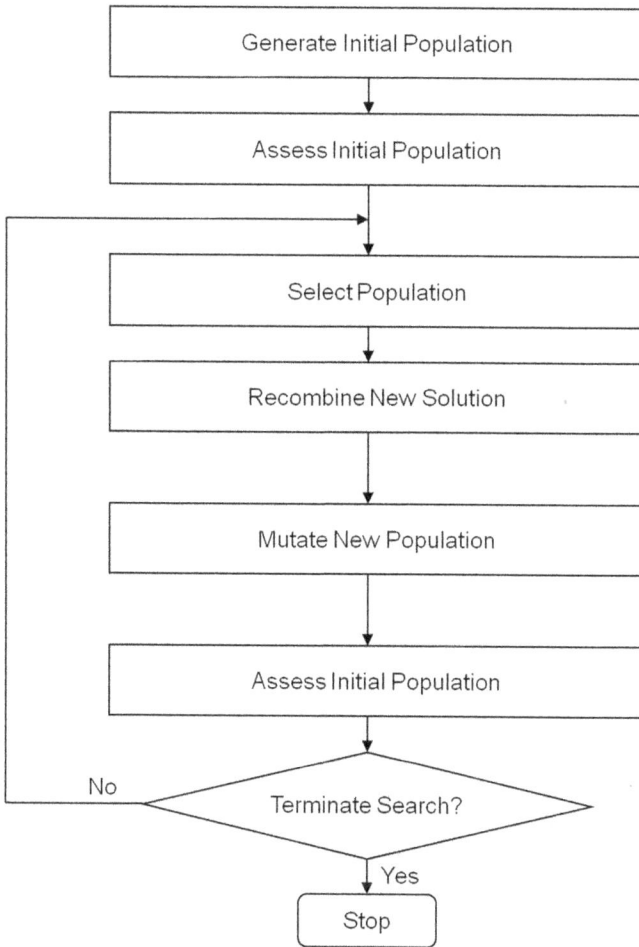

FIGURE 11.6   The basic structure of a genetic algorithm.

11.8.8.1 Examples

Crossover: This is a generic operator applied to two randomly selected parent solutions in the mating pool to generate two offspring solutions. Crossover operation is not performed on all pairs of parent solutions selected from the mating pool. Crossover is performed according to a probability, p (usually $p = 0.6$). If GA decides by this probability not to perform crossover on a pair of parent solutions, they are simply copied to the new population. If crossover is to take place, then one or two random splicing points are chosen in a string. The two strings are spliced, and the spliced regions are mixed together to create two (potentially) new strings. These child-strings are then placed in the new generation.

For example, using strings 10010111 and 00101010, suppose the GA chooses at random to perform crossover at point 5,

```
1   0   0   1   0 | 1   1   1     crossover     1   0   0   1   0   0   1   0

0   0   1   0   1 | 0   1   0                   0   0   1   0   1   1   1   1

parent strings                                 child - strings  (after crossover)

                              the new strings are :   1   0   0   1   0   0   1   0

                                                      0   0   1   0   1   1   1   1
```

For two crossover points at points 2 and 6, the crossing looks like,

```
1   0 | 0   1   0   1 | 1   1     crossover     1   0 | 1   0   1   0 | 1   1

0   0 | 1   0   1   0 | 1   0                   0   0 | 0   1   0   1 | 1   0

parent strings                                 child - strings  (after crossover)

                              the new strings are :   1   0   1   0   1   0   1   1

                                                      0   0   1   0   1   0   1   1   0
```

Mutation: Selection and crossover alone can obviously generate a staggering amount of differing strings. However, depending on the initial population chosen, there may not be enough variety of strings to ensure that GA sees the entire problem space in the event of which GA may find itself converging on strings that are not close to the optimum it seeks due to a bad initial population. Some of these problems may be overcome by introducing a mutation operator into the GA. The GA has a mutation probability, m, which dictates the frequency at which mutation occurs. Mutation can be performed either during selection or crossover (though crossover is more usual). For each string, GA checks if it should perform a mutation. If it should, it randomly changes the element value to a new one. In a binary string, 1s are changed to 01s and 0s to 1s. For example, given the string 10101111, if the GA determines to mutate this string at point 3 the 1 in that position is changed to 0, that is,

$$10101111 \xrightarrow{\text{mutate}} 10001111$$

The mutation probability is kept as low as possible (usually about 0.01), as high mutation rate will destroy fit strings and degenerate the GA algorithm into a random walk, with all the associated problems. GA has been applied successfully to a wide variety of systems. We can only highlight a small subset of the myriad of applications, including Packing and Scheduling, Design of Engineering Systems, Robots and Transport Systems.

For a number of reasons, MetaHeuristic Optimization Techniques differ from classical search and Optimization methods.

1. MetaHeuristic Optimization Techniques work with a coding of decision variables, a matter, which is very uncommon in classical methods. Coding discretizes the search spaces and allows MetaHeuristic Optimization Techniques to be applied to both discrete and discontinuous problems. MetaHeuristic Optimization Techniques also exploit coding similarities to make faster and parallel searches.
2. Since no gradient information is used in MetaHeuristic Optimization Techniques, they can also be applied to non-differentiable functions. This makes MetaHeuristic Optimization Techniques robust in the sense that they can be applied to a wide variety of problems.
3. Unlike many classical optimization methods, MetaHeuristic Optimization Techniques, like GA, work with a population of points. This increases the possibility of obtaining the global optimal solution even in ill-behaved problems.
4. MetaHeuristic Optimization Techniques use probabilistic transition rules, instead of fixed rules. For example, in early GA iterations, this randomness in GA operators makes the search unbiased toward any particular region in the search space and has an affect not making a hasty wrong decision. Use of stochastic transition rules also increases the chance of recovering from a mistake.
5. MetaHeuristic Optimization Techniques use only pay-off information to guide them through the problem space. Many search techniques need a variety of information to guide them. Hill climbing methods require derivatives, for example. The only information a MetaHeuristic needs is some measure of fitness about a point in the space (sometimes known as objective function value). Once the MetaHeuristic knows the current measure of "goodness" about a point, it can use this to continue searching for the optimum.
6. MetaHeuristic Optimization Techniques allow procedure-based function declaration. Most classical search methods do not permit such declarations. Thus, where procedures of optimization need to be declared, MetaHeuristic Optimization Techniques prove a better optimization tool.

The coding of decision variables in MetaHeuristic Optimization Techniques also makes it proficient in solving optimization problems involving equality and inequality constraints.

11.8.8.2　Applications of Heuristics to Intelligent Systems

Some of these techniques have been successfully applied to model and simulate some intelligent systems. These include Application of Genetic Algorithms to Bi-modal Transport Scheduling problem. In this application, GAs are applied using a bi-level programming approach to obtain transit schedules for a bi-modal transfer station in an urban transit network with multiple dispatching stations. The arrival rates of passengers are captured as fuzzy numbers in order to confer intelligibility in the system. The model results in comparably better schedules with better levels of service (LOS) in the transit network (in terms of total waiting and transfer times for passengers) than is obtained otherwise.

Another application is on an intelligent path planner for autonomous mobile robots (Asaolu, 2002). For this robotic motion planner, a real-time optimal path planner was developed for autonomous mobile robots navigation. MetaHeuristic Optimization Techniques were used to guide the robot within a workspace with both static and roving obstacles. With the introduction of arbitrarily moving obstacles, the dynamic obstacle avoidance problem was re-cast into a dynamic graph-search. The instantaneous graph is made of the connected edges of the rectangles boxing the ellipses swept out by the robot, goal, and moving obstacles. This was easily transversed in an optimal fashion.

We are presently also looking at the problems of machine translation as another application of intelligent MetaHeuristic Optimization Techniques.

11.8.8.3   High-Performance Optimization Programming

Future developments in the field of optimization will undoubtedly be influenced by recent interest and rapid developments in new technologies – powerful vector and parallel machines. Indeed, their exploitation for algorithm design and solution of "grand challenge" applications is expected to bring new advances in many fields, such as computational chemistry and computational fluid dynamics. Supercomputers can provide speedup over traditional architectures by optimizing both scalar and vector computations. This can be accomplished by pipelining data as well as offering special hardware instructions for calculating intrinsic functions (e.g., $\left(\exp(x), \sqrt{x}\right)$, arithmetic, and array operations. In addition, parallel computers can execute several operations concurrently. Communication among processors is crucial for efficient algorithm design so that the full parallel apparatus is exploited. These issues will only increase in significance, as massively parallel networks enter into regular use. In general, one of the first steps in optimizing codes for these architectures is implementation of standard basic linear algebra subroutines (BLAS). These routines, continuously being improved, expanded, and adapted optimally to more machines, perform operations such as dot products $(x^T y)$ and vector manipulations $(ax + y)$, as well as matrix/vector and matrix/matrix operations.

Specific strategies for optimization algorithms have been quite recent and are not yet unified. For parallel computers, natural improvements may involve the following ideas:

(1) Performing multiple minimization procedures concurrently from different starting points;
(2) Evaluating function and derivatives concurrently at different points (e.g., for a finite-difference approximation of gradient or Hessian or for an improved line search);
(3) Performing matrix operations or decompositions in parallel for special structured systems (e.g., Cholesky factorizations of block-band preconditioned).

With increased computer storage and speed, the feasible methods for solution of very large [e.g., $O(10^5)$ or more variable] nonlinear optimization problems arising in important applications (macromolecular structure, meteorology, economics) will undoubtedly expand considerably and make possible solution of larger and far more complex problems in all fields of science and engineering.

REFERENCES

Asaolu, O. S., (2002). An Intelligent Path Planner for Autonomous Mobile Robots. PhD Thesis in Engineering Analysis, University of Lagos, Lagos, Nigeria.

Badiru, A. B., O. Ibidapo-Obe and B. J. Ayeni (2012). *Industrial Control Systems: Mathematical and Statistical Models and Techniques*. Taylor & Francis Group/CRC Press, Boca Raton, FL.

Cadenas, M. and F. Jimenez (1994). Genetic Algorithm for the Multi-objective Solid Transportation Problem: A Fuzzy Approach. In *International Symposium on Automotive Technology and Automation, Proceedings for the dedicated conferences on Mechatronics and Supercomputing Applications in the Transportation Industries*, pp. 327–334, Aachen, Germany.

Darwin, C. (1964). *On the Origin of Species* (1st edition; facsimile, 1964). Harvard University Press, Cambridge, MA.

Davis, L. (1991). *Handbook of Genetic Algorithms*. Van Nostrand Reinhold, New York.

Gill, P. E., W. Murray and M. H. Wright (1983). *Practical Optimization*. Academic Press, New York.

Glover, F. and C. Macmillan, (1986). The General Employee Scheduling Problem: An Integration of Management Science and Artificial Intelligence. *Computers & Operations Research* 15: 563–593.

Gray, P., W. Hart, L. Painton, C. Phillips, M. Trahan and J. Wagner "A Survey of Global Optimization Methods". *Sandia National Laboratories Albuquerque, NM 87185*. http://www.cs.sandia.gov/opt/survey/, Accessed Dec 2, 2022.

Haykin, S. (1993). *Neural Networks: A Comprehensive Foundation*, Prentice Hall International Inc., Trenton, NJ, ISBN: 0-13-908385-5.

Ingber, L. A., (1993). "Simulated Annealing: Practice versus Theory". *Journal of Mathematical Computer Modelling* 18(11): 29–57.

Luenberger, D. G. (1984). *Linear and Nonlinear Programming* (2nd edition). Addison-Wesley, Reading, MA.

Marti, K. (1996). "Stochastic Optimization Methods in Engineering," in *System Modeling and Optimization* (J. Dolezal and J. Fiedler, Eds.), Chapman and Hall, London–New York.

Powell, M. J. D. (1964). "A Monte Carlo Method for Finding the Minimum of a Function of Several Variables without Calculating Derivatives". *The Computer Journal* 7: 155–162.

Rayward-Smith, V. J. (1995). *Applications of Modern Heuristic Methods*, pp. 145–156, Alfred Walter Limited Publishers in association with UNICOM, New York, NY.

Siarry, P., G. Berthiau, F. Durbin and J. Haussy (1997). "Enhanced Simulated Annealing for Globally Minimizing Functions of Many-Continuous Variables". *ACM Transactions on Mathematical Software* 23(2): 209–228.

Wasan, M. T. (1969). *Stochastic Approximation*, Cambridge University Press, London, UK.

12 Mathematical Modeling and Control of Resource Constraints

12.1 INTRODUCTION

The premise of this chapter is the development of a generic project scheduling tool that incorporates (1) resource characteristics, such as preferences, time-effective capabilities, costs and availability of project resources, and (2) performance interdependencies among different resource groups, and proceeds to map the most adequate resource units to each newly scheduled project activity. The chapter is based on the work of Milatovic and Badiru (2004). The principal challenge in this generic model development is to make it applicable to realistic project environments, which often involve multifunctional resources, whose capabilities or other characteristics may cross activities, as well as within a single activity relative to specific interactions among resources themselves. The scope of this research challenge further increases when the actual duration, cost, and successful completion of a project activity is assumed as resource driven and dependent on the choice of particular resource units assigned to it (Badiru et al., 2012).

The proposed methodology dynamically executes two alternative procedures: the *activity scheduler* and *resource mapper*. The *activity scheduler* prioritizes and schedules activities based on their attributes and may also attempt to centralize selected resource loading graphs based on activity resource requirements. bThe *resource mapper* considers resource characteristics, incorporates interdependencies among resource groups or types, and maps the available resource units to newly scheduled activities according to a project manager's pre-specified mapping (objective) function.

12.2 THE NATURE OF RESOURCE CONSTRAINTS

Project resources, generally limited in quantity, are the most important constraints in scheduling of activities. In cases when resources have pre-specified assignments and responsibilities toward one or more activities, their allocation is concurrently performed with the scheduling of applicable activities. In other cases, an activity may only require a certain number of (generic) resource units of a particular type(s), which are assigned after the scheduling of the particular activity. These two approaches coarsely represent the dominant paradigms in project scheduling. The objective of this research is to propose a new strategy that will shift these paradigms to facilitate more refined guidance for allocation and assignment of project resources. In other words, there is a need for tools, which will provide for more effective resource tracking, control, interaction, and, most importantly, resource-activity mapping.

The main assumption in the methodology of this chapter is that project environments often involve multi-capable resource units with different characteristics. This is especially the case in knowledge-intensive settings and industries, which are predominantly staffed with highly trained personnel. The specific characteristics considered were resource preferences, time-effective capabilities, costs, and availability. Each resource unit's characteristics may not only further vary across project activities, but also within a single activity relative to interaction among resource units. Finally, resource preferences, cost, and time-effective capabilities may also independently vary with time due to additional factors, such as learning, forgetting, weather, type of work, etc. Therefore, although we do not exclude a possibility that an activity duration is independent of resources assigned to it, in this research, we assume that it is those resource units assigned to a particular activity that determine how long it will take for the activity to be completed.

The scheduling strategy, as illustrated above, promotes a more balanced and integrated activity-resource mapping approach. Mapping the most qualified resources to each project activity, and thus preserving the values of resource, is achieved by proper consideration or resource time-effective capabilities and costs. By considering resource preferences and availability, which may be entered in either crisp or fuzzy form, the model enables consideration of personnel's voice and its influence on a project schedule and quality. Furthermore, resource interactive dependencies may also be evaluated for each of the characteristics and their effects incorporated into resource-activity mapping. Finally, by allowing flexible and dynamic modifications of scheduling objectives, the model permits managers or analysts to incorporate some of their tacit knowledge and discretionary input into project schedules.

12.3   LITERATURE BACKGROUND

Literature presents extensive work on scheduling projects by accounting for worker (resource) preferences, qualifications, and skills, as decisive factors to their allocation. Yet, a recent survey of some 400 top contractors in construction showed that 96.2% of them still use basic CPM for project scheduling (Mattila & Abraham, 1998). Roberts (1992) argued that information sources for project planners and schedulers are increasingly non-human and stressed that planners must keep computerized tools for project management and scheduling in line and perspective with human resources used by projects. In other words, the author warns that too much technicalities may prompt and mislead the managers into ignoring human aspects of management.

Franz and Miller (1993) considered a problem of scheduling medical residents to rotations, and approached it as a large-scale multi-period staff assignment problem. The objective of the problem was to maximize residents' schedule preferences while meeting hospital's training goals and contractual commitments for staffing assistance.

Gray et al. (1993) discussed the development of an expert system to schedule nurses according to their scheduling preferences. Assuming consistency in nurses' preferences, an expert system was proposed and implemented to produce feasible schedules considering nurses' preferences, but also accounting for overtime needs, desirable staffing levels, patient acuity, etc. A similar problem was also addressed by Yura (1994), where the objective was to satisfy worker's preferences for time off as well as overtime, but under due date constraints.

Campbell (1999) further considered allocation of cross-trained resources in multi-departmental service environment. Employers generally value more resource units with various skills and capabilities for performing a greater number of jobs. It is in those cases when managers face challenges of allocating these workers such that the utility of their assignment to a department is maximized. The results of experiments showed that the benefits of cross-training utilization may be significant. In most cases, only a small degree of cross-training captured the most benefits, and tests also showed that beyond a certain amount, the additional cross-training adds little additional benefits.

Finally, many companies face problems of continuous reassignment of people in order to facilitate new projects (Cooprider, 1999). Cooprider suggests a seven-step procedure to help companies consider a wide spectrum of parameters when distributing people within particular projects or disciplines.

Badiru (1993) proposed *Critical Resource Diagramming* (*CRD*), which is a simple extension to traditional *CPM* graphs. In other words, criticalities in project activities may also be reflected on resources. Different resource types or units may vary in skills, supply, or be very expensive. This discrimination in resource importance should be accounted for when carrying out their allocation in scheduling activities.

Unlike activity networks, the *CRD*s use nodes to represent each resource unit. Also, unlike activities, a resource unit may appear more than once in a *CRD* network, specifying all different tasks for which a particular unit is assigned to. Similar to *CPM*, the same backward and forward computations may be performed to *CRD*s.

12.4 METHODOLOGY

The methodology of this chapter represents an analytical extension of *CRDs* discussed in the preceding chapter. As previously mentioned, the design considerations of the proposed model consist of two distinct procedures: activity scheduling and resource mapping. At each decision instance during a scheduling process, the *activity scheduler* prioritizes and schedules some or all candidate activities, and then, the *resource mapper* iteratively assigns the most adequate resource units to each of the newly scheduled activities.

12.5 MATHEMATICAL NOTATIONS

i $\quad\equiv$ project activity i, such that $i = 1, \ldots, I$

I $\quad\equiv$ number of activities in project network.

t_c $\quad\equiv$ decision instance, that is, time moment at which one or more activities qualify to be scheduled since their predecessor activities have been completed.

$PR(i)$ $\equiv$ Set of predecessor activities of activity i.

$Q(t_c)$ $\equiv$ Set of activities qualifying to be scheduled at t_c, that is, $Q(t_c) = \{i \mid PR(i) = \varnothing\}$.

j	$\equiv$ resource type j, $j = 1, \ldots, J$.
J	$\equiv$ number of resource types involved in the project.
$R_j^{\grave{}}$	$\equiv$ number of units of resource type j available for the project.
$<j,k>$	$\equiv$ notation for k-th unit of type j.
ρ_i^j	$\equiv$ number of resource units type j required by activity i.
$u_{t_c}^{j,k}$	$\equiv$ a binary variable with a value of one if k-th unit of type j is engaged in one of the project activities that are in progress at the decision instance t_c, and zero otherwise. All $u_{t_c}^{j,k}$'s are initially set to zero.
$t_i^{j,k}$	$\equiv$ time-effective executive capability of kth unit of resource type j if assigned to work on activity i.
$p_i^{j,k}$	$\equiv$ preference of k-th unit of resource type j to work on activity i.
$c_i^{j,k}$	$\equiv$ estimated cost of k-th unit of resource type j if assigned to work on activity i.
$\alpha_i^{j,k}\left(t_c\right)$	$\equiv$ desired start time or interval availability of k-th unit of type j to work on activity i at the decision instance t_c. In many cases, this parameter is invariant across activities, and the subscript i may often be dropped.

12.6 REPRESENTATION OF RESOURCE INTERDEPENDENCIES AND MULTIFUNCTIONALITY

This study is primarily focused on renewable resources. In addition, resources are not necessarily categorized into types or groups according to their similarities (i.e., into personnel, equipment, space, etc.), but more according to hierarchy of their interdependencies. In other words, we assume that time-effective capabilities, preferences, or even cost of any particular resource unit assigned to work on an activity may be dependent on other resource units also assigned to work on the same activity. Some or all of these other resource units may, in a similar fashion, be also dependent on a third group of resources, and so on. Based on the above assumptions, we model competency of project resources in terms of following four resource characteristics: time-effective capabilities, preferences, cost, and availability. Time-effective capability of a resource unit with respect to a particular activity is the amount of time the unit needs to complete its own task if assigned to that particular activity. Preferences are relative numerical weights that indicate personnel's degree of desire to be assigned to an activity, or manager's perception on assigning certain units to particular activities. Similarly, each resource unit may have different costs associated with it relative to which activities it gets assigned to. Finally, not all resource units may be available to some or all activities at all times during project execution. Thus, times during which a particular unit is available to some or all activities are also incorporated into the mapping methodology. Each of the characteristics described may vary across different project activities. In addition, some or all of these characteristics (especially time-effective capabilities and preferences) may also vary within a particular activity relative to resource interaction with other resources that are also assigned to work on the same activity.

Those resources whose performance is totally independent of their interaction with other units are grouped together and referred to as the type or group "one" and allocated first to scheduled activities. Resource units whose performance or competency is affected by their interaction with the type or group "one" units are grouped into type or group "two" and assigned (mapped) next. Resource units whose competency or performance is a function of type "two" or both types "one" and "two" are grouped into type "three" and allocated to scheduled activities after the units of the first two types have been assigned to them.

A project manager may consider one, more than one, or all of the four characteristics when performing activity-resource mapping. For example, a manager may wish to keep project costs as low as possible, while at the same time attempting to use resources with the best time-effective capabilities, consider their availability, and even incorporate their voice (in case of humans) or his/her own perception (in cases of human or non-human resources) in the form of preferences. This objective may be represented as follows:

$$\mathcal{U}_i^{j,k} = f\left(t_i^{j,k}, c_i^{j,k}, p_i^{j,k}, \alpha_i^{j,k}\left(t_c\right)\right)$$

(see Appendix for a detailed notation)

Mapping units of all resource types according to the same mapping function may often be impractical and unrealistic. Cost issues may be of greater importance in mapping some, while inferior to time-effective capabilities of other resource types. To accommodate the need for a resource-specific mapping function as mapping objective, we formulated the mapping function as additive utility function (Keeney & Raiffa, 1993). In such a case, each of its components pertains to a particular resource type and is multiplied by a *Kronecker's delta* function (Bracewell, 1978). Kronecker's delta then detects resource type whose units are currently being mapped and filters out all mapping function components, except the one that pertains to the currently mapped resource type.

As an example, consider again a case where all resource types would be mapped according to their time-effective capabilities, except in the case of resource types "two" and "three" where costs would also be of consideration, and in the case of type "five," resource preferences and availabilities would be considered:

$$\mathcal{U}_i^{j,k} = f\left(t_i^{j,k}\right) + f_2\left(c_i^{j,k}\right)\cdot\delta\left(j,2\right) + f_3\left(c_i^{j,k}\right)\cdot\delta\left(j,3\right) + f_5\left(p_i^{j,k}, \alpha_i^{j,k}\left(t_c\right)\right)\cdot\delta\left(j,5\right)$$

The above example illustrates a case where mapping of resource units is performed according to filtered portions of a manager's objective (mapping) function, which may in turn be dynamically adaptive and varying with project scheduling time. As previously indicated, some resource characteristics may be of greater importance to a manager in the early scheduling stages of a project rather than in the later stages. Such a mapping function may be modeled as follows:

$$\mathcal{U}_i^{j,k} = f_g\left(t_i^{j,k}, c_i^{j,k}, p_i^{j,k}, a_i^{j,k}\left(t_c\right)\right) + \sum_{s\in T} f_s\left(t_i^{j,k}, c_i^{j,k}, p_i^{j,k}, a_i^{j,k}\left(t_c\right)\right)\cdot w\left(t_{LO}^s, t_{HI}^s, t_c\right)$$

where

$f_g \equiv$ Component of the mapping function that is common to all resource types.

$f_s \equiv$ Component of the mapping function that pertains to a *specific* project scheduling interval.

$t_{LO}^s, t_{HI}^s \equiv$ Specific time interval during which resource mapping must be performed according to a unique function.

$T \equiv$ Set of above defined time intervals for a particular project.

$w\left(t_{LO}^s, t_{HI}^s, t_c\right) \equiv$ Window function with a value of one if t_c falls within the interval $[t_{LO}^s, t_{HI}^s]$, and zero otherwise.

Finally, it is also possible to map different resource types according to different objectives <u>and</u> at different times simultaneously, by simply combining the two concepts above. For example, assume again that a manager forms his objective in the early stage of the project based on resources' temporal capabilities, costs, and preferences. Then, at a later stage, the manager wishes to drop the costs and preferences and consider only resource capabilities, with the exception of resource type "three" whose costs should still remain in consideration for mapping. An example of a mapping function that would account for this scenario may be as follows:

$$\mathcal{U}_i^{j,k} = f\left(c_i^{j,k}, p_i^{j,k}, t_i^{j,k}\right) \cdot w\left(0, 30, t_c\right) + \left(f\left(t_i^{j,k}\right) + f\left(c_i^{j,k}\right) \cdot \delta\left(j, 3\right)\right) \cdot w\left(30, 90, t_c\right)$$

12.7　MODELING OF RESOURCE CHARACTERISTICS

For resource units whose performance on a particular activity is independent of their interaction with other units, that is, for the *drivers*, $t_i^{j,k}$ is defined as the time, t, it takes kth unit of resource type j to complete its own task or process when working on activity i. Thus, different resource units, if multi-capable, can be expected to perform differently on different activities. Each *dependent* unit, on the other hand, instead of $t_i^{j,k}$, generally has a set of interdependency functions associated with it.

In this research, we consider two types of interactive dependencies among resources, which due to their simplicity, are expected to be the most commonly used ones: *additive* and *percentual*. *Additive* interaction between a *dependent* and each of its *driver* resource unit indicates the amount of time that the *dependent* will need to complete its own task if assigned to work in conjunction with a particular driver. This is in addition to the time the driver itself needs to spend working on the same activity:

$$\left(T_i^{j,k}\right)_z \equiv \left(t_i^{j_D, k_D} + \tilde{t}_i^{j,k}\right) \cdot y_i^{j_D, k_D}$$

where

$<j_D, k_D> \in D^{j,k}$, where $D^{j,k}$ is a set of *driver* units (each defined by an indexed pair $<j_D, k_D>$) for a particular resource unit $<j, k>$.

$\left(T_i^{j,k}\right)_z \equiv z$-th interactive time-effective dependency of k-th unit of type j on its *driver* $<j_D, k_D>$, $z = 1, \ldots,$ size$(D^{j,k})$. The actual number of these dependencies will depend on a manager's knowledge and familiarity with his/her resources.

$\tilde{t}_i^{j,k} \equiv$ time needed in addition to $t_i^{j_D,k_D}$ for k-th *dependent* unit of type j to complete its task on activity i if it interacts with its *driver* unit j_D, k_D.

$y_i^{j_D,k_D} \equiv$ binary (zero-one) variable indicating mapping status of the *driver* unit $<j_D, k_D>$. It equals one if the unit $<j_D, k_D>$ is assigned to activity i, and zero if the unit $<j_D, k_D>$ has been assigned to activity i. Therefore, each $\left(T_i^{j,k}\right)_z$ will have a nonzero value only if $y_i^{j_D,k_D}$ is also nonzero (i.e., if the *driver* resource unit $<j_D, k_D>$ has been previously assigned to activity i).

The percentual interactive dependency is similarly defined as:

$$\left(T_i^{j,k}\right)_z = t_i^{j_D,k_D} \cdot \left(1 + \tilde{t}_i^{j,k}\%\right) \cdot y_i^{j_D,k_D}$$

where $\tilde{t}_i^{j,k}\%$ is the percentage of time by which $t_i^{j_D,k_D}$ will be prolonged if the unit k of type j interacts with its *driver* $<j_D, k_D>$.

Modeling cost characteristics follow a similar logic used for representation of temporal capabilities and interdependencies. In place of $t_i^{j,k}$, we now define a variable $c_i^{j,k}$, which represents the cost (say, in Dollars) of k-th unit of resource type j if it gets assigned to work on activity i. This value of $c_i^{j,k}$ may be invariant regardless of a unit's interaction with other resources, or it may vary relative to interaction among resources, and thus, implying cost interdependencies, which need to be evaluated before any mapping is performed (provided that the cost considerations are, indeed, a part of a manager's utility or objective for mapping).

In cases when a cost of a resource unit for an activity varies depending on its interaction with units of other (lower indexed) types, we define cost dependencies as:

$$\left(C_i^{j,k}\right)z = \tilde{c}_i^{j,k} \cdot y_i^{j_D,k_D}$$

where

$y_i^{j_D,k_D} \equiv$ a binary variable indicating the status of the particular *driver* resource unit $<j_D, k_D>$, as defined in the previous section.

$\tilde{c}_i^{j,k} \equiv$ interactive cost of k-th unit of type j on its *driver* $<j_D, k_D>$, with respect to activity i.

$\left(C_i^{j,k}\right)z \equiv z$-th evaluated interactive cost dependency of k-th unit of type j on its *driver* $<j_D, k_D>$, $z = 1, \ldots, \text{size}(D^{j,k})$. The values of each $\left(C_i^{j,k}\right)_z$ equals $\tilde{c}_i^{j,k}$ when $y_i^{j_D,k_D}$ equals one, and zero otherwise. The actual number of these interactive cost dependencies will again depend on a manager's knowledge and information about available resources.

Given a set of cost dependencies, we compute the overall $c_i^{j,k}$ as a sum of all evaluated $\left(C_i^{j,k}\right)z$'s as follows:

$$c_i^{j,k} = \sum_{z=1}^{\left| D^{j,k} \right|} \left(C_i^{j,k} \right)_z$$

In many instances, due to political, environmental, safety, or community standards, aesthetics, or other similar non-monetary reasons, pure monetary factors may not necessarily prevail in decision making. It is those other non-monetary factors that we wish to capture by introducing preferences in resource mapping to newly scheduled activities. The actual representation of preferences is almost identical to those of the costs:

$$\left(P_i^{j,k} \right)z = \tilde{p}_i^{j,k} \cdot y_i^{j_D,k_D}$$

where $\tilde{p}_i^{j,k}$ is an interactive preference of k-th unit of type j on its *driver* $<j_D, k_D>$, with respect to activity i. $\left(P_i^{j,k} \right)z$ is z-th evaluated interactive preference dependency of k-th unit of type j, with respect to activity i. Finally, again identically to modeling costs, $p_i^{j,k}$ is computed as:

$$p_i^{j,k} = \sum_{z=1}^{\left| D^{j,k} \right|} \left(P_i^{j,k} \right)_z$$

Having a certain number of resource units of each type available for a project does not necessarily imply that all of the units are available all the time for the project or any of its activities in particular. Due to transportation, contracts, learning, weather conditions, logistics, or other factors, some units may only have *time preferences* for when they are available to start working on a project activity or the project as a whole. Others may have *strict time intervals* during which they are allowed to start working on a particular activity or the project as a whole. This latter, strictly constrained availability may be easily accommodated by the previously considered *window* function, $w(t_{LO}, t_{HI}, t_c)$.

In many cases, especially for humans, resources may have a desired or "ideal" time when to start their work or be available in general. This flexible availability can simply be represented by fuzzifying the specified desired times using the following function:

$$\alpha_i^{j,k}\left(t_c \right) = \frac{1}{1 + a\left(t_c - \tau_i^{j,k} \right)^b}$$

where

$\tau_i^{j,k} \equiv$ desired time for k-th unit of resource type j to start its task on activity i. This desirability may either represent the voice of project personnel (as in the case of preferences), or manager's perception on resource's readiness and availability to take on a given task.

$\alpha_i^{j,k}(t_c)$ ≡ fuzzy membership function indicating a degree of desirability of <j, k>-th unit to start working on activity i, at the decision instance t_c.

a ≡ parameter that adjusts for the width of the membership function.

b ≡ parameter that defines the extent of *start time* flexibility.

12.8   RESOURCE MAPPER

At each scheduling time instance, t_c, available resource units are mapped to newly scheduled activities. This is accomplished by solving J number of zero-one linear integer problems (i.e., one for each resource type), where the coefficients of the decision vector correspond to evaluated mapping function for each unit of the currently mapped resource type:

$$\max \sum_{h\in\Omega(t_c)} \sum_{k=1}^{R_j} u_h^{j,k} \cdot y_h^{j,k} \qquad \text{for } j = 1,\dots,J$$

where:

$y_h^{j,k}$ ≡ binary variable of the decision vector.

$\Omega(t_c)$ ≡ set of newly scheduled activities at decision instance t_c.

A $y_i^{j,k}$ resulting in a value of one would mean that k-th unit of resource type j is mapped to i-th ($i\in\Omega(t_c)$) newly scheduled activity at t_c. The above objective in each of J number of problems is subjected to four types of constraints, as illustrated below.

I)   The first type of constraints ensure that each newly scheduled activity receives its required number of units of each project resource type:

$$\sum_{k=1}^{R_j} y_i^{j,k} = \rho_i^j \qquad \text{for } i \in \Omega(t_c) \qquad \text{for } j = 1,\dots,J$$

II) The second type of constraints prevent mapping of any resource units to more than one activity at the same time at t_c:

$$\sum_{i\in\Omega(t_c)} y_i^{j,k} \le 1 \qquad \text{for } k = 1, \dots, R_j \qquad \text{for } j = 1,\dots,J$$

III) The third type of constraints prevent mapping of those resource units that are currently in use by activities in progress at time t_c:

$$\sum_{k=1}^{R_j} u_{t_c}^{j,k} \cdot y_i^{j,k} = 0 \quad \text{for } i \in \Omega(t_c) \qquad \text{for } j = 1,\dots,J$$

IV) The fourth type of constraints ensure that the variables in the decision vector $y_i^{j,k}$ take on binary values:

$$y_i^{j,k} = 0 \text{ or } 1 \qquad \text{for } k = 1,\ldots,R_j, i \in \Omega\left(t_c\right), \quad \text{for } j = 1,\ldots,J$$

Therefore, in the first of the total of J runs at each decision instance t_c, available
units of resource type "one" compete (based on their characteristics and pre-specified
mapping function) for their assignments to newly scheduled activities. In the second
run, resources of type "two" compete for their assignments. Some of their characteristics, however, may vary depending on the "winners" from the first run. Thus, the
information from the first run is used to refine the mapping of type or group "two"
resources. Furthermore, the information from either or both of the first two runs is
then used in tuning the coefficients of the objective function for the third run when
resources of type "three" are mapped.

Due to the nature of linear programming, zeros in the coefficients of the objective
do not imply that corresponding variables in the solution will also take the value of
zero. In our case, that would mean that although we flagged off a resource unit as
unavailable, the solution may still map it to an activity. Thus, we need to strictly
enforce the interval (un)availability by adding information into constraints. Thus, we
perturbed the third mapping constraint, which was previously set to prohibit mapping
of resource units at time t_c, which are in use by activities in progress at that time. The
constraint was originally defined as:

$$\sum_{k=1}^{R_j} u_{t_c}^{j,k} \cdot y_i^{j,k} = 0 \qquad \text{for } i \in \Omega\left(t_c\right) \qquad \text{for } j = 1,\ldots,J$$

To now further prevent mapping of resource units whose $\alpha_i^{j,k}\left(t_c\right)$ equals zero at t_c,
we modify the above constraint as follows:

$$\sum_{k=1}^{R_j} \left(u_{t_c}^{j,k} + \left(1-\alpha_i^{j,k}\left(t_c\right)\right)\right) \cdot y_i^{j,k} = 0 \quad \text{for } i \in \Omega\left(t_c\right) \qquad \text{for } j = 1,\ldots,J$$

This modified constraint now not only filters out those resource units that are
engaged in activities in progress at t_c, but also those units which were flagged as
unavailable at t_c due to any other reasons.

12.9 ACTIVITY SCHEDULER

Traditionally, a project manager estimates duration of each project activity first, and
then assigns resources to it. In this study, although we do not exclude a possibility
that an activity duration is independent of resources assigned to it, we assume that it
is those resource units and their skills or competencies assigned to a particular activity that determine how long it will take for the activity to be completed. Normally,
more capable and qualified resource units are likely to complete their tasks faster,

and vice versa. Thus, activity duration in this research is considered a *resource driven activity attribute*.

At each decision instance t_c (in resource constrained non-preemptive scheduling as investigated in this study), activities whose predecessors have been completed enter the set of qualifying activities, $Q(t_c)$. In cases of resource conflicts, we often have to prioritize activities in order to decide which ones to schedule. In this methodology, we prioritize activities based on two (possibly conflicting) objectives:

1. Basic *Activity Attributes*, such as the *current amount of depleted slack*, number of successors, and initially estimated optimistic activity duration, d_i.
2. Degree of manager's desire to *centralize* (or balance) *the loading* of one or more pre-selected project resource types.

Amount of Depleted Slack, $S_i(t_c)$, is defined in this research as a measure of how much total slack of an activity from unconstrained *CPM* computations has been depleted each time the activity is delayed in resource-constrained scheduling due to lack of available resource units. The larger the $S_i(t_c)$ of an activity, the more it has been delayed from its unconstrained schedule, and the greater probability that it will delay the entire project.

Before resource-constrained scheduling of activities (as well as resource mapping which is performed concurrently) starts, we perform a single run of *CPM* computations to determine initial unconstrained *Latest Finish Time*, LFT_i of each activity. Then, as the resource-constrained activity scheduling starts, at each decision instance t_c, we calculate $S_i(t_c)$ for each candidate activity [from the set $Q(t_c)$] as follows:

$$S_i\left(t_c\right) = \frac{t_c + d_i}{LFT_i} = \frac{t_c + d_i}{LST_i + d_i} \qquad \text{for } i \in Q\left(t_c\right)$$

$S_i(t_c)$, as a function of time, is always a positive real number. The value of its magnitude is interpreted as follows:

- when $S_i(t_c) < 1$, the activity i still has some slack remaining and it may be safely delayed;
- when $S_i(t_c) = 1$, the activity i has depleted all of its resource unconstrained slack and any further delay to it will delay its completion as initially computed by conventional unconstrained *CPM*;

when $S_i(t_c) > 1$, *the* activity i has exceeded its slack and its completion will be delayed beyond its unconstrained *CPM* duration.

Once calculated at each t_c, the current *amount of depleted*, $S_i(t_c)$, is then used in combination with the other two activity attributes for assessing activity priority for scheduling. (These additional attributes are *the number of activity successors*, as well as its *initially estimated duration d_i*.) The number of successors is an important determinant in prioritizing, because if an activity with many successors is delayed, chances are that any of its successors will also be delayed, thus eventually prolonging the entire project itself. Therefore, the prioritizing weight, w_p^t, pertaining to basic activity attributes is computed as follows:

$$w_i^p = S_i\left(t_c\right)\cdot\left(\frac{\varsigma_i}{\max\left(\varsigma_i\right)}\right)\cdot\left(\frac{d_i}{\max\left(d_i\right)}\right)$$

where

$w_i^p \equiv$ activity prioritizing weight that pertains to basic activity attributes.
$\varsigma_i \equiv$ number of successors activities of current candidate activity i.
$\max(\varsigma_i) \equiv$ maximum number of activity successors in project network.
$\max(d_i) \equiv$ maximum of the most optimistic activity durations in a project network.

The second *objective* that may influence activity prioritizing is a manager's desire for a somewhat centralized (i.e., balanced) resource loading graph for one or more resource groups or types. This is generally desirable in cases when a manager does not wish to commit all of the available project funds or resources at the very beginning of the project (Dreger, 1992), or to avoid frequent hiring and firing or project resources (Badiru & Simin Pulat, 1995).

In this research, we attempt to balance (centralize) loading of pre-specified resources by scheduling those activities whose resource requirements will minimize the increase in loading graph's stairstep size of the early project stages, and then minimize the decrease in the step size in the later stages. A completely balanced resource loading graph contains no depression regions as defined by Konstantinidis (1998), that is, it is a nondecreasing graph up to a certain point at which it becomes nonincreasing.

The activity prioritizing weight that pertains to attempting to centralize resource loading is computed in this research as follows:

$$w_i^r = \sum_{j=1}^{J}\frac{\rho_i^j}{R_j}$$

where

$w_i^r \equiv$ prioritizing weight that incorporates activity resource requirements.
$\rho_i^j \equiv$ number of resource type j units required by activity i.
$R_j \equiv$ total number of resource type j units required for the project.

Notice that w_i^p and w_i^r are weights of possibly conflicting objectives in prioritization of candidate activities for scheduling.

To further limit the range of w_i^r between zero and one, we scale it as follows:

$$w_i^r = \frac{w_i^r}{\max\left(w_i^r\right)}$$

With the two weights w_i^p and w_i^r defined and computed, we further use them as the coefficients of activity scheduling objective function:

$$\max \left(\sum_{i \in Q(t_c)} w_i^p \cdot x_i \right) + W \left(\sum_{i \in Q(t_c)} \left(1 - w_i^r \cdot x_i \right) \right)$$

where

$x_i \equiv$ binary variable whose value becomes one if a candidate activity $i \in Q(t_c)$ is scheduled at t_c, and zero if the activity i is not scheduled at t_c.

$W \equiv$ Decision Maker's supplied weight that conveys the importance of resource centralization (balancing) in project schedule.

Notice that W is a parameter that allows a manager to further control the influence of w_i^p. Large values of W will place greater emphasis on the importance of resource balancing. However, to again localize the effect of W to the early stages of a project, we dynamically decrease its value at each subsequent decision instance, t_c according to the following formula:

$$W_{new} = W_{old} \left(\frac{\sum_{i=1}^{I} d_i - \sum_{i \in H(t_c)} d_i}{\sum_{i=1}^{I} d_i} \right)$$

where

$\sum_{i=1}^{I} d_i \equiv$ The sum of all the most optimistic activity durations (as determined by conventional resource unconstrained CPM computations) for all activities in project network.

$H(t_c) \equiv$ set of activities that have been so far scheduled by the time t_c.

Figure 12.1 shows a Gantt chart and resource loading graphs of sample project with seven activities and two resource types. Clearly, neither of the two resource types are balanced. The same project has been re-run using the above reasoning and is shown in Figure 12.2. The same project has been re-run using the above methodology, and is shown in Figure 12.3. Notice that the loading of resource type two is now fully balanced. The loading of resource type one still contains depression regions, but to a considerably lesser extent than in Figure 12.1.

With the two weights w_i^p and w_i^r defined and computed, we further use them as the coefficients of activity scheduling objective function:

$$\max \left(\sum_{i \in Q(t_c)} w_i^p \cdot x_i \right) + W \left(\sum_{i \in Q(t_c)} \left(1 - w_i^r \cdot x_i \right) \right)$$

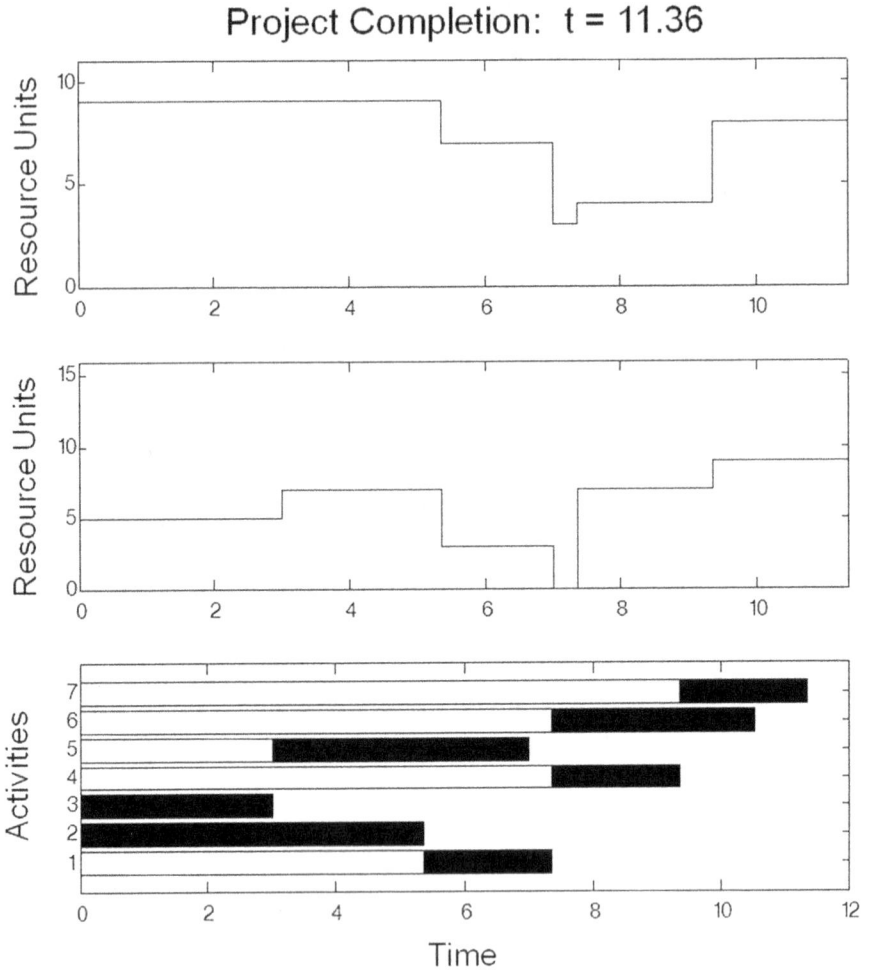

FIGURE 12.1   Gantt chart and resource loading graphs.

where

$x_i \equiv$ binary variable whose value becomes one if a candidate activity $i \in Q(t_c)$ is scheduled at t_c, and zero if the activity i is not scheduled at t_c.

$W \equiv$ Decision Maker's supplied weight that conveys the importance of resource centralization (balancing) in project schedule.

Large values of W will place greater emphasis on the importance of resource balancing. However, to again localize the effect of W to the early stages of a project, we

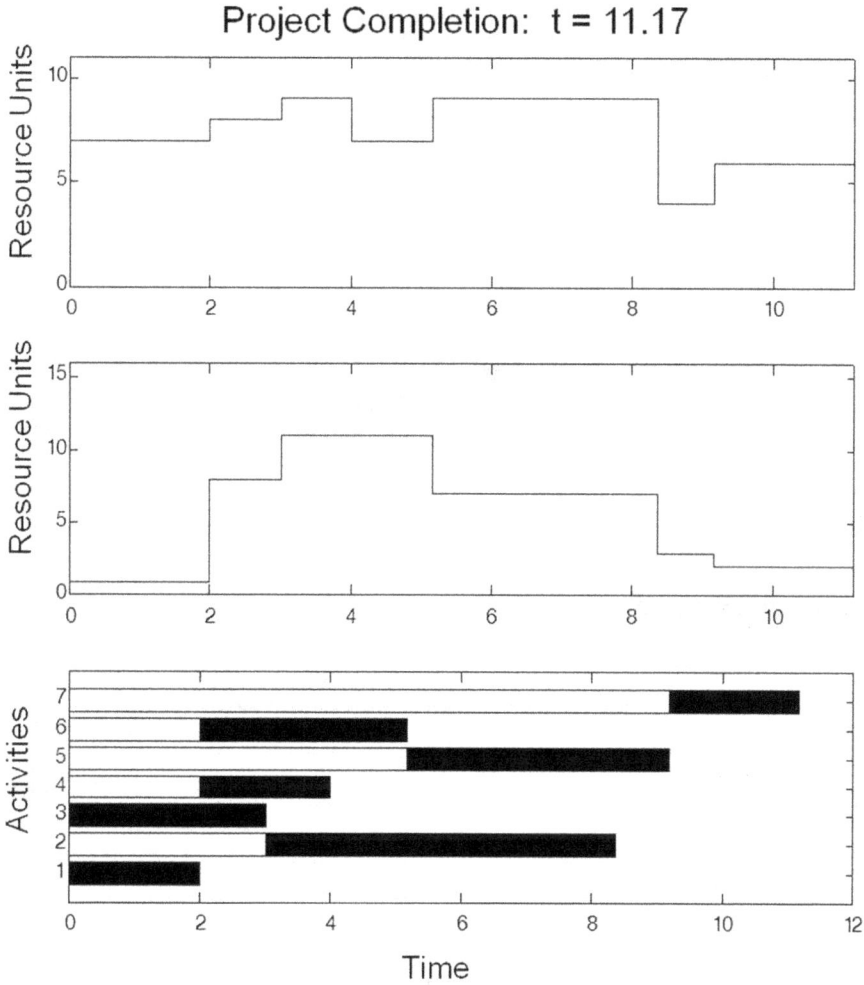

FIGURE 12.2 Rerun of Gantt chart and resource loading graphs.

dynamically decrease its value at each subsequent decision instance, t_c according to the following formula:

$$W_{new} = W_{old} \left(\frac{\sum_{i=1}^{I} d_i - \sum_{i \in H(t_c)} d_i}{\sum_{i=1}^{I} d_i} \right)$$

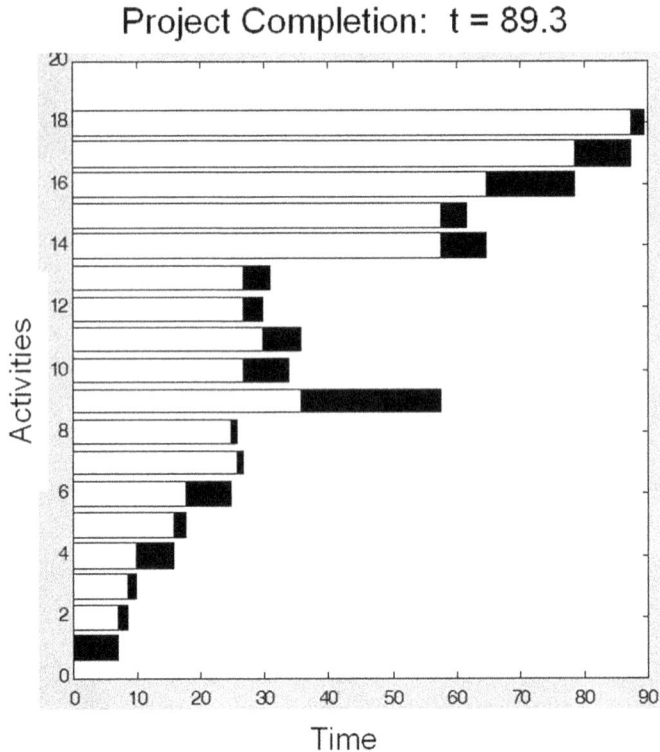

FIGURE 12.3   Project completion at Time 89.3.

where

$$\sum_{i=1}^{I} d_i \equiv \text{The sum of all the most optimistic activity durations (as determined by}$$
conventional resource unconstrained CPM computations) for all activities in project network.

$H(t_c) \equiv$ set of activities that have been so far scheduled by the time t_c.

Previously, it was proposed that one way of balancing resource loading was to keep minimizing the increase in the stairstep size of the loading graph in the early project stages, and then minimize the decrease in the step size in the later stages. The problem with this reasoning is that a continuous increase in the loading graph in early stages may eventually lead to scheduling infeasibility due to limiting constraints in resource availability. Therefore, an intelligent mechanism is needed that will detect the point when resource constraints become binding and force the scheduling to proceed in a way that will start the decrease in resource loading, as shown in Figure 12.3. In other words, we need to formulate a linear programming model whose constraints will drive the increase in resource stairstep shaped loading function up to a point when limits in resource availability are reached. At that point, the model must adjust the objective function and modify (relax) the constraints, in order to start minimizing the stairstep decrease of resource loading.

To ensure this, the constraints are formulated such that at each decision instance t_c, maximal number of candidate activities are scheduled, while satisfying activity precedence relations, preventing the excess of resource limitations, and most importantly, flag off the moment when resource limitations are reached. To facilitate a computer implementation and prevent the strategy from crashing, we introduce an auxiliary zero-one variable, $\hat{x}$, in this study referred to as the *peak flag*. The value of $\hat{x}$ in the decision vector is zero as long as current constraints are capable of producing a feasible solution. Once that is impossible, all variables in the decision vector must be forced to zero, except $\hat{x}$, which will then take a value of one and indicate that the peak of resource loading is reached. At that moment, the constraints that force the increase in resource loading are relaxed (eliminated).

The *peak flag* is appended to the previous objective function as follows:

$$\max\left(\sum_{i\in Q(t_c)} w_i^p \cdot x_i\right) + W\left(\sum_{i\in Q(t_c)}\left(1 - w_i^r \cdot x_i\right)\right) - b\hat{x}$$

where

$b \equiv$ arbitrary large positive number (in computer implementation of this study,

b was taken as $b = \sum_{i=1}^{I} d_i$).

There are two types of constraints associated with the above objective of scheduling project activities. The first type simply serves to prevent scheduling of activities, which would overuse available resource units:

$$\sum_{i\in Q(t_c)} \rho_i^j \cdot x_i + \left(R_j - \sum_{i\in G(t_c)} \rho_i^j\right)\hat{x} \le \left(R_j - \sum_{i\in G(t_c)} \rho_i^j\right), \qquad j = 1, \ldots, J$$

where

$x_i \equiv$ candidate activity qualified to be scheduled at t_c

$G(t_c) \equiv$ set of activities that are in progress at time t_c.

$\left(R_j - \sum_{i\in G(t_c)} \rho_i^j\right) \equiv$ difference between the total available units of resource type j

(denoted as R_j) and the number of units of the same resource type being currently consumed by the activities in progress during the scheduling instant t_c.

The second type of constraints serves to force the gradual increase in the stairstep resource loading graphs. In other words, at each scheduling instant t_c, this group of constraints will attempt to force the model to schedule those candidate activities whose total resource requirements are greater or than equal the total requirements of the activities that have just finished at t_c. The constraints are formulated as follows:

$$\sum_{i \in Q(t_c)} \rho_i^j x_i + \left(\sum_{i \in F(t_c)} \rho_i^j \right) \hat{x} \geq \left(\sum_{i \in F(t_c)} \rho_i^j \right), \qquad j \in \mathcal{D}$$

where

$F(t_c) \equiv$ Set of activities that have been just completed at t_c,

$\mathcal{D} \equiv$ set of manager's pre-selected resource types whose loading graphs are to be centralized (i.e., balanced).

$\left(\sum_{i \in F(t_c)} \rho_i^j \right) \equiv$ total resource type j requirements by all activities that have been completed at the decision instance t_c.

Finally, to ensure an integer zero-one solution, we impose the last type of constraints as follows:

$$x_i = 0 \text{ or } 1, \qquad \text{for } i \in Q(t_c)$$

As previously discussed, once $\hat{x}$ becomes unity, we adjust the objective function and modify the constraints that will, from that point on, allow a decrease in resource loading graph(s). Objective function for activity scheduling is modified such that the product $w_i^t \cdot x_i$ is not being subtracted from one anymore, while the second type of constraint is eliminated completely:

$$\min \left(- \sum_{i \in Q(t_c)} w_i^t \cdot x_i \right) - W \left(\sum_{i \in Q(t_c)} w_i^r \cdot x_i \right)$$

subject to:

$$\sum_{i \in Q(t_c)} \rho_i^j \cdot x_i \leq \left(R_j - \sum_{i \in G(t_c)} \rho_i^j \right), \qquad j = 1, \dots, J$$

$$x_i = 0 \text{ or } 1$$

Since the second type of constraint is eliminated, resource loading function is now allowed to decrease. The first type of constraint still remains in place to prevent any overuse of available resources.

12.10   MODEL IMPLEMENTATION AND GRAPHICAL ILLUSTRATIONS

The model as described above has been implemented in a software prototype *Project Resource Mapper* (*PROMAP*), with its code, input format, and sample outputs illustrated in the appendices. The output consists of five types of charts. The more traditional ones include project *Gantt char* (Figure 12.3) and *resource loading graphs* (Figure 12.4) for all resource groups or types involved in a project. More specific graphs include *resource-activity mapping grids* (Figure 12.5), *resource utilization* (Figure 12.6), and *resource cost* (Figure 12.7) bar charts. Based on the imported resource characteristics, their interdependencies, and the form of the objective, the *resource-activity mapping grid* provides a decision support in terms of which units of each specified resource group should be assigned to which particular project activity. Therefore, the *resource-activity grids* are, in effect, the main contributions of this study. *Unit utilization charts* track the resource assignments and provide a relative resource usage of each unit relative to the total project duration. The bottom (darker shaded) bars indicate the total time it takes each unit to complete all of its own project tasks. The upper (lighted shaded) bars indicate the total additional time a unit

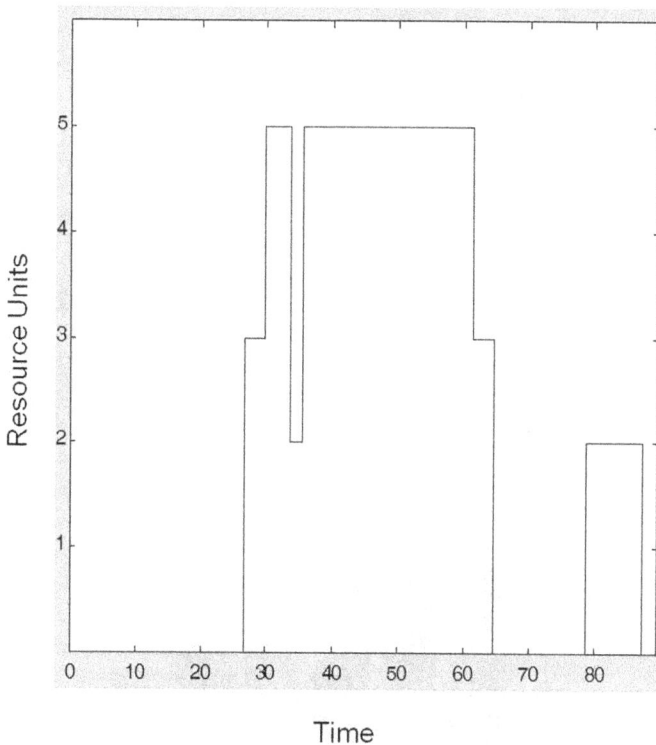

FIGURE 12.4   Resource Type 3 loading graph.

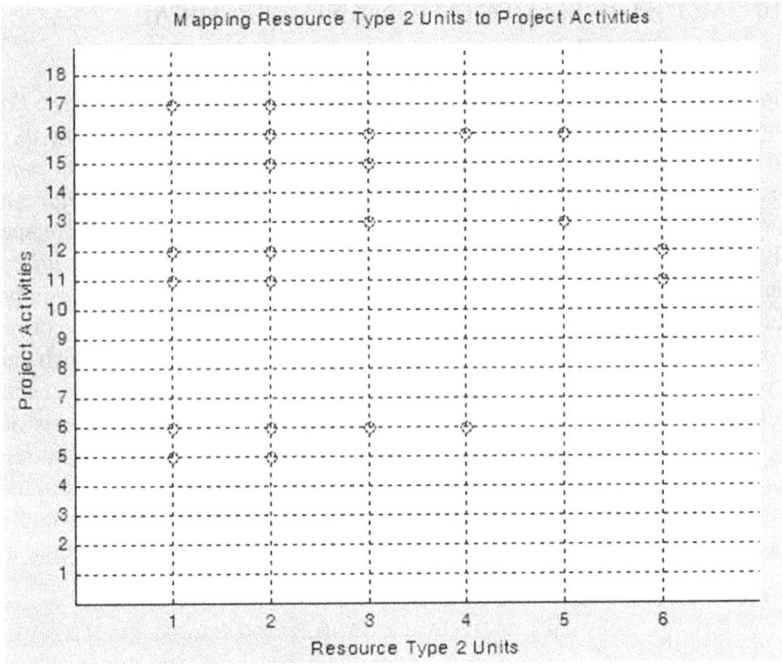

FIGURE 12.5   Mapping of resource Type 2 units to project activities.

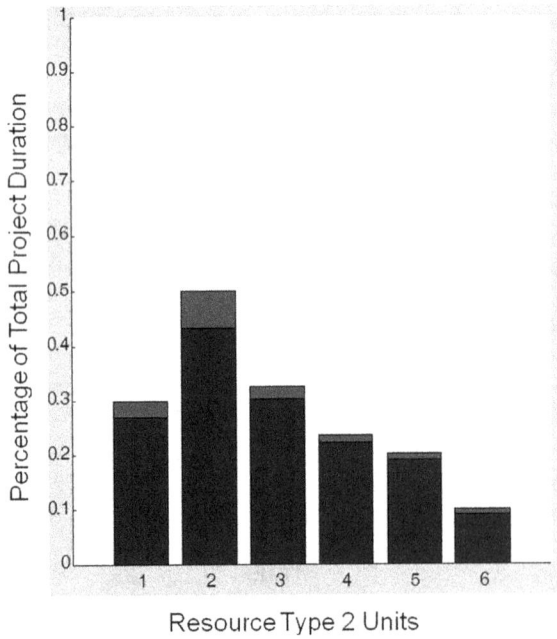

FIGURE 12.6   Time percentage of resource Type 2 units engagement versus total project duration.

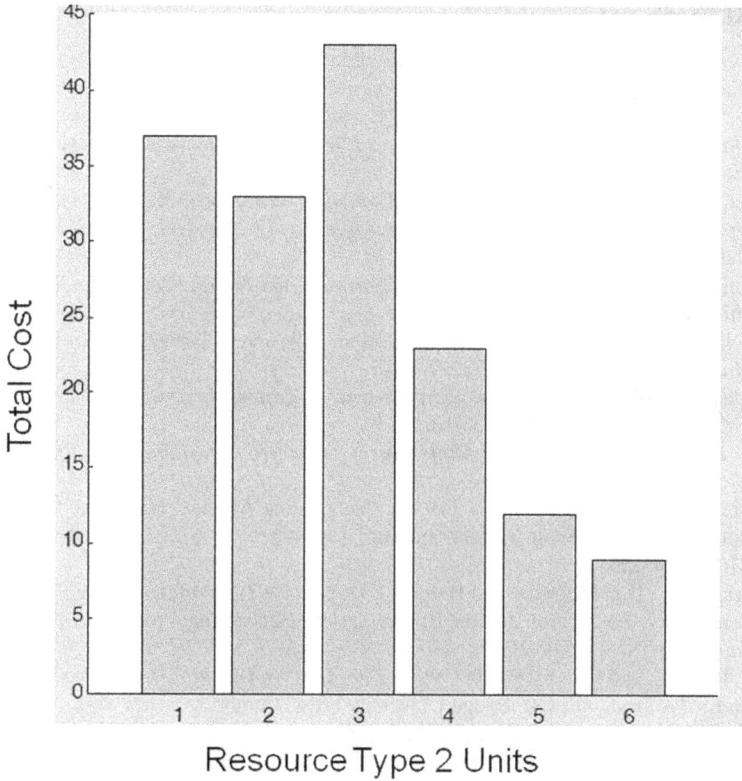

FIGURE 12.7   Project cost for Type 2 resource units.

may be locked in or engaged in an activity by waiting for other units to finish their tasks. In other words, the upper bars indicate the total possible resource idle time during which it cannot be reassigned to other activities because it is blocked waiting for other units to finish their own portions of work. This information is very useful in non-preemptive scheduling as assumed in this study, as well as in contract employment of resources. *Resource cost* charts compare total project resource expenditures for each resource unit.

The model developed in this chapter represents an initial step toward a more comprehensive resource-activity integration in project scheduling and management. It provides for both effective activity scheduling based on dynamically updated activity attributes and intelligent iterative mapping of resources to each activity based on resource characteristics and pre-selected shape of project manager's objectives. The model consists of two complementary procedures: an *activity scheduler* and *resource mapper*. The procedures are alternatively being executed throughout the scheduling process at each newly detected decision instance, such that the final output is capable of providing decision support and recommendations with respect to both scheduling project activities and resource assignments. This approach allows human, social, as well as technical resources to interact and be utilized in value creating ways, while facilitating effective resource tracking and job distribution control.

REFERENCES

Badiru, Adedeji B. (1993). "Activity Resource Assignments Using Critical Resource Diagramming". *Project Management Journal* 14(3): 15–21.

Badiru, A. B., Oye Ibidapo-Obe and B. J. Ayeni (2012). *Industrial Control Systems: Mathematical and Statistical Models and Techniques*, Taylor & Francis Group/CRC Press, Boca Raton, FL.

Badiru, Adedeji B. and P. Simin Pulat (1995). *Comprehensive Project Management: Integrating Optimization Models, Management Principles, and Computers*, pp. 162–209, Prentice Hall, Hoboken, NJ.

Bracewell, Ronald N. (1978). *The Fourier Transform and Its Applications*, p. 97, McGraw-Hill, Inc., New York.

Campbell, Gerard M. (1999). "Cross-Utilization of Workers Whose Capabilities Differ". *Management Science* 45(5): 722–732.

Cooprider, C. (1999). "Solving a Skill Allocation Problem". *Production and Inventory Management Journal* Third Quarter: 1–6.

Dreger, J. Brian (1992). *Project Management: Effective Scheduling*, pp. 202–231, Van Nostrand Reinhold, New York.

Franz, Lori S. and Janis L. Miller (1993). "Scheduling Medical Residents to Rotations: Solving the Large Scale Multiperiod Staff Assignment Problem". *Operations Research* 41(2): 269–279.

Gray, Jennifer J., Don McIntire, and Herbert J. Doller (1993). "Preferences for Specific Work Schedulers: Foundation for an Expert-System Scheduling Program". *Computers in Nursing* 11(3): s115–s121.

Keeney, Ralph L. and Howard Raiffa (1993). *Decisions with Multiple Objectives: Preferences and Value Tradeoffs*, Cambridge University Press, Cambridge, New York.

Konstantinidis, P. D. (1998). "A Model to Optimize Project Resource Allocation by Construction of a Balanced Histogram". *European Journal of Operational Research* 104: 559–571.

Mattila, K. G. and D. M. Abraham (1998). "Resource Leveling of Linear Schedules Using Integer Linear Programming". *Journal of Construction Engineering and Management* 124(3): 232–244.

Milatovic, M. and A. B. Badiru (2004). "Applied Mathematics Modeling of Intelligent Mapping and Scheduling of Interdependent and Multifunctional Project Resources". *Applied Mathematics and Computation* 149(3): 703–721.

Roberts, Stephen M. (1992). "Human Skills – Keys to Effectiveness". *Cost Engineering* 34(2): 17–19.

Yura, Kenji (1994). "Production Scheduling to Satisfy Worker's Preferences for Days Off and Overtime Under Due-Date Constraints". *International Journal of Production Economics* 33: 265–270.

Appendix A
Mathematical Expressions and Collections (Series, Patterns, and Formulae)

GREEK ALPHABET

Alpha	$= A, \alpha = A$, a
Beta	$= B, \beta = B$, b
Gamma	$= \Gamma, \gamma = G$, g
Delta	$= \Delta, \delta = D$, d
Epsilon	$= E, e = E$, e
Zeta	$= Z, \zeta = Z$, z
Eta	$= H, \eta = E$, e
Theta	$= \theta, \theta = $ Th,th
Iota	$= I, \iota = I$, i
Kappa	$= K, \kappa = K$, k
Lambda	$= \wedge, \lambda = L$, 1
Mu	$= M, \mu = M$, m
Nu	$= N, \nu = N$, n
Xi	$= \Xi, \xi = X$, x
Omicron	$= O, o = O$, o
Pi	$= \Pi, \pi = P$, p
Rho	$= P, p = R$, r
Sigma	$= \sum, \sigma = S$, s
Tau	$= T, \tau = T$, t
Upsilon	$= T, \upsilon = U$,u
Phi	$= \Phi, \phi = $ Ph ph
Chi	$= X, x = $ Ch,ch
Psi	$= \Psi, \psi = $ Ps,ps
Omega	$= \Omega, \omega = O$, o

NUMBER SEQUENCE AND PATTERNS

$$1 \times 8 + 1 = 9$$
$$12 \times 8 + 2 = 98$$
$$123 \times 8 + 3 = 987$$
$$1234 \times 8 + 4 = 9876$$
$$12345 \times 8 + 5 = 98765$$
$$123456 \times 8 + 6 = 987654$$
$$1234567 \times 8 + 7 = 9876543$$
$$12345678 \times 8 + 8 = 98765432$$
$$123456789 \times 8 + 9 = 987654321$$

$$1 \times 9 + 2 = 11$$
$$12 \times 9 + 3 = 111$$
$$123 \times 9 + 4 = 1111$$
$$1234 \times 9 + 5 = 11111$$
$$12345 \times 9 + 6 = 111111$$
$$123456 \times 9 + 7 = 1111111$$
$$1234567 \times 9 + 8 = 11111111$$
$$12345678 \times 9 + 9 = 111111111$$
$$123456789 \times 9 + 10 = 1111111111$$

$$9 \times 9 + 7 = 88$$
$$98 \times 9 + 6 = 888$$
$$987 \times 9 + 5 = 8888$$
$$9876 \times 9 + 4 = 88888$$
$$98765 \times 9 + 3 = 888888$$
$$987654 \times 9 + 2 = 8888888$$
$$9876543 \times 9 + 1 = 88888888$$
$$98765432 \times 9 + 0 = 888888888$$

$$1 \times 1 = 1$$
$$11 \times 11 = 121$$
$$111 \times 111 = 12321$$
$$1111 \times 1111 = 1234321$$
$$11111 \times 11111 = 123454321$$
$$111111 \times 111111 = 12345654321$$
$$1111111 \times 1111111 = 1234567654321$$
$$11111111 \times 11111111 = 123456787654321$$
$$111111111 \times 111111111 = 12345678987654321$$

$$1 \times 8 + 1 = 9$$
$$12 \times 8 + 2 = 98$$
$$123 \times 8 + 3 = 987$$
$$1234 \times 8 + 4 = 9876$$
$$12345 \times 8 + 5 = 98765$$
$$123456 \times 8 + 6 = 987654$$
$$1234567 \times 8 + 7 = 9876543$$
$$12345678 \times 8 + 8 = 98765432$$
$$123456789 \times 8 + 9 = 987654321$$

$$9 \times 9 + 7 = 88$$
$$98 \times 9 + 6 = 888$$
$$987 \times 9 + 5 = 8888$$
$$9876 \times 9 + 4 = 88888$$
$$98765 \times 9 + 3 = 888888$$
$$987654 \times 9 + 2 = 8888888$$
$$9876543 \times 9 + 1 = 88888888$$
$$98765432 \times 9 + 0 = 888888888$$

CLOSED FORM MATHEMATICAL EXPRESSIONS

$$\sum_{n=0}^{\infty} \frac{x^n}{n!} = e^x$$

$$\sum_{n=0}^{\infty} \frac{x^n}{n} = \ln\left(\frac{1}{1-x}\right)$$

$$\sum_{n=0}^{k} x^n = \frac{x^{k+1} - 1}{x - 1}, \qquad x \neq 1$$

$$\sum_{n=1}^{k} x^n = \frac{x - x^{k+1}}{1 - x}, \qquad x \neq 1$$

$$\sum_{n=2}^{k} x^n = \frac{x^2 - x^{k+1}}{1 - x}, \qquad x \neq 1$$

$$\sum_{n=0}^{\infty} p^n = \frac{1}{1 - p}, \qquad \text{if } |p| < 1$$

$$\sum_{n=0}^{\infty} n x^n = \frac{x}{(1-x)^2}, \qquad x \neq 1$$

$$\sum_{n=0}^{\infty} n^2 x^n = \frac{2x^2}{(1-x)^3} + \frac{x}{(1-x)^2} = \frac{x(1+x)}{(1-x)^3}, \quad |x| < 1$$

$$\sum_{n=0}^{\infty} n^3 x^n = \frac{6x^3}{(1-x)^4} + \frac{6x^2}{(1-x)^3} = \frac{x}{(1-x)^2}, \quad |x| < 1$$

$$\sum_{n=0}^{M} nx^n = \frac{x\left[1 - (M+1)x^M + Mx^{M+1}\right]}{(1-x)^2}, \quad |x| < 1$$

$$\sum_{x=0}^{\infty} \binom{r+x-1}{x} u^x = (1-u)^{-r}, \quad \text{if } |u| < 1$$

$$\sum_{k=1}^{\infty} (-1)^{k+1} \frac{1}{k} = 1 - \frac{1}{2} + \frac{1}{3} - \frac{1}{4} + \frac{1}{5} - \frac{1}{6} + \ldots = \ln 2$$

$$\sum_{k=1}^{\infty} (-1)^{k+1} \frac{1}{(2k-1)} = 1 - \frac{1}{3} + \frac{1}{5} - \frac{1}{7} + \frac{1}{9} - \ldots = \frac{\pi}{4}$$

$$\sum_{k=0}^{\infty} (-1)^k x^k = \frac{1}{1+x}, \quad -1 < x < 1$$

$$\sum_{k=1}^{n} (-1)^k \binom{n}{k} = 1, \quad \text{for } n \geq 2$$

$$\sum_{k=0}^{n} \binom{n}{k}^2 = \binom{2n}{n}$$

$$\sum_{k=1}^{n} k = 1 + 2 + 3 + \ldots + n = \frac{n(n+1)}{2}$$

$$\sum_{k=1}^{n} k^2 = 1 + 4 + 9 + \ldots + n^2 = \frac{n(n+1)(2n+1)}{6}$$

$$\sum_{k=0}^{n-1} k^2 x^k = \frac{(x-1)^2 n^2 x^n - 2(x-1)nx^{n+1} + x^{n+2} - x^2 + x^{n+1} - x}{(x-1)^3}$$

$$\sum_{k=1}^{n} k^3 = 1 + 8 + 27 + \ldots + n^3 = \left(\frac{n(n+1)}{2}\right)^2$$

$$\sum_{k=1}^{n} (2k) = 2 + 4 + 6 + \ldots + 2n = n(n+1)$$

$$\sum_{k=1}^{n} (2k-1) = 1 + 3 + 5 + \ldots + (2n-1) = n^2$$

$$\sum_{k=0}^{\infty} (a+kd)r^k = a + (a+d)r + (a+2d)r^2 + \ldots = \frac{a}{1-r} + \frac{rd}{(1-r)^2}$$

$$\sum_{k=1}^{n} k^3 = 1 + 8 + 27 + \ldots + n^3 = \frac{n^2(n+1)^2}{4} = \left[\frac{n(n+1)}{2}\right]^2 = \left[\sum_{k=1}^{n} k\right]^2$$

$$\sum_{x=1}^{\infty} \frac{1}{x} = 1 + \frac{1}{2} + \frac{1}{3} + \ldots (\text{does not converge})$$

$$\sum_{m=0}^{k} m a^m = \frac{a}{(1-a)^2}\left[1 - (k+1)a^k + k a^{k+1}\right] = \sum_{m=1}^{k} m a^m$$

$$\sum_{k=0}^{n} (1) = n$$

$$\sum_{k=0}^{n} \binom{n}{k} = 2^n$$

$$(a+b)^n = \sum_{k=0}^{n} \binom{n}{k} a^k b^{n-k}$$

$$\prod_{n=1}^{\infty} a_n = e^{\left(\sum_{n=1}^{\infty} \ln(a_n)\right)}$$

$$\ln\left(\prod_{n=1}^{\infty} a_n\right) = \sum_{n=1}^{\infty} \ln a_n$$

$$\ln(x) = \sum_{k=1}^{\infty} \frac{1}{k}\left(\frac{x-1}{x}\right)^k, \quad x \geq \frac{1}{2}$$

$$\lim_{h \to \infty} (1+h)^{1/h} = e$$

$$\lim_{n \to \infty} \left(1 + \frac{x}{n}\right)^n = e^{-x}$$

$$\lim_{n \to \infty} \sum_{k=0}^{n} \frac{e^{-n} n^r}{K!} = \frac{1}{2}$$

$$\lim_{k \to \infty} \left(\frac{x^k}{k!}\right) = 0$$

$$|x + y| \leq |x| + |y|$$

$$|x - y| \geq |x| - |y|$$

$$\ln(1+x) = \sum_{k=1}^{\infty} (-1)^{k+1}\left(\frac{x^k}{k}\right), \quad \text{if } -1 < x \leq 1$$

$$\Gamma(\alpha + 1) = \alpha \Gamma(\alpha)$$

$$\Gamma\left(\frac{n}{2}\right) = \frac{\sqrt{\pi}\,(n-1)!}{2^{n-1}\left(\dfrac{n-1}{2}\right)!}, \qquad n \ \text{odd}$$

$$\Gamma(n) = \int_0^\infty e^{-x} x^{n-1} dx$$

$$\binom{n}{2} = \frac{1}{2}\left(n^2 - n\right) = \sum_{k=1}^{n-1} k$$

$$\binom{n+1}{2} = \binom{n}{2} + n$$

$$2.4.6.8\ldots 2n = \prod_{k=1}^{n} 2k = 2^n n!$$

$$1.3.5.7\ldots(2n-1) = \frac{(2n-1)!}{2^{2n-2}(2n-2)!} = \frac{2n-1}{2^{2n-2}}$$

Derivation of closed form expression for $\displaystyle\sum_{k=1}^{n} k x^k$

$$\sum_{k=1}^{n} k x^k = x \sum_{k=1}^{n} k x^{k-1}$$

$$= x \sum_{k=1}^{n} \frac{d}{dx}\left[x^k\right]$$

$$= x\,\frac{d}{dx}\left[\sum_{k=1}^{n} x^k\right]$$

$$= x\,\frac{d}{dx}\left[\frac{x\left(1-x^n\right)}{1-x}\right]$$

$$= x\left[\frac{\left(1-(n+1)x^n\right)(1-x) - x\left(1-x^n\right)(-1)}{(1-x)^2}\right]$$

$$= \frac{x\left[1-(n+1)x^n + nx^{n+1}\right]}{(1-x)^2}, \qquad x \neq 1$$

DERIVATION OF THE QUADRATIC FORMULA

Formula:

$$ax^2 + bx + c = 0$$

Solution

$$x = \frac{-b \pm \sqrt{b^2 - 4ac}}{2a}$$

If $b^2 - 4ac < 0$, the roots are complex,
If $b^2 - 4ac > 0$, the roots are real,
If $b^2 - 4ac = 0$, the roots are real and repeated.

Formula

$$ax^2 + bx + c = 0$$

Dividing both sides by 'a', ($a \neq 0$)

$$x^2 + \frac{b}{a}x + \frac{c}{a} = 0$$

Note if $a = 0$, the solution to $ax^2 + bx + c = 0$ is $x = -\frac{c}{b}$.
Rewrite

$$x^2 + \frac{b}{a}x + \frac{c}{a} = 0$$

as

$$\left(x + \frac{b}{2a}\right)^2 - \frac{b^2}{4a^2} + \frac{c}{a} = 0$$

$$\left(x + \frac{b}{2a}\right)^2 = \frac{b^2}{4a^2} - \frac{c}{a} = \frac{b^2 - 4ac}{4a^2}$$

$$x + \frac{b}{2a} = \pm\sqrt{\frac{b^2 - 4ac}{4a^2}} = \pm\frac{\sqrt{b^2 - 4ac}}{2a}$$

$$x = -\frac{b}{2a} = \pm\sqrt{\frac{b^2 - 4ac}{2a}}$$

$$x = \frac{-b \pm \sqrt{b^2 - 4ac}}{2a}$$

Numbers, Prefixes, and Power

Notation	Expansion
quecca (10^{30})	1, 000, 000, 000, 000, 000, 000, 000, 000, 000, 000
ronna (10^{27})	1, 000, 000, 000, 000, 000, 000, 000, 000, 000
yotta (10^{24})	1, 000, 000, 000, 000, 000, 000, 000, 000
zetta (10^{21})	1, 000, 000, 000, 00,0 000, 000, 000
exa (10^{18})	1, 000, 000, 000, 000, 000, 000
peta (10^{15})	1, 000, 000, 000, 000, 000
tera (10^{12})	1, 000, 000, 000, 000
giga (10^{9})	1, 000, 000, 000
mega (10^{6})	1, 000, 000
kilo (10^{3})	1, 000
hecto (10^{2})	100
deca (10^{1})	10
deci (10^{-1})	0.1
centi (10^{-2})	0.01
milli (10^{-3})	0.001
micro (10^{-6})	0.000 001
nano (10^{-9})	0.000 000 001
pico (10^{-12})	0.000 000 000 001
femto (10^{-15})	0.000 000 000 000 001
atto (10^{-18})	0.000 000 000 000 000 001
zepto (10^{-21})	0.000 000 000 000 000 000 001
yocto (10^{-24})	0.000 000 000 000 000 000 000 001
ronto (10^{-27})	0.000 000 000 000 000 000 000 000 001
quecto (10^{-30})	0.000 000 000 000 000 000 000 000 000 001
stringo (10^{-35})	0.000 000 000 000 000 000 000 000 000 000 000 01

Units of Measurement

English System

1 foot (ft)	12 inches (in)
1 yard (yd)	3 feet
1 mile (mi)	1760 yards
1 sq. foot	144 sq. inches
1 sq. yard	9 sq. feet
1 acre	4840 sq. yards

Metric System

mm	millimeter	.001 m
cm	centimeter	.01 m
dm	decimeter	.1 m
m	meter	1 m
dam	decameter	10 m
hm	hectometer	100 m
km	kilometer	1000 m

Constants	
speed of light	$2.997,925 \times 10^{10}$ cm/sec
	983.6×10^{6} ft/sec
	186,284 miles/sec
velocity of sound	340.3 meters/sec
	1116 ft/sec
Gravity (acceleration)	9.80665 m/sec square
	32.174 ft/sec square
	386.089 inches/sec square

Area Conversion

Multiply	By	To Obtain
Acres	43,560	sq feet
Acres	4,047	sq meters
Acres	4,840	sq yards
Acres	0.405	hectare
sq cm	0.155	sq inches
sq feet	144	sq inches
sq feet	0.09290	sq meters
sq feet	0.1111	sq yards
sq inches	645.16	sq millimeters
sq kilometers	0.3861	sq miles
sq meters	10.764	sq feet
sq meters	1.196	sq yards
sq miles	640	acres
sq miles	2.590	sq kilometers

Volume Conversion

Multiply	By	To Obtain
acre-foot	1233.5	cubic meters
cubic cm	0.06102	cubic inches
cubic feet	1728	cubic inches
	7.480	gallons (US)
	0.02832	cubic meters
	0.03704	cubic yards
Liter	1.057	liquid quarts
	0.908	dry quarts
	61.024	Cubic inches
Gallons (US)	231	cubic inches
	3.7854	liters
	4	quarts
	0.833	British gallons
	128	U.S. fluid ounces
quarts (US)	0.9463	liters

Energy, Heat, and Power Conversion

Multiply	By	To Obtain
BTU	1055.9	joules
BTU	0.2520	kg-calories
watt-hour	3600	joules
watt-hour	3.409	BTU
HP (electric)	746	watts
BTU/second	1055.9	watts
Watt-second	1.00	joules

Mass Conversion

Multiply	By	To Obtain
Carat	0.200	cubic grams
Grams	0.03527	ounces
Kilograms	2.2046	pounds
Ounces	28.350	grams
Pound	16	ounces
Pound	453.6	grams
stone (UK)	6.35	kilograms
stone (UK)	14	pounds
ton (net)	907.2	kilograms
ton (net)	2000	pounds
ton (net)	0.893	gross ton
ton (net)	0.907	metric ton
ton (gross)	ton (gross)	pounds
ton (gross)	1.12	net tons
ton (gross)	1.016	metric tons
tonne (metric)	2,204.623	pounds
tonne (metric)	0.984	gross pound
tonne (metric)	1000	kilograms

Temperature Conversion

Celsius to Kelvin	$K = C + 273.15$
Celsius to Fahrenheit	$F = (9/5)C + 32$
Fahrenheit to Celsius	$C = (5/9)(F - 32)$
Fahrenheit to Kelvin	$K = (5/9)(F + 459.67)$
Fahrenheit to Rankin	$R = F + 459.67$
Rankin to Kelvin	$K = (5/9)R$

Velocity conversion

Multiply	By	To Obtain
feet/minute	5.080	mm/second
feet/second	0.3048	meters/second
inches/second	0.0254	meters/second
km/hour	0.6214	miles/hour
meters/second	3.2808	feet/second

Velocity conversion

Multiply	By	To Obtain
meters/second	2.237	miles/hour
miles/hour	88.0	feet/minute
miles/hour	0.44704	meters/second
miles/hour	1.6093	km/hour
miles/hour	0.8684	knots
Knot	1.151	miles/hour

Pressure Conversion

Multiply	By	To Obtain
Atmospheres	1.01325	bars
	33.90	feet of water inches of mercury mm of mercury
	29.92	
	760.0	
Bars	75.01	cm of mercury pounds/sq inch
	14.50	
dyne/sq cm	0.1	N/sq meter
newtons/sq cm	1.450	pounds/sq inch
pounds/sq inch	0.06805	atmospheres inches of mercury inches of water
	2.036	millibars
	27.708	mm of mercury
	68.948	
	51.72	

Distance conversion

Multiply	By	To Obtain
Angstrom	10^{-10}	meters
Feet	0.30480	meters
Feet	12	meters
Inches	25.40	millimeters
Inches	0.02540	meters
Inches	0.08333	feet
Kilometers	3,280.8	feet
Kilometers	0.6214	miles
Kilometers	1,094	yards
Meters	39.370	inches
Meters	3.2808	feet
Meters	1.094	yards
Miles	5,280	feet
Miles	1.6093	kilometers
Miles	0.8694	nautical miles
Millimeters	0.03937	inches
nautical miles	6,076	feet
nautical miles	1.852	kilometers
Yards	0.9144	meters
Yards	3	feet
Yards	36	inches

Common Measurements

1 pinch	1/8 teaspoon or less
3 teaspoons	1 tablespoon
2 tablespoons	1/8 cup
4 tablespoons	1/4 cup
8 tablespoons	1/2 cup
12 tablespoons	3/4 cup
16 tablespoons	1 cup
5 tablespoons + 1 teaspoon	1/3 cup
4 oz	1/2 cup
8oz	1 cup
16 oz	1 lb
1 oz	2 tablespoons fat or liquid
1 cup of liquid	1/2 pint
2 cups	1 pint
2 pints	1 quart
4 cup of liquid	1 quart
4 quarts	1 gallon
8 quarts	1 peck (such as apples, pears, etc.)
1 jigger	1 ½ fluid oz
1 jigger	3 tablespoons

MATHEMATICAL SIGNS AND SYMBOLS

$\pm(\mp)$	plus or minus (minus or plus)
:	divided by, ratio sign
::	proportional sign
<	less than
$\not<$	not less than
>	greater than
$\not>$	not greater than
$\cong$	approximately equals, congruent
~	similar to
$\equiv$	equivalent to
$\neq$	not equal to
$\doteq$	approaches, is approximately equal to
$\propto$	varies as
∞	infinity
$\therefore$	therefore
$\sqrt{}$	square root
$\sqrt[3]{}$	cube root
$\sqrt[n]{}$	nth root
$\angle$	angle
$\perp$	perpendicular to
—	parallel to
$|x|$	numerical value of x

log *or* log$_{10}$	common logarithm or Briggsian logarithm
log$_e$ *or In*	natural logarithm or hyperbolic logarithm or Napierian logarithm
e	base (2.718) of natural system of logarithms
$a°$	an angle a degrees
a'	a prime, an angle a minutes
a''	a double prime, an angle a seconds, a second
sin	sine
cos	cosine
tan	tangent
ctn or cot	cotangent
sec	secant
csc	cosecant
vers	versed sine
covers	coversed sine
exsec	exsecant
sin^{-1}	anti-sine or angle whose sine is
sinh	hyperbolic sine
cosh	hyperbolic cosine
tanh	hyperbolic tangent
sinh^{-1}	anti-hyperbolic sine or angle whose hyperbolic sine is
$f(x)$ or $\phi(x)$	function of x
Δx	increment of x
Σ	summation of
dx	differential of x
dy/dx or y'	derivative of y with respect to x
d^2y/dx^2 or y''	second derivative of y with respect to x
$d^n y/dx^n$	nth derivative of y with respect to x
$\partial y/\partial x$	partial derivative of y with respect to x
$\partial^n y/\partial x^n$	nth partial derivative of y with respect to x
$\dfrac{\partial^n y}{\partial x \partial y}$	nth partial derivative with respect to x and y
$\int$	integral of
$\int_a^b$	integral between the limits a and b
$\dot{y}$	first derivative of y with respect to time
$\ddot{y}$	second derivative of y with respect to time
Δ or ∇^2	the "Laplacian"

$$\left(\frac{\partial^2}{\partial x^2} + \frac{\partial^2}{\partial y^2} + \frac{\partial^2}{\partial x^2} \right)$$

δ	sign of a variation
Ξ	sign of integration around a closed path

OVERALL MEAN

$$\bar{x} = \frac{n_1\bar{x}_1 + n_2\bar{x}_2 + n_3\bar{x}_3 + \ldots + n_k\bar{x}_k}{n_1 + n_2 + n_3 + \ldots + n_k} = \frac{\sum n\bar{x}}{\sum n}$$

Chebyshev's Theorem

$$1 - 1/k^2$$

Permutations

A permutation of m elements from a set of n elements is any arrangement, without repetition, of the m elements. The total number of all the possible permutations of n distinct objects taken m times is

$$P(n,m) = \frac{n!}{(n-m)!}, \quad (n \geq m)$$

Example

Find the number of ways a president, vice-president, secretary, and a treasurer can be chosen from a committee of eight members.

Solution

$$P(n,m) = \frac{n!}{(n-m)!} = P(8,4) = \frac{8!}{(8-4)!} = \frac{8.7.6.5.4.3.2.1}{4.3.2.1} = 1680$$

There are 1680 ways of choosing the four officials from the committee of eight members.

Combinations

The number of combination of n distinct elements taken is given by

$$C(n,\ m) = \frac{n!}{m!(n-m)!}, \quad (n \geq m)$$

Example

How many poker hands of five cards can be dealt from a standard deck of 52 cards?

Solution

Note: The order in which the 5 cards care dealt is not important.

$$C(n,m) = \frac{n!}{m!(n-m)!} = C(52,5) = \frac{52!}{5!(52-5)!} = \frac{52!}{5!47!}$$

$$= \frac{52.51.50.49.48}{5.4.3.2.1} = 2,598,963$$

Failure

$$q = 1 - p = \frac{n-s}{n}$$

Probability

$$P(X \leq x) = F(x) = \int_{-\infty}^{x} f(x)dx$$

Expected Value

$$\mu = \Sigma\left(xf(x)\right)$$

Variance

$$\sigma^2 = \Sigma(x-\mu)^2 f(x) \quad \text{or} \quad \sigma^2 = \int_{-\infty}^{\infty} (x-\mu)^2 f(x)dx$$

Binomial Distribution

$$f(x) = {}^{n}c_x p^x (1-p)^{n-x}$$

Poisson Distribution

$$f(x) = \frac{(np)^x e^{-np}}{x!}$$

Mean of a Binomial Distribution

$$\mu = np$$

Variance

$$\sigma^2 = npq$$

where $q = 1 - p$ and is the probability of obtaining x failures in the n trials.

Normal Distribution

$$f(x) = \frac{1}{\sigma\sqrt{2\pi}} e^{\frac{-(x-\mu)^2}{2\sigma^2}}$$

Cumulative Distribution Function

$$f(x) = P(X \le x) = \frac{1}{\sigma\sqrt{2\pi}} \int_{-\infty}^{x} e^{\frac{-(x-\mu)^2}{2\sigma^2}} dx$$

Population Mean

$$\mu_{\bar{x}} = \mu$$

Standard Error of the Mean

$$\sigma_{\bar{x}} = \frac{\sigma}{\sqrt{n}}$$

t-Distribution

$$\bar{x} - t_{\alpha/2}\left(\frac{s}{\sqrt{n}}\right) \le \mu \le \bar{x} + t_{\alpha/2}\left(\frac{s}{\sqrt{n}}\right)$$

where

$\bar{x}$ = sample mean
μ = population mean
s = sample standard deviation

Chi-squared Distribution

$$\frac{(n-1)s^2}{\chi^2_{\alpha/2}} \le \sigma^2 \le \frac{(n-1)s^2}{\chi^2_{1-\alpha/2}}$$

DEFINITION OF SET AND NOTATION

A set is a collection of object called elements. In mathematics we write a set by putting its elements between the curly brackets. { }.

Set A which containing numbers 3, 4, and 5 is written

$$A = \{3,4,5\}$$

a) Empty set
   A set with no elements is called an empty set and it denoted by

$$\{\} = \Phi$$

b) Subset
   Sometimes every element of one set also belongs to another set:

$$A = \{3,4,5\} \text{ and } B = \{1,2,3,4,5,6,7,\},$$

A set A is a subset of a set B because every elements of set A is also an element of set B, and it is written as

$$A \subseteq B$$

a) Set equality
   The sets A and B are equal if and only if they have exactly the same elements, and the equality is written as

$$A = B$$

b) Set union
   The union of a set A and Set B is the set of all elements that belong to either A or B or both, and is written as

$$A \cup b = \{x \mid x \in A \text{ or } x \in B \text{ or both}\}$$

Set Terms and Symbols

{ } set brace
$\in$ is an element of
$\notin$ is not an element of
$\subseteq$ is a subset of
$\not\subset$ is not a subset of
A' complement of set A
$\cap$ set intersection
$\cup$ set union

Operations on Sets

If A, B, and C are arbitrary subsets of universal set U, then the following rules govern the operations on sets:

1) Commutative law for union

$$A \cup B = B \cup A$$

2) Commutative law for intersection

$$A \cap B = B \cap A$$

3) Associative law for union

$$A \cup (B \cup C) = (A \cup B) \cup C$$

4) Associative law for intersection

$$A \cap (B \cap C) = (A \cap B) \cap C$$

5) Distributive law for union

$$A \cup (B \cap C) = (A \cup B) \cap (A \cup C)$$

6) Distributive law for intersection

$$A \cap (B \cap C) = (A \cap B) \cup (A \cap C)$$

De Morgan's Laws

$$(A \cup B)' = A' \cap B' \quad (1)$$
$$(A \cap B)' = A' \cup B' \quad (2)$$

The complement of the union of two sets is equal to the intersection of their complements. The complement of the intersection of two sets is equal to the union of their complements.

Counting the Elements is a Set

The number of the elements in a finite set is determined by simply counting the elements in the set.

If A and B are disjoint sets, then

$$n(A \cup B) = n(A) + n(B)$$

In general, A and B need not to be disjoint, so

$$n(A \cup B) = n(A) + n(B) - n(A \cap B)$$

where
n = number of the elements in a set

Permutations
A permutation of m elements from a set of n elements is any arrangement, without repetition, of the m elements. The total number of all the possible permutations of n distinct objects taken m times is

$$P(n,m) = \frac{n!}{(n-m)!}, \quad (n \geq m)$$

Example
Find the number of ways a president, vice-president, secretary, and a treasurer can be chosen from a committee of eight members.

Solution

$$P(n,m) = \frac{n!}{(n-m)!}, = P(8,4) = \frac{8!}{(8-4)!} = \frac{8.7.6.5.4.3.2.1}{4.3.2.1} = 1680$$

There are 1680 ways of choosing the four officials from the committee of eight members.

Combinations
The number of combination of n distinct elements taken is given by

$$C(n,m) = \frac{n!}{m!(n-m)!}, \quad (n \geq m)$$

Example
How many poker hands of five cards can be dealt from a standard deck of 52 cards?

Solution
Note: The order in which the 5 cards care dealt is not important.

$$C(n,m) = \frac{n!}{m!(n-m)!}, = C(52,5) = \frac{52!}{5!(52-5)!} = \frac{52!}{5!47!} =$$

$$\frac{52.51.50.49.48}{5.4.3.2.1} = 2,598,963$$

PROBABILITY TERMINOLOGY

A number of specialized terms are used in the study of probability.

Experiment: An experiment is an activity or occurrence with an observable result.
Outcome: The result of the experiment.
Sample point: An outcome of an experiment.
Event: An event is a set of outcomes (a subset of the sample space) to which a probability is assigned.

Basic Probability Principles

Consider a random sampling process in which all the outcomes solely depend on chance, that is, each outcome is equally likely to happen. If S is a uniform sample space and the collection of desired outcomes is E, the probability of the desired outcomes is

$$P(E) = \frac{n(E)}{n(S)}$$

where

$n(E)$ = number of favorable out comes in E
$n(S)$ = number of possible outcomes in S

Since E is a subset of S,

$$0 \le n(E) \le n(S),$$

the probability of the desired outcome is

$$0 \le P(E) \le 1$$

Random Variable

A random variable is a rule that assigns a number to each outcome of a chance experiment.
Example:

1. A coin is tossed six times. The random variable X is the number of tails that are noted. X can only take the values 1,2,..., 6, so X is a discrete random variable.

2. A light bulb is burned until it burns out. The random variable Y is its lifetime in hours. Y can take any positive real value, so Y is a continuous random variable.

Mean Value $\hat{x}$ or Expected Value μ

The mean value or expected value of a random variable indicates its average or central value. It is a useful summary value of the variable's distribution.

1. If random variable X is a discrete mean value,

$$\hat{x} = x_1 p_1 + x_2 p_2 + \ldots + x_n p_n = \sum_{i=1}^{n} x_1 p_1$$

where
    p_i = probability densities

2. If X is a continuous random variable with probability density function $f(x)$, then the expected value of X is

$$\mu = E(X) = \int_{-\infty}^{+\infty} x f(x) \, dx$$

where
    $f(x)$ = probability densities

Series Expansions
(a) Expansions of Common Functions

$$e = 1 + \frac{1}{1!} + \frac{1}{2!} + \frac{1}{3!} + \ldots$$

$$e^x = 1 + x + \frac{x^2}{2!} + \frac{x^3}{3!} + \ldots$$

$$a^x = 1 + x \ln a + \frac{(x \ln a)^2}{2!} + \frac{(x \ln a)^3}{3!} + \ldots$$

$$e^{-x^2} = 1 - x^2 + \frac{x^4}{2!} - \frac{x^6}{3!} + \frac{x^8}{4!} - \dots$$

$$\ln x = (x-1) - \frac{1}{2}(x-1)^2 + \frac{1}{3}(x-1)^3 - \dots, \qquad 0 < x \le 2$$

$$\ln x = \frac{x-1}{x} + \frac{1}{2}\left(\frac{x-1}{x}\right)^2 + \frac{1}{3}\left(\frac{x-1}{x}\right)^3 + \dots, \qquad x > \frac{1}{2}$$

$$\ln x = 2\left[\frac{x-1}{x+1} + \frac{1}{3}\left(\frac{x-1}{x+1}\right)^3 + \frac{1}{5}\left(\frac{x-1}{x+1}\right)^5 + \dots\right], \qquad x > 0$$

$$\ln(1+x) = x - \frac{x^2}{2} + \frac{x^3}{3} - \frac{x^4}{4} + \dots, \qquad |x| \le 1$$

$$\ln(a+x) = \ln a + 2\left[\frac{x}{2a+x} + \frac{1}{3}\left(\frac{x}{2a+x}\right)^3 + \frac{1}{5}\left(\frac{x}{2a+x}\right)^5 + \dots\right],$$

$$a > 0, \qquad -a < x < +\infty$$

$$\ln\left(\frac{1+x}{1-x}\right) = 2\left(x + \frac{x^3}{3} + \frac{x^5}{5} + \frac{x^7}{7} + \dots\right), \qquad x^2 < 1$$

$$\ln\left(\frac{1+x}{1-x}\right) = 2\left[\frac{1}{x} + \frac{1}{3}\left(\frac{1}{x}\right)^3 + \frac{1}{5}\left(\frac{1}{x}\right)^5 + \left(\frac{1}{x}\right)^7 + \dots\right], \qquad x^2 > 1$$

$$\ln\left(\frac{1+x}{x}\right) = 2\left[\frac{1}{2x+1} + \frac{1}{3(2x+1)^3} + \frac{1}{5(2x+1)^5} + \dots\right], \qquad x > 0$$

$$\sin x = x - \frac{x^3}{3!} + \frac{x^5}{5!} - \frac{x^7}{7!} + \dots$$

$$\cos x = 1 - \frac{x^2}{2!} + \frac{x^4}{4!} - \frac{x^6}{6!} + \dots$$

$$\tan x = x + \frac{x^3}{3} + \frac{2x^5}{15} + \frac{17x^7}{315} + \frac{62x^9}{2835} + \dots, \qquad x^2 < \frac{\pi^2}{4}$$

$$\sin^{-1} x = x + \frac{x^3}{6} + \frac{1}{2}\cdot\frac{3}{4}\cdot\frac{x^3}{5} + \frac{1}{2}\cdot\frac{3}{4}\cdot\frac{5}{6}\cdot\frac{x^7}{7} + \dots, \qquad x^2 < 1$$

$$\tan^{-1} x = x - \frac{1}{3}x^3 + \frac{1}{5}x^5 - \frac{1}{7}x^7 + \dots, \qquad x^2 < 1$$

$$\tan^{-1} x = \frac{\pi}{2} - \frac{1}{x} + \frac{1}{3x^5} - \frac{1}{5x^5} + \dots, \qquad x^2 > 1$$

$$\sinh x = x + \frac{x^3}{3!} + \frac{x^5}{5!} + \frac{x^7}{7!} + \dots$$

$$\cosh x = 1 + \frac{x^2}{2!} + \frac{x^4}{4!} + \frac{x^6}{6!} + \dots$$

$$\tanh x = x - \frac{x^3}{3} + \frac{2x^5}{15} - \frac{17x^7}{315} + \dots$$

$$\sinh^{-1} x = x - \frac{1}{2} \cdot \frac{x^3}{3} + \frac{1 \cdot 3}{2 \cdot 4} \cdot \frac{x^5}{5} - \frac{1 \cdot 3 \cdot 5}{2 \cdot 4 \cdot 6} \cdot \frac{x^7}{7} + \dots, \qquad x^2 < 1$$

$$\sinh^{-1} x = \ln 2x + \frac{1}{2} \cdot \frac{1}{2x^2} - \frac{1 \cdot 3}{2 \cdot 4} \cdot \frac{1}{4x^4} + \frac{1 \cdot 3 \cdot 5}{2 \cdot 4 \cdot 6} \cdot \frac{1}{6x^6} - \dots, \qquad x > 1$$

$$\cosh^{-1} x = \ln 2x - \frac{1}{2} \cdot \frac{1}{2x^2} - \frac{1 \cdot 3}{2 \cdot 4} \cdot \frac{1}{4x^4} - \frac{1 \cdot 3 \cdot 5}{2 \cdot 4 \cdot 6} \cdot \frac{1}{6x^6} - \dots$$

$$\tanh^{-1} x = x + \frac{x^3}{3} + \frac{x^5}{5} + \frac{x^7}{7} + \dots, \qquad x^2 < 1$$

(b) Binomial Theorem

$$(a+x)^n = a^n + na^{n-1}x + \frac{n(n-1)}{2!} a^{n-2}x^2 + \frac{n(n-1)(n-2)}{3!}$$

$$a^{n-3}x^3 + \dots, \qquad x^2 < a^2$$

(c) Taylor Series Expansion

A function $f(x)$ may be expanded about $x = a$ if the function is continuous, and its derivatives exist and are finite at $x = a$.

$$f(x) = f(a) + f'(a)\frac{(x-a)}{1!} + f''(a)\frac{(x-a)^2}{2!} + f'''(a)\frac{(x-a)^3}{3!} + \dots$$

$$+ f^{n-1}(a)\frac{(x-a)^{n-1}}{(n-1)!} + R_n$$

(d) Maclaurin Series Expansion

The Maclaurin series expansion is a special case of the Taylor series expansion for $a = 0$.

$$f(x) = f(0) + f'(0)\frac{x}{1!} + f''(0)\frac{x^2}{2!} + f'''(0)\frac{x^3}{3!} + \dots + f^{(n-1)}(0)\frac{x^{n-1}}{(n-1)!} + R_n$$

(e) Arithmetic Progression

The sum to n terms of the arithmetic progression

$$S = a + (a+d) + (a+2d) + \dots + \left[a + (n-1)d\right]$$

is (in terms of the last number l)

$$S = \frac{n}{2}(a+l)$$

where $l = a + (n - 1)\,d$.

(f) Geometric Progression
The sum of the geometric progression to n terms is

$$S = a + ar + ar^2 + \cdots + ar^{n-1} = a\left(\frac{1-r^n}{1-r}\right)$$

(g) Sterling's Formula for Factorials

$$n! \approx \sqrt{2\pi}\, n^{n+1/2} e^{-n}$$

ALGEBRA

Laws of Algebraic Operations
(a) Commutative law: $a + b = b + a$, $ab = ba$;
(b) Associative law: $a + (b + c) = (a + b) + c$, $a\,(bc) = (ab)\,c$;
(c) Distributive law: $c\,(a + b) = ca + cb$.

Special Products and Factors

$$\left(x+y\right)^2 = x^2 + 2xy + y^2$$
$$\left(x-y\right)^2 = x^2 - 2xy + y^2$$
$$\left(x+y\right)^3 = x^3 - 3x^2y + 3xy^2 + y^3$$
$$\left(x-y\right)^3 = x^3 - 3x^2y + 3xy^2 - y^3$$
$$\left(x+y\right)^4 = x^4 + 4x^3y + 6x^2y^2 + 4xy^3 + y^4$$
$$\left(x-y\right)^4 = x^4 - 4x^3y + 6x^2y^2 - 4xy^3 + y^4$$
$$\left(x+y\right)^5 = x^5 + 5x^4y + 10x^3y^2 + 10x^2y^3 + 5xy^4 + y^5$$
$$\left(x-y\right)^5 = x^5 - 5x^4y + 10x^3y^2 - 10x^2y^3 + 5xy^4 - y^5$$
$$\left(x+y\right)^6 = x^6 + 6x^5y + 15x^4y^2 + 20x^3y^3 + 15x^2y^4 + 6xy^5 + y^6$$
$$\left(x-y\right)^6 = x^6 - 6x^5y + 15x^4y^2 - 20x^3y^3 + 15x^2y^4 - 6xy^5 + y^6$$

The results above are special cases of the binomial formula.

$$x^2 - y^2 = (x - y)(x + y)$$
$$x^3 - y^3 = (x - y)(x^2 + xy + y^2)$$
$$x^3 + y^3 = (x + y)(x^2 - xy + y^2)$$
$$x^4 - y^4 = (x - y)(x + y)(x^2 + y^2)$$
$$x^5 - y^5 = (x - y)(x^4 + x^3 y + x^2 y^2 - xy^3 + y^4)$$
$$x^5 + y^5 = (x + y)(x^4 - x^3 y + x^2 y^2 - xy^3 + y^4)$$
$$x^6 - y^6 = (x - y)(x + y)(x^2 + xy + y^2)(x^2 - xy + y^2)$$
$$x^4 + x^2 y^2 + y^4 = (x^2 + xy + y^2)(x^2 - xy + y^2)$$
$$x^4 + 4y^4 = (x^2 + 2xy + 2y^2)(x^2 - 2xy + 2y^2)$$

Some generalizations of the above are given by the following results where n is a positive integer.

$$x^{2n+1} - y^{2n+1} = (x - y)(x^{2n} + x^{2n-1} y + x^{2n-2} y^2 + \cdots + y^{2n})$$
$$= (x - y)\left(x^2 - 2xy\cos\frac{2\pi}{2n+1} + y^2\right)\left(x^2 - 2xy\cos\frac{4\pi}{2n+1} + y^2\right)$$
$$\cdots\left(x^2 - 2xy\cos\frac{2n\pi}{2n+1} + y^2\right)$$

$$x^{2n+1} + y^{2n+1} = (x + y)(x^{2n} - x^{2n-1} y + x^{2n-2} y^2 - \cdots + y^{2n})$$
$$= (x + y)\left(x^2 + 2xy\cos\frac{2\pi}{2n+1} + y^2\right)\left(x^2 + 2xy\cos\frac{4\pi}{2n+1} + y^2\right)$$
$$\cdots\left(x^2 + 2xy\cos\frac{2n\pi}{2n+1} + y^2\right)$$

$$x^{2n} - y^{2n} = (x - y)(x + y)(x^{n-1} + x^{n-2} y + x^{n-3} y^2 + \cdots)(x^{n-1} - x^{n-2} y + x^{n-3} y^2 - \cdots)$$
$$= (x - y)(x + y)\left(x^2 - 2xy\cos\frac{\pi}{n} + y^2\right)\left(x^2 - 2xy\cos\frac{2\pi}{n} + y^2\right)$$
$$\cdots\left(x^2 - 2xy\cos\frac{(n-1)\pi}{n} + y^2\right)$$

$$x^{2n} + y^{2n} = \left(x^2 + 2xy\cos\frac{\pi}{2n} + y^2\right)\left(x^2 + 2xy\cos\frac{3\pi}{2n} + y^2\right)$$

$$\cdots\left(x^2 + 2xy\cos\frac{(2n-1)\pi}{2n} + y^2\right)$$

Powers and Roots

$a^x \times a^y = a^{(x+y)}$ $\qquad\qquad$ $a^0 = 1$ [if $a \neq 0$] $\qquad\qquad$ $(ab^x) = a^x b^x$

$\dfrac{a^x}{a^y} = a^{(x-y)}$ $\qquad\qquad$ $a^{-x} = \dfrac{1}{a^x}$ $\qquad\qquad$ $\left(\dfrac{a}{b}\right)^x = \dfrac{a^x}{b^x}$

$(a^x)^y = a^{xy}$ $\qquad\qquad$ $a^{\frac{1}{x}} = \sqrt[x]{a}$ $\qquad\qquad$ $\sqrt[x]{ab} = \sqrt[x]{a}\sqrt[x]{b}$

$\sqrt[x]{\sqrt[y]{a}} = \sqrt[xy]{a}$ $\qquad\qquad$ $a^{\frac{x}{y}} = \sqrt[y]{a^x}$ $\qquad\qquad$ $\sqrt[x]{\dfrac{a}{b}} = \dfrac{\sqrt[x]{a}}{\sqrt[x]{b}}$

Proportion

$$\text{If} \quad \frac{a}{b} = \frac{c}{d}, \qquad \text{then} \qquad \frac{a+b}{b} = \frac{c+d}{d}$$

$$\frac{a-b}{b} = \frac{c-d}{d} \qquad\qquad \frac{a-b}{a+b} = \frac{c-d}{c+d}$$

Sum of Arithmetic Progression to n Terms[1]

$$a + (a+d) + (a+2d) + \ldots + \left(a + (n-1)d\right)$$

$$= na + \frac{1}{2}n(n-1)d = \frac{n}{2}(a+l),$$

$$\text{last term in series} = l = a + (n-1)d$$

Sum of Geometric Progression to n Terms

$$S_n = a + ar + ar^2 + \ldots + ar^{n-1} = \frac{a(1-r^n)}{1-r}$$

$$\lim_{n \to \infty} 8_n = a!(1-r) \qquad (-1 < r < 1)$$

Arithmetic Mean of *n* Quantities A

$$A = \frac{a_1 + a_2 + \ldots + a_n}{n}$$

Geometric Mean of n Quantities G

$$G = \left(a_1 a_2 \ldots a_n\right)^{1/n}$$
$$(a_k > 0, k = 1, 2, \ldots, n)$$

Harmonic Mean of n Quantities H

$$\frac{1}{H} = \frac{1}{n}\left(\frac{1}{a_1} + \frac{1}{a_2} + \ldots + \frac{1}{a_n}\right)$$
$$(a_k > 0, k = 1, 2, \ldots, n)$$

Generalized Mean

$$M(t) = \left(\frac{1}{n}\sum_{k=1}^{n} a_k^t\right)^{1/t}$$
$$M(t) = 0 (t < 0, \text{some } a_k \text{ zero})$$
$$\lim_{t \to \infty} M(t) = \max. \quad (a_1, a_2, \ldots, a_n) = \max.a$$
$$\lim_{t \to -\infty} M(t) = \min. \quad (a_1, a_2, \ldots, a_n) = \min.a$$
$$\lim_{t \to 0} M(t) = G$$
$$M(1) = A$$
$$M(-1) = H$$

Solution of Quadratic Equations
Given $az^2 + bz + c = 0$

$$z_{1,2} = -\left(\frac{b}{2a}\right) \pm \frac{1}{2a}q^{\frac{1}{2}}, q = b^2 - 4ac,$$

$$z_1 + z_2 = -b/a, z_1 z_2 = c/a$$

If

$q < 0, \text{two real roots},$
$q = 0, \text{two equal roots},$
$q < 0, \text{pair of complex conjugate roots}.$

Solution of Cubic Equations

Given $z^2 + a_2 z^2 + a_1 z + a_0 = 0$, let

$$q = \frac{1}{3} a_1 - \frac{1}{9} a_2^2;$$

$$r = \frac{1}{6}(a_1 a_2 - 3a_0) - \frac{1}{27} a_2^3.$$

If

$q^3 + r^2 > 0$, one real root and a pair of complex conjugate roots,

$q^3 + r^2 = 0$, all roots real and atleast two are equal,

$q^3 + r^2 < 0$, all roots real $(\text{irreducible case})$.

Let

$$s_1 = \left[r + \left(q^3 + r^2 \right)^{\frac{1}{2}} \right]^{\frac{1}{2}},$$

$$s_2 = \left[r - \left(q^3 + r^2 \right)^{\frac{1}{2}} \right]^{\frac{1}{2}}$$

then

$$z_1 = \left(s_1 + s_2 \right) - \frac{a_2}{3}$$

$$z_2 = -\frac{1}{2}\left(s_1 + s_2 \right) - \frac{a_2}{3} + \frac{i\sqrt{3}}{2}\left(s_1 - s_2 \right)$$

$$z_3 = -\frac{1}{2}\left(s_1 + s_2 \right) - \frac{a_2}{3} - \frac{i\sqrt{3}}{2}\left(s_1 - s_2 \right).$$

If z_1, z_2, z_3 are the roots of the cubic equation

$$z_1 + z_2 + z_3 = -a_2$$

$$z_1 z_2 + z_1 z_3 + z_2 z_3 = a_1$$

$$z_1 z_2 z_3 = a_0$$

Trigonometric Solution of the Cubic Equation

The form $x^3 + ax + b = 0$ with $ab \neq 0$ can always be solved by transforming it to the trigonometric identity

$$4\cos^3\theta - 3\cos\theta - \cos(3\theta) \equiv 0$$

Let $x = m\cos\theta$, then

$$x^3 + ax + b = m^3\cos^3\theta + am\cos\theta + b = 4\cos^3\theta - 3\cos\theta - \cos(3\theta) \equiv 0$$

Hence

$$\frac{4}{m^3} = -\frac{3}{am} = \frac{-\cos(3\theta)}{b},$$

from which follows that

$$m = 2\sqrt{-\frac{a}{3}}, \cos(3\theta) = \frac{3b}{am}$$

Any solution θ_1 which satisfies $\cos(3\theta) = \dfrac{3b}{am}$, will also have the solutions

$$\theta_1 + \frac{2\pi}{3} \text{ and } \theta_1 + \frac{4\pi}{3}$$

The roots of the cubic $x^3 + ax + b = 0$ are

$$2\sqrt{-\frac{a}{3}}\cos\theta_1,$$

$$2\sqrt{-\frac{a}{3}}\cos\left(\theta_1 + \frac{2\pi}{3}\right),$$

$$2\sqrt{-\frac{a}{3}}\cos\left(\theta_1 + \frac{4\pi}{3}\right)$$

Given $z^4 + a_3z^3 + a_2z^2 + a_1z + a_0 = 0$, find the real root u_1 of the cubic equation

$$u^3 - a_2u^2 + (a_1a_3 - 4a_0)u - (a_1^2 + a_0a_3^2 - 4a_0a_2) = 0$$

and determine the four roots of the quadric as solutions of the two quadratic equations

$$v^2 + \left[\frac{a_3}{2} \mp \left(\frac{a_3^2}{4} + u_1 - a_2\right)^{\frac{1}{2}}\right]v + \frac{u_1}{2} \mp \left[\left(\frac{u_1}{2}\right)^2 - a_0\right]^{\frac{1}{2}} = 0$$

If all roots of the cubic equation are real, use the value of u_1 which gives real coefficients in the quadratic equation and select signs so that if

$$z^4 + a_3 z^3 + a_2 z^3 + a_1 z + a_0 = \left(z^2 + p_1 z + q_1\right)\left(z^2 + p_2 z + q_2\right),$$

then

$$p_1 + p_2 = a_3, \; p_1 p_2 + q_1 + q_2 = a_2, \; p_1 q_2 + p_2 q_1 = a_1, \; q_1 q_2 = a_0.$$

If z_1, z_2, z_3, z_4 are the roots,

$$\sum z_t = -a_3, \sum z_t z_j z_k = -a_1,$$
$$\sum z_t z_j = a_2, z_1 z_2 z_3 z_4 = a_0.$$

Partial Fractions

This section applies only to rational algebraic fractions with numerator of lower degree than the denominator. Improper fractions can be reduced to proper fractions by long division.

Every fraction may be expressed as the sum of component fractions whose denominators are factors of the denominator of the original fraction.

Let $N(x)$ = numerator, a polynomial of the form

$$N(x) = n_0 + n_1 x + n_2 x^2 + \cdots + n_1 x^1$$

Non-Repeated Linear Factors

$$\frac{N(x)}{(x-a)G(x)} = \frac{A}{x-a} + \frac{F(x)}{G(x)}$$

$$A = \left[\frac{N(x)}{G(x)}\right]_{x=a}$$

$F(x)$ determined by methods discussed in the following sections.

Repeated Linear Factors

$$\frac{N(x)}{x^m G(x)} = \frac{A_0}{x^m} + \frac{A_1}{x^{m-1}} + \cdots + \frac{A_{m-1}}{x} + \frac{F(x)}{G(x)}$$

$$N(x) = n_o + n_1 x + n_2 x^2 + n_3 x^3 + \cdots$$

$$F(x) = f_0 + f_1 x + f_2 x^2 + \cdots,$$

$$G(x) = g_0 + g_1 x + g_2 x^2 + \cdots$$

$$A_0 = \frac{n_0}{g_0}, A_1 = \frac{n_1 - A_0 g_1}{g_0}$$

$$A_0 = \frac{n_2 - A_0 g_2 - A_1 g_1}{g_0}$$

General Terms

$$A_0 = \frac{n_0}{g_0}, A_k = \frac{1}{g_0}\left[n_k - \sum_{t=0}^{k-1} A_t g_k - t \right] k \geq 1$$

$$m^* = 1 \begin{cases} f_0 = n_1 - A_0 g_1 \\ f_1 = n_2 - A_0 g_2 \\ f_1 = n_{j+1} - A_0 g_{t+1} \end{cases}$$

$$m = 2 \begin{cases} f_0 = n_2 - A_0 g_2 - A_1 g_1 \\ f_1 = n_3 - A_0 g_3 - A_1 g_2 \\ f_1 = n_{j+2} - \left[A_0 g_{1+2} + A_1 g_1 + 1 \right] \end{cases}$$

$$m = 3 \begin{cases} f_0 = n_3 - A_0 g_3 - A_1 g_2 - A_2 g_1 \\ f_1 = n_3 - A_0 g_4 - A_1 g_3 - A_2 g_2 \\ f_1 = n_{j+3} - \left[A_0 g_{j+3} + A_1 g_{j+2} + A_2 g_{j+1} \right] \end{cases}$$

$$\text{any } m : f_1 = n_{m+1} - \sum_{i=0}^{m-1} A_1 g_{m+j-1}$$

$$\frac{N(x)}{(x-a)^m G(x)} = \frac{A_0}{(x-a)^m} + \frac{A_1}{(x-a)^{m-1}} + \cdots + \frac{A_{m-1}}{(x-a)} + \frac{F(x)}{G(x)}$$

Change to form $\dfrac{N'(y)}{y^m G'(y)}$ by substitution of $x = y + a$. Resolve into partial fractions in terms of y as described above. Then express in terms of x by substitution $y = x - a$.

Repeated Linear Factors

Alternative method of determining coefficients:

$$\frac{N(x)}{(x-a)^m G(x)} = \frac{A_0}{(x-a)^m} + \cdots + \frac{A_k}{(x-a)^{m-k}} + \cdots + \frac{A_{m-1}}{x-a} + \frac{F(x)}{G(x)}$$

$$A_k = \frac{1}{k!}\left\{D_x^k\left[\frac{N(x)}{G(x)}\right]\right\}_{x-G}$$

where D_x^k is the differentiating operator, and the derivative of zero order is defined as

$$D_x^0 u = u.$$

Factors of Higher Degree

Factors of higher degree have the corresponding numerators indicated.

$$\frac{N(x)}{\left(x^2 + h_1 x + h_0\right)G(x)} = \frac{a_1 x + a_0}{x^2 + h_1 x + h_0} + \frac{F(x)}{G(x)}$$

$$\frac{N(x)}{\left(x^2 + h_1 x + h_0\right)^2 G(x)} = \frac{a_1 x + a_0}{\left(x^2 + h_1 x + h_0\right)^2} + \frac{b_1 x + b_0}{\left(x^2 + h_1 x + h_0\right)} + \frac{F(x)}{G(x)}$$

$$\frac{N(x)}{\left(x^3 + h_2 x^2 + h_1 x + h_0\right)G(x)} = \frac{a_2 x^2 + a_1 x + a_0}{x^3 + h_2 x^2 + h_1 x + h_0} + \frac{F(x)}{G(x)} \quad \text{etc.}$$

   Problems of this type are determined first by solving for the coefficients due to linear factors as shown above, and then determining the remaining coefficients by the general methods given below.

GEOMETRY

Mensuration Formulas are used for measuring angles and distances in geometry. Examples are presented below.

Triangles

Let K = area, r = radius of the inscribed circle, R = radius of circumscribed circle.

Right Triangle

$A + B = C = 90°$

   $c^2 = a^2 + b^2$ (Pythagorean relations)

$$a = \sqrt{(c+b)(c-b)}$$

$$K = \frac{1}{2}ab$$

$$r = \frac{ab}{a+b+c}, \quad R = \frac{1}{2}c$$

$$h = \frac{ab}{c}, \quad m = \frac{b^2}{c}, \quad n = \frac{a^2}{c}$$

Equilateral Triangle

$$A = B = C = 60^0$$

$$K = \frac{1}{4}a^2\sqrt{3}$$

$$r = \frac{1}{6}a\sqrt{3}, \quad R = \frac{1}{3}a\sqrt{3}$$

$$h = \frac{1}{2}a\sqrt{3}$$

General Triangle

Let $s = \frac{1}{2}(a+b+c), h_c = $ length of altitude on side c, $t_c = $ length of bisector of angle C, $m_c = $ length of median to side c.

$$A + B + C = 180^0$$

$$c^2 = a^2 + b^2 - 2ab\cos C \quad \text{(law of cosines)}$$

$$K = \frac{1}{2}h_c c = \frac{1}{2}ab \sin C$$

$$= \frac{c^2 \sin A \sin B}{2\sin C}$$

$$= rs = \frac{abc}{4R}$$

$$= \sqrt{s(s-a)(s-b)(s-c)} \,\text{(Heron's formula)}$$

$$r = c\sin\frac{A}{2}\sin\frac{B}{2}\sec\frac{C}{2} = \frac{ab\sin C}{2s} = (s-c)\tan\frac{C}{2}$$

$$= \sqrt{\frac{(s-a)(s-b)(s-c)}{s}} = \frac{K}{s} = 4R\sin\frac{A}{2}\sin\frac{B}{2}\sin\frac{C}{2}$$

$$R = \frac{c}{2\sin C} = \frac{abc}{4\sqrt{s(s-a)(s-b)(s-c)}} = \frac{abc}{4K}$$

$$h_c = a\sin B = b\sin A = \frac{2K}{c}$$

$$t_c = \frac{2ab}{a+b}\cos\frac{C}{2} = \sqrt{ab\left\{1 - \frac{c^2}{(a+b)^2}\right\}}$$

$$m_c = \sqrt{\frac{a^2}{2} + \frac{b^2}{2} - \frac{c^2}{4}}$$

Menelaus' Theorem

A necessary and sufficient condition for points D, E, F on the respective side lines BC, CA, AB of a triangle ABC to be collinear is that

$$BD \cdot CE \cdot AF = -DC \cdot EA \cdot FB,$$

where all segments in the formula are directed segments.

Ceva's Theorem

A necessary and sufficient condition for AD, BE, CF, where D, E, F are points on the respective side lines BC, CA, AB of a triangle ABC, to be concurrent is that

$$BD \cdot CE \cdot AF = +DC \cdot EA \cdot FB,$$

where all segments in the formula are directed segments.

Quadrilaterals
Let K = area, p and q are diagonals.

Rectangle

$$A = B = C = D = 90°$$

$$K = ab, \quad p = \sqrt{a^2 + b^2}$$

Parallelogram

$$A = C, \quad B = D, \quad A + B = 180°$$

$$K = bh = ab\sin A = ab\sin B$$

$$h = a\sin A = a\sin B$$

$$p = \sqrt{a^2 + b^2 - 2ab\cos A}$$

$$q = \sqrt{a^2 + b^2 - 2ab\cos B} = \sqrt{a^2 + b^2 - 2ab\cos A}$$

Rhombus

$$p^2 + q^2 = 4a^2$$

$$K = \frac{1}{2}pq$$

Trapezoid

$$m = \frac{1}{2}(a+b)$$

$$K = \frac{1}{2}(a+b)h = mh$$

General Quadrilateral
Let

$$s = \frac{1}{2}(a+b+c+d).$$

$$K = \frac{1}{2}pq\sin\theta$$

$$= \frac{1}{4}\left(b^2 + d^2 - a^2 - c^2\right)\tan\theta$$

$$= \frac{1}{4}\sqrt{2p^2q^2 - \left(b^2 + d^2 - a^2 - c^2\right)^2}$$

(Bretschneider's formula)

$$= \sqrt{(s-a)(s-b)(s-c)(s-d) - abcd\cos^2\left(\frac{A+B}{2}\right)}$$

Theorem

The diagonals of a quadrilateral with consecutive sides a, b, c, d are perpendicular if and only if $a^2 + c^2 = b^2 + d^2$.

Regular Polygon of n Sides Each of Length b

$$\text{Area} = \frac{1}{4}nb^2 \cot\frac{\pi}{n} = \frac{1}{4}nb^2\frac{\cos(\pi/n)}{\sin(\pi/n)}$$

Perimeter $= nb$
Circle of Radius r
Area $= \pi r^2$
Perimeter $= 2\pi r$

Regular Polygon of n sides inscribed in a Circle of Radius r

$$\text{Area} = \frac{1}{2}nr^2\sin\frac{2\pi}{n} = \frac{1}{2}nr^2\sin\frac{360°}{n}$$

$$\text{Perimeter} = 2nr\sin\frac{\pi}{n} = 2nr\sin\frac{180°}{n}$$

Regular Polygon of n Sides Circumscribing a Circle of Radius r

$$\text{Area} = nr^2\tan\frac{\pi}{n} = nr^2\tan\frac{180°}{n}$$

$$\text{Perimeter} = 2nr\sin\frac{\pi}{n} = 2nr\tan\frac{180°}{n}$$

Cyclic Quadrilateral

Let R = radius of the circumscribed circle.

$$A + C = B + D = 180^0$$

$$K = \sqrt{(s-a)(s-b)(s-c)(s-d)} = \frac{\sqrt{(ac+bd)(ad+bc)(ab+cd)}}{4R}$$

$$p = \sqrt{\frac{(ac+bd)(ab+cd)}{ad+bc}}$$

$$q = \sqrt{\frac{(ac+bd)(ad+bc)}{ab+cd}}$$

$$R = \frac{1}{2}\sqrt{\frac{(ac+bd)(ad+bc)(ab+cd)}{(s-a)(s-b)(s-c)(s-d)}}$$

$$\sin\theta = \frac{2K}{ac+bd}$$

Prolemy's Theorem

A convex quadrilateral with consecutive sides a, b, c, d and diagonals p and q is cyclic if and only if $ac + bd = pq$.

Cyclic-Inscriptable Quadrilateral

Let r = radius of the inscribed circle,

R = radius of the circumscribed circle,

m = distance between the centers of the inscribed and the circumscribed circles.

$$A + C = B + D = 180°$$

$$a + c = b + d$$

$$K = \sqrt{abcd}$$

$$\frac{1}{(R-m)^2} + \frac{1}{(R+m)^2} = \frac{1}{r^2}$$

$$r = \frac{\sqrt{abcd}}{s}$$

$$R = \frac{1}{2}\sqrt{\frac{(ac+bd)(ad+bc)(ab+cd)}{abcd}}$$

Sector of Circle of Radius r

$$\text{Area} = \frac{1}{2}r^2\theta\left[\theta \text{ in radians}\right]$$

Arc length $s = r\theta$

Radius of Circle Inscribed in a Triangle of Sides a, b, c

$$r = \frac{\sqrt{s(s-a)(s-b)(s-b)}}{s}$$

where

$$s = \frac{1}{2}(a+b+c) = \text{semiperimeter}$$

Radius of Circle Circumscribing a Triangle of Sides a, b, c

$$R = \frac{abc}{4\sqrt{s(s-a)(s-b)(s-c)}}$$

where

$$s = \frac{1}{2}(a+b+c) = \text{semiperimeter}$$

Segment of Circle of Radius r

$$\text{Area of shaded part} = \frac{1}{2}r^2(\theta - \sin\theta)$$

Ellipse of Semi-Major Axis a and Semi-Minor Axis b
Area $= \pi ab$

$$\text{Perimeter} = 4a \int_0^{\pi/2} \sqrt{1 - k^2 \sin^2\theta}\ d\theta$$

$$= 2\pi \sqrt{\frac{1}{2}(a^2 + b^2)}\left[\text{approximately}\right]$$

where

$$k = \sqrt{a^2 - b^2}\ /\ a.$$

Segment of a Parabola

$$\text{Area} = \frac{2}{8}ab$$

$$\text{Arc length ABC} = \frac{1}{2}\sqrt{b^2 + 16a^2} + \frac{b^2}{8a}\ln\left(\frac{4a + \sqrt{b^2 + 16a^2}}{b}\right)$$

PLANAR AREAS BY APPROXIMATION

Divide the planar area K into n strips by equidistant parallel chords of lengths y_0, y_1, y_2, ..., y_n (where y_0 and/or y_n may be zero), and let h denote the common distance between the chords.

Then, approximately:

Trapezoidal Rule

$$K = h\left(\frac{1}{2}y_0 + y_1 + y_2 + \cdots + y_{n-1} + \frac{1}{2}y_n\right)$$

Durand's Rule

$$K = h\left(\frac{4}{10}y_0 + \frac{11}{10}y_1 + y_2 + y_3 + \cdots + y_{n-2} + \frac{11}{10}y_{n-1} + \frac{4}{10}y_n\right)$$

Simpson's Rule (*n* even)

$$K = \frac{1}{3}h\left(y_0 + 4y_1 + 2y_2 + 4y_3 + 2y_4 + \cdots + 2y_{n-2} + 4y_{n-1} + y_n\right)$$

Weddle's Rule (*n* = 6)

$$K = \frac{3}{10}h\left(y_0 + 5y_1 + y_2 + 6y_3 + y_4 + 5y_5 + y_6\right)$$

SOLIDS BOUNDED BY PLANES

In the following: S = lateral surface, T = total surface, V = volume.

Cube

Let a = length of each edge.

$$T = 6a^2, \text{diagonal of face} = a\sqrt{2}$$
$$V = a^3, \text{diagonal of cube} = a\sqrt{3}$$

Rectangular Parallelepiped (or Box)
Let a, b, c, be the lengths of its edges.

$$T = 2(ab + bc + ca), \quad V = abc$$
$$\text{diagonal} = \sqrt{a^2 + b^2 + c^2}$$

Prism

S = (perimeter of right section) × (lateral edge)
V = (area of right section) × (lateral edge)
= (area of base) × (altitude)

Truncated Triangular Prism

$$V = \left(\text{area of right section}\right) \times \frac{1}{3}\left(\text{sum of the three lateral edges}\right)$$

Pyramid

$$S \text{ of regular pyramid} = \frac{1}{2}(\text{perimeter of base}) \times (\text{slant height})$$

$$V = \frac{1}{3}(\text{area of base}) \times (\text{altitude})$$

Frustum of Pyramid
  Let B_1 = area of lower base, B_2 = area of upper base, h = altitude.

$$S \text{ of regular figure} = \frac{1}{2}(\text{sum of perimeters of base}) \times (\text{slant height})$$

$$V = \frac{1}{3}h(B_1 + B_2 + \sqrt{B_1 B_2})$$

Prismatoid

A prismatoid is a polyhedron having for bases two polygons in parallel planes, and for lateral faces triangles or trapezoids with one side lying in one base, and the opposite vertex or side lying in the other base, of the polyhedron. Let B_1 = area of lower base, M = area of midsection, B_2 = area of upper base, h = altitude.

$$V = \frac{1}{6}h(B_1 + 4M + B_2)(\text{the prismoidal formula})$$

Note: Since cubes, rectangular parallelepipeds, prisms, pyramids, and frustums of pyramids are all examples of prismatoids, the formula for the volume of a prismatoid subsumes most of the above volume formulae.

REGULAR POLYHEDRA

Let:

  v = number of vertices
  e = number of edges
  f = number of faces
  α = each dihedral angle
  a = length of each edge
  r = radius of the inscribed sphere
  R = radius of the circumscribed sphere
  A = area of each face
  T = total area
  V = volume.

$$v - e + f = 2$$

$$T = fA$$

$$V = \frac{1}{3}rfA = \frac{1}{3}rT$$

Name	Nature of Surface	T	V
Tetrahedron	4 equilateral triangles	1.73205 a^2	0.11785 a^3
Hexahedron (cube)	6 squares	6.00000 a^2	1.00000 a^3
Octahedron	8 equilateral triangles	3.46410 a^2	0.47140 a^3
Dodecahedron	12 regular pentagons	20.64573 a^2	7.66312 a^3
Icosahedron	20 equilateral triangles	8.66025 a^2	2.18169 a^2

Name	v	e	f	α	a	r
Tetrahedron	4	6	4	$70^\circ\ 32'$	1.633R	0.333R
Hexahedron	8	12	6	90°	1.155R	0.577R
Octahedron	6	12	8	$190^\circ\ 28'$	1.414R	0.577R
Dodecahedron	20	30	12	$116^\circ\ 34'$	0.714R	0.795R
Icosahedron	12	30	20	$138^\circ\ 11'$	1.051R	0.795R

Name	A	r	R	V
Tetrahedron	$\frac{1}{4}a^2\sqrt{3}$	$\frac{1}{12}a\sqrt{6}$	$\frac{1}{4}a\sqrt{6}$	$\frac{1}{12}a^2\sqrt{2}$
Hexahedron (cube)	a^2	$\frac{1}{2}a$	$\frac{1}{2}a\sqrt{3}$	a^3
Octahedron	$\frac{1}{4}a^2\sqrt{3}$	$\frac{1}{6}a\sqrt{6}$	$\frac{1}{2}a\sqrt{2}$	$\frac{1}{3}a^3\sqrt{2}$
Dodecahedron	$\frac{1}{4}a^2\sqrt{25+10\sqrt{5}}$	$\frac{1}{20}a\sqrt{250+110\sqrt{5}}$	$\frac{1}{4}a\left(\sqrt{15}+\sqrt{3}\right)$	$\frac{1}{4}a^3\left(15+7\sqrt{5}\right)$
Icosahedron	$\frac{1}{4}a^2\sqrt{3}$	$\frac{1}{12}a\sqrt{42+18\sqrt{5}}$	$\frac{1}{4}a\sqrt{10+2\sqrt{5}}$	$\frac{1}{12}a^3\left(3+\sqrt{5}\right)$

Sphere of Radius r

Volume $= \frac{3}{4}\pi r^3$

Surface area $= 4\pi r^2$

Right Circular Cylinder of Radius r and Height h

Volume $= \pi r^2 h$
Lateral surface area $= 2\pi rh$

Circular Cylinder of Radius r and Slant Height ℓ

Volume $= \pi r^2 h = \pi r^2 \ell \sin \theta$
Lateral surface area $= p\ell$

Cylinder of Cross-Sectional Area A and Slant Height ℓ
Volume $= Ah = A\ell \sin \theta$
Lateral Surface area $= p\ell$

Right Circular Cone of Radius r and Height h

Volume $\quad = \dfrac{1}{3}\pi r^2 h$
Lateral surface area $= \pi r \sqrt{r^2 + h^2} = \pi rl$

Spherical Cap of Radius r and Height h

Volume $\left(\text{shaded in figure}\right) = \dfrac{1}{3}\pi h^2 \left(3r - h\right)$
Surface area $= 2\pi rh$

Frustum of Right Circular Cone of Radii a, b and Height h

$$\text{Volume} = \frac{1}{3}\pi h\left(a^2 + ab + b^2\right)$$

$$\text{Lateral surface area} = \pi\left(a + b\right)\sqrt{h^2 + \left(b - a\right)^2}$$
$$= \pi\left(a + b\right)l$$

Zone and Segment of Two Bases

$S = 2\pi Rh = \pi Dh$

$$V = \frac{1}{6}\pi h\left(3a^2 + 3b^2 + h^2\right)$$

Lune
$S = 2R^3\theta, \quad \theta \text{ in radians}$

Spherical Sector

$$V = \frac{2}{3}\pi R^2 h = \frac{1}{6}\pi D^2 h$$

SPHERICAL TRIANGLE AND POLYGON

Let A, B, C be the angles, in radians, of the triangle; let θ = sum of angles, in radians, of a spherical polygon on n sides.

$$S = \left(A + B + C - \pi\right)R^2$$
$$S = \left[\theta - (n-2)\pi\right]R^2$$

Spheroids
Ellipsoid
Let a, b, c be the lengths of the semi-axes.

$$V = \frac{4}{3}\pi abc$$

Oblate Spheroid

An oblate spheroid is formed by the rotation of an ellipse about its minor axis. Let a and b be the major and minor semi-axes, respectively, and $\in$ the eccentricity, of the revolving ellipse.

$$S = 2\pi a^2 + \pi \frac{b^2}{\in} \log_e \frac{1+\in}{1-\in}$$

$$V = \frac{4}{3}\pi a^2 b$$

Prolate Spheroid

A prolate spheroid is formed by the rotation of an ellipse about its major axis. Let a and b be the major and minor semi-axes, respectively, and $\in$ the eccentricity, of the revolving ellipse.

$$S = 2\pi b^2 + 2\pi \frac{ab}{\in} \sin^{-1} \in$$

$$V = \frac{3}{4}\pi ab^2$$

Circular Torus

A circular torus is formed by the rotation of a circle about an axis in the plane of the circle and not cutting the circle. Let r be the radius of the revolving circle and let R be the distance of its center from the axis of rotation.

$$S = 4\pi^2 Rr$$
$$V = 2\pi^2 Rr^2$$

Formulas from Plane Analytic Geometry
Distance d between Two Points

$$P_1(x_1, y_1) \quad \text{and} \quad P_2(x_2, y_2)$$

$$d = \sqrt{(x_2 - x_1)^2 + (y_2 - y_1)^2}$$

Slope m of Line Joining Two Points

$$P_1(x_1, y_1) \quad \text{and} \quad P_2(x_2, y_2)$$

$$m = \frac{y_2 - y_1}{x_2 - x_1} = \tan\theta$$

EQUATION OF LINE JOINING TWO POINTS

$$P_1(x_1, y_1) \quad \text{and} \quad P_2(x_2, y_2)$$

$$\frac{y - y_1}{x - x_1} = \frac{y_2 - y_1}{x_2 - x_1} = m \quad \text{or} \quad y - y_1 = m(x - x_1)$$

$$y = mx + b$$

where $b = y_1 - mx_1 = \dfrac{x_2 y_1 - x_1 y_2}{x_2 - x_1}$ is the intercept on the y axis, i.e. the y intercept

Equation of Line in Terms of x Intercept a ≠ 0 and y intercept b ≠ 0

$$\frac{x}{a} + \frac{y}{b} = 1$$

Normal Form for Equation of Line

$$x\cos\alpha + y\sin\alpha = p$$

where p = perpendicular distance from origin O to line
and α = angle of inclination of perpendicular with positive x axis.

General Equation of Line

$$Ax + By + C = 0$$

Distance From Point (x_1, y_1) to Line $Ax + By + C = 0$

$$\frac{Ax_1 + By_1 + C}{\pm\sqrt{A^2 + B^2}}$$

where the sign is chosen so that the distance is nonnegative.

Angle ψ between Two Lines Having Slopes m_1 and m_2

$$\tan\psi = \frac{m_2 - m_1}{1 + m_1 m_2}$$

Lines are parallel or coincident if and only if $m_1 = m_2$
Lines are perpendicular if and only if $m_2 = -1/m_1$.

Area of Triangle with Vertices

At

$$\left(x_1, y_1\right), \quad \left(x_2, y_2\right), \quad \left(x_3, y_3\right)$$

$$\text{Area} = \pm\frac{1}{2}\begin{vmatrix} x_1 & y_1 & 1 \\ x_2 & y_2 & 1 \\ x_3 & y_3 & 1 \end{vmatrix}$$

$$= \pm\frac{1}{2}\left(x_1 y_2 + y_1 x_3 + y_3 x_2 - y_2 x_3 - y_1 x_2 - x_1 y_3\right)$$

where the sign is chosen so that the area is nonnegative.
If the area is zero the points all lie on a line.

Transformation of Coordinates Involving Pure Translation

$$\begin{cases} x = x' + x_0 \\ y = y' + y_0 \end{cases} \quad \text{or} \quad \begin{cases} x' = x + x_0 \\ y' = y + y_0 \end{cases}$$

Where x, y are old coordinates [i.e. coordinates relative to xy system], (x', y') are new coordinates [relative to $x'y'$ system] and $(x_0\ y_0)$ are the coordinates of the new origin O' relative to the old xy coordinate system.

Transformation of Coordinates Involving Pure Rotation

$$\begin{cases} x = x'\cos\alpha - y'\sin\alpha \\ y = x'\sin\alpha + y'\cos\alpha \end{cases} \quad \text{or} \quad \begin{cases} x' = x\cos\alpha + y\sin\alpha \\ y' = y\cos\alpha - x\sin\alpha \end{cases}$$

where the origins of the old [xy] and new [$x'y'$] coordinate systems are the same but the x' axis makes an angle α with the positive x axis.

Transformation of Coordinates Involving Translation and Rotation

$$\begin{cases} x = x'\cos\alpha - y'\sin\alpha + x_0 \\ y = x'\sin\alpha + y'\cos\alpha + y_0 \end{cases}$$

or $\begin{cases} x' = (x - x_0)\cos\alpha + (y - y_0)\sin\alpha \\ y' = (y - y_0)\cos\alpha - (x - x_0)\sin\alpha \end{cases}$

where the new origin O' of $x'y'$ coordinate system has coordinates (x_0, y_0) relative to the old xy coordinate system and the x' axis makes an angle α with the positive x axis.

Polar Coordinates (r, θ)

A point P can be located by rectangular coordinates (x, y) or polar coordinates (r, θ).
The transformation between these coordinates is

$$\begin{cases} x = r\cos\theta \\ y = r\sin\theta \end{cases} \quad \text{or} \quad \begin{cases} r = \sqrt{x^2 + y^2} \\ \theta = \tan^{-1}(y/x) \end{cases}$$

Plane Curves

$$\left(x^2 + y^2\right)^2 = ax^2 y$$

$$r = a\sin\theta\cos^2\theta$$

Catenary, Hyperbolic Cosine

$$y = \frac{a}{2}\left(e^{x/e} + e^{-x/e}\right) = a\cosh\frac{x}{a}$$

Cardioid

$$\left(x^2 + y^2 - ax\right)^2 = a^2\left(x^2 + y^2\right)$$

$$r = a(\cos\theta + 1)$$

or

$$r = a(\cos\theta - 1)$$

$$\left[P'A = AP = a\right]$$

Circle

$$x^2 + y^2 = a^2$$

$$r = a$$

Cassinian Curves

$$x^2 + y^2 = 2ax$$

$$r = 2a\cos\theta$$

$$x^2 + y^2 = ax + by$$

$$r = a\cos\theta + b\sin\theta$$

Cotangent Curve

$$y = \cot x$$

Cubical Parabola

$$y = ax^3, \quad a > 0$$

$$r^2 = \frac{1}{a}\sec^2\theta\tan\theta, \quad a > 0$$

Cosecant Curve

$$y = \csc x$$

Cosine Curve

$$y = \cos x$$

Ellipse

$$x^2/a^2 + y^2/b^2 = 1$$

$$\begin{cases} x = a\cos\phi \\ y = b\sin\phi \end{cases}$$

Gamma Function

$$\Gamma(n) = \int_0^\infty x^{n-1}e^{-x}dx \quad (n > 0)$$

$$\Gamma(n) = \frac{\Gamma(n+1)}{n}(0 > n \neq -1, -2, -3, \ldots)$$

Hyperbolic Functions

$$\sinh x = \frac{e^x - e^{-x}}{2} \quad \csc hx = \frac{2}{e^x - e^{-x}}$$

$$\cosh x = \frac{e^x - e^{-x}}{2} \quad \csc hx = \frac{2}{e^x - e^{-x}}$$

$$\tanh x = \frac{e^x - e^{-x}}{e^x + e^{-x}} \quad \coth x = \frac{e^x + e^{-x}}{e^x - e^{-x}}$$

Inverse Cosine Curve

$$y = \arccos x$$

Inverse Sine Curve

$$y = \arcsin x$$

Inverse Tangent Curve

$$y = \arctan x$$

Logarithmic Curve

$$y = \log_a x$$

Parabola

$$y = x^2$$

Cubical Parabola

$$y = x^3$$

Tangent Curve

$$y = \tan x$$

Ellipsoid

$$\frac{x^2}{a^2} + \frac{y^2}{b^2} + \frac{z^2}{c^2} = 1$$

Elliptic Cone

$$\frac{x^2}{a^2} + \frac{y^2}{b^2} - \frac{z^2}{c^2} = 0$$

Elliptic Cylinder

$$\frac{x^2}{a^2} + \frac{y^2}{b^2} = 1$$

Hyperboloid of One Sheet

$$\frac{x^2}{a^2} + \frac{y^2}{b^2} - \frac{z^2}{c^2} = 1$$

Elliptic Paraboloid

$$\frac{x^2}{a^2} + \frac{y^2}{b^2} = cz$$

Hyperboloid of Two Sheets

$$\frac{z^2}{c^2} - \frac{x^2}{a^2} - \frac{y^2}{b^2} = 1$$

Hyperbolic Paraboloid

$$\frac{x^2}{a^2} - \frac{y^2}{b^2} = cz$$

Sphere

$$x^2 + y^2 + z^2 = a^2$$

Distance d between Two Points

$$P_1\left(x_1, y_1, z_1\right) \quad \text{and} \quad P_2\left(x_2, y_2, z_2\right)$$

$$d = \sqrt{\left(x_2 - x_1\right)^2 + \left(y_2 - y_1\right)^2 + \left(z_2 - z_1\right)^2}$$

Equations of Line Joining $P_1(x_1, y_1, z_1)$ and $P_2(x_2, y_2, z_2)$ in Standard Form

$$\frac{x - x_1}{x_2 - x_1} = \frac{y - y_1}{y_2 - y_1} = \frac{z - z_1}{z_2 - z_1} \quad \text{or}$$

$$\frac{x - x_1}{l} = \frac{y - y_1}{m} = \frac{z - z_1}{n}$$

Equations of Line Joining $P_1(x_1, y_1, z_1)$ and $P_2(x_2, y_2, z_2)$ in Parametric Form

$$x = x_1 + lt, \; y = y_1 + mt, \; z = z_1 + nt$$

Angle ϕ between Two Lines with Direction Cosines l_1, m_1, n_1 and l_2, m_2, n_2

$$\cos\phi = l_1 l_2 + m_1 m_2 + n_1 n_2$$

General Equation of a Plane

$$Ax + By + Cz + D = 0$$

where A, B, C, D are constants

Equation of Plane Passing through Points

$$\left(x_1, y_1, z_1\right), \quad \left(x_2, y_2, z_2\right), \quad \left(x_3, y_3, z_3\right)$$

$$\begin{vmatrix} x - x_1 & y - y_1 & z - z_1 \\ x_2 - x_1 & y_2 - y_1 & z_2 - z_1 \\ x_3 - x_1 & y_3 - y_1 & z_3 - z_1 \end{vmatrix} = 0$$

or

$$\begin{vmatrix} y_2 - y_1 & z_2 - z_1 \\ y_3 - y_1 & z_3 - z_1 \end{vmatrix} \left(x - x_1\right) + \begin{vmatrix} z_2 - z_1 & x_2 - x_1 \\ z_3 - z_1 & x_3 - x_1 \end{vmatrix} \left(y - y_1\right)$$

$$+ \begin{vmatrix} x_2 - x_1 & y_2 - y_1 \\ x_3 - x_1 & y_3 - y_1 \end{vmatrix} \left(z - z_1\right) = 0$$

Equation of Plane in Intercept Form

$$\frac{x}{a} + \frac{y}{b} + \frac{z}{c} = 1$$

where a, b, c are the intercepts on the x, y, z axes respectively.

Equations of Line through (x_0, y_0, z_0) and Perpendicular to Plane

$$Ax + By + Cz + D = 0$$

$$\frac{x - x_0}{A} = \frac{y - y_0}{B} = \frac{z - z_0}{C}$$

or

$$x = x_0 + At, \, y = y_0 + Bt, \, z = z_0 + Ct$$

Distance from Point (x, y, z) to Plane $Ax + By + D = 0$

$$\frac{Ax_0 + By_0 + Cz_0 + D}{\pm\sqrt{A^2 + B^2 + C^2}}$$

where the sign is chosen so that the distance is nonnegative.

Normal form for Equation of Plane

$$x \cos \alpha + y \cos \beta + z \cos \gamma = p$$

where p = perpendicular distance from O to plane at P and α, β, γ are angles between OP and positive x, y, z axes.

Transformation of Coordinates Involving Pure Translation

$$\begin{cases} x = x' + x_0 \\ y = y' + y_0 \\ z = z' + z_0 \end{cases} \text{ or } \begin{cases} x' = x + x_0 \\ y' = y + y_0 \\ z' = z + z_0 \end{cases}$$

where (x, y, z) are old coordinates [i.e. coordinates relative to system], (x', y', z') are new coordinates [relative to (x', y', z') system and (x_0, y_0, z_0) are the coordinates of the new origin O' relative to the old xyz coordinate system.

Transformation of Coordinates Involving Pure Rotation

$$\begin{cases} x = l_1 x' + l_2 y' + l_3 z' \\ y = m_1 x' + m_2 y' + m_3 z' \\ z = n_1 x' + n_2 y' + n_3 z' \end{cases} \text{ or } \begin{cases} x' = l_1 x + m_1 y + n_1 z \\ y' = l_2 x + m_2 y + n_3 z \\ z' = l_3 x + m_3 y + n_3 z \end{cases}$$

where the origins of the xyz and x', y', z' systems are the same and l_1, m_1, n_1; l_2, m_2, n_2; l_3, m_3, n_3 are the direction cosines of the x', y', z' axes relative to the x, y, z axes respectively.

Transformation of Coordinates Involving Translation and Rotation

$$\begin{cases} x = l_1 x' + l_2 y' + l_3 z' + x_0 \\ y = m_1 x' + m_2 y' + m_3 z' + y_0 \\ z = n_1 x' + n_2 y' + n_3 z' + z_0 \end{cases}$$

or

$$\begin{cases} x' = l_1 (x - x_0) + m_1 (y - y_0) + n_1 (z - z_0) \\ y' = l_2 (x - x_0) + m_2 (y - y_0) + n_2 (z - z_0) \\ z' = l_3 (x - x_0) + m_3 (y - y_0) + n_3 (z - z_0) \end{cases}$$

where the origin O' of the $x'\,y'\,z'$ system has coordinates (x_0, y_0, z_0) relative to the xyz system and l_1, m_1, n_1; l_2, m_2, n_2; l_3, m_3, n_3 are the direction cosines of the $x'\,y'\,z'$ axes relative to the x, y, z axes respectively.

Cylindrical Coordinates (r, θ, z)

A point P can be located by cylindrical coordinates (r, θ, z) as well as rectangular coordinates (x, y, z). The transformation between these coordinates is

$$\begin{cases} x = r\cos\theta \\ y = r\sin\theta \\ z = z \end{cases} \quad \text{or} \quad \begin{cases} r = \sqrt{x^2 + y^2} \\ \theta = \tan^{-1}(y/x) \\ z = z \end{cases}$$

Spherical Coordinates (r, θ, ϕ)

A point P can be located by cylindrical coordinates (r, θ, ϕ) as well as rectangular coordinates (x, y, z). The transformation between these coordinates is

$$\begin{cases} x = r\cos\theta\cos\phi \\ y = r\sin\theta\sin\phi \\ z = r\cos\theta \end{cases} \quad \text{or} \quad \begin{cases} r = \sqrt{x^2 + y^2 + z^2} \\ \phi = \tan^{-1}(y/x) \\ \theta = \cos^{-1}\left(z/\sqrt{x^2 + y^2 + z^2}\right) \end{cases}$$

Equation of Sphere in Rectangular Coordinates

$$(x - x_0)^2 + (y - y_0)^2 + (z - z_0)^2 = R^2$$

where the sphere has cent (x_0, y_0, z_0) and radius R.

Equation of Sphere in Cylindrical Coordinates

$$r^2 - 2r_0 r\cos(\theta - \theta_0) + r_0^2 + (z - z_0)^2 = R^2$$

where the sphere has center (r_0, θ_0, z_0) in cylindrical coordinates and radius R. If the center is at the origin, the equation is

$$r^2 + z^2 = R^2$$

Equation of Sphere in Spherical Coordinates

$$r^2 + r_0^2 - 2r_0 r\sin\theta\sin\theta_0\cos(\phi - \phi_0) = R^2$$

where the sphere has center (r_0, θ_0, ϕ_0) in spherical coordinates and radius R. If the center is at the origin, the equation is

$r = R$

LOGARITHMIC IDENTITIES

$$\text{Ln}(z_1 z_2) = \text{Ln}\, z_1 + \text{Ln}\, z_2.$$

$$\ln(z_1 z_2) = \ln z_1 + \ln z_2 \quad (-\pi < \arg z_1 + \arg z_2 \leq \pi)$$

$$\text{Ln}\,\frac{z_1}{z_2} = \text{Ln}\,z_1 - \text{Ln}\,z_2$$

$$\ln\frac{z_1}{z_2} = \ln z_1 - \ln z_2 \quad (-\pi < \arg z_1 - \arg z_2 \le \pi)$$

$$\ln z^n = n\ln z \quad (n\,\text{integer})$$

$$\ln z^n = n\ln z \quad (n\,\text{integer}, \quad -\pi < n\arg z \le \pi)$$

Special Values

$$\ln 1 = 0$$

$$\ln 0 = -\infty$$

$$\ln(-1) = \pi i$$

$$\ln(\pm i) = \pm\frac{1}{2}\pi i$$

$\ln e = 1$, e is the real number such that

$$\int_1^e \frac{dt}{t} = 1$$

$$e = \lim_{n\to\infty}\left(1+\frac{1}{n}\right)^n = 2.71828\quad 18284\dots$$

Logarithms to General Base

$$\log_a z = \ln z\,/\ln a$$

$$\log_a z = \frac{\log_b z}{\log_b a}$$

$$\log_a b = \frac{1}{\log_b a}$$

$$\log_e z = \ln z$$

$$\log_{10} z = \ln z\,/\ln 10 = \log_{10} e\,\ln z = (.43429\quad 44819\dots)\ln z$$

$$\ln z = \ln 10\,\log_{10} z = (2.30258\quad 50929\dots)\log_{10} z$$

$$\left(\begin{array}{l}\log_e x = \ln x,\,\text{called natural, Napierian, or hyperbolic logarithms;}\\ \log_{10} x,\,\text{called common or Briggs logarithms.}\end{array}\right)$$

Series Expansions

$$\ln\left(1+z\right) = z - \frac{1}{2}z^2 + \frac{1}{3}z^3 - \ldots \left(\left|z\right| \le 1 \text{ and } z \ne -1\right)$$

$$\ln z = \left(\frac{z-1}{z}\right) + \frac{1}{2}\left(\frac{z-1}{z}\right)^2 + \frac{1}{3}\left(\frac{z-1}{z}\right)^3 + \ldots \left(\Re z \ge \frac{1}{2}\right)$$

$$\ln z = \left(z-1\right) - \frac{1}{2}\left(z-1\right)^2 + \frac{1}{3}\left(z-1\right)^3 - \ldots \left(\left|z-1\right| \le 1,\ z \ne 0\right)$$

$$\ln z = 2\left[\left(\frac{z-1}{z+1}\right) + \frac{1}{3}\left(\frac{z-1}{z+1}\right)^3 + \frac{1}{5}\left(\frac{z-1}{z+1}\right)^5 + \ldots\right]\left(\Re z \ge 0,\ z \ne 0\right)$$

$$\ln\left(\frac{z+1}{z-1}\right) = 2\left(\frac{1}{z} + \frac{1}{3z^3} + \frac{1}{5z^5} + \ldots\right)\left(\left|z\right| \ge 1, \qquad z \ne \pm 1\right)$$

$$\ln\left(z+a\right) = \ln a + 2\left[\left(\frac{z}{2a+z}\right) + \frac{1}{3}\left(\frac{z}{2a+z}\right)^3\right.$$

$$\left. + \frac{1}{5}\left(\frac{z}{2a+z}\right)^5 + \ldots\right]\left(a > 0, \qquad \Re z \ge -a \ne z\right)$$

Limiting Values

$$\lim_{x \to \infty} x^{-\alpha} \ln x = 0 \qquad (\alpha \text{ constant},\ \Re\alpha > 0)$$

$$\lim_{x \to 0} x^{-\alpha} \ln x = 0 \qquad (\alpha \text{ constant},\ \Re\alpha > 0)$$

$$\lim_{m \to \infty}\left(\sum_{k=1}^{m}\frac{1}{k} - \ln m\right) = \gamma \quad \left(\text{Euler's Constant}\right) = .57721\ 56649\ldots$$

Inequalities

$$\frac{x}{1+x} < \ln\left(1+x\right) < x \quad (x > -1,\quad x \ne 0)$$

$$x < -\ln\left(1-x\right) < \frac{x}{1+x} \qquad (x < 1,\quad x \ne 0)$$

$$\left|\ln\left(1-x\right)\right| < \frac{3x}{2} \quad (0 < x \le .5828)$$

$$\ln x \le x - 1 \qquad (x > 0)$$

$$\ln x \le n(x^{1/n}-1) \text{ for any positive } n \ (x>0)$$

$$\left|\ln(1-z)\right| \le -\ln(1-|z|) \quad (|z|<1)$$

Continued Fractions

$$\ln(1+z) = \frac{z}{1+}\frac{z}{2+}\frac{z}{3+}\frac{4z}{4+}\frac{4z}{5+}\frac{9z}{6+}\cdots$$

$$\left(z \text{ in the plance cut from } -1 \text{ } to -\infty\right)$$

$$\ln\left(\frac{1+z}{1-z}\right) = \frac{2z}{1-}\frac{z^2}{3-}\frac{4z^2}{5-}\frac{9z^2}{7-}\cdots$$

Polynomial Approximations

$$\frac{1}{\sqrt{10}} \le x \le \sqrt{10}$$

$$\log_{10} x = a_1 t + a_3 t^3 + \varepsilon(x), \quad t = (x-1)/(x+1)$$

$$\left|\varepsilon(x)\right| \le 6 \times 10^{-4}$$

$$a_1 = .86304 \qquad a_3 = .36415$$

$$\frac{1}{\sqrt{10}} \le x \le \sqrt{10}$$

$$\log_{10} x = a_1 t + a_3 t^3 + a_5 t^5 + a_7 t^7 + a_9 t^9 + \varepsilon(x)$$

$$t = (x-1)/(x+1)$$

$$\left|\varepsilon(x)\right| \le 10^{-7}$$

$$a_1 = .86859\ 1718$$
$$a_3 = .28933\ 5524$$
$$a_5 = .17752\ 2071$$
$$a_7 = .09437\ 6476$$
$$a_9 = .19133\ 7714$$

$$0 \le x \le 1$$

$$\ln(1+x) = a_1 x + a_2 x^2 + a_3 x^3 + a_4 x^4 + a_5 x^5 + \varepsilon(x)$$

$$\left|\varepsilon(x)\right| \le 1 \times 10^{-5}$$

$$a_1 = .99949\ 556$$
$$a_2 = .49190\ 896$$
$$a_3 = .28947\ 478$$
$$a_4 = .13606\ 275$$
$$a_5 = .03215\ 845$$

$$0 \le x \le 1$$
$$\ln(1+x) = a_1 x + a_2 x^2 + a_3 x^3 + a_4 x^4 + a_5 x^5 + a_6 x^6 + a_7 x^7 + a_8 x^8 + \varepsilon(x)$$
$$|\varepsilon(x)| \le 3X10^{-8}$$

$$a_1 = \ \ \ .99999\ \ \ 64239$$
$$a_2 = -.49987\ \ \ 41238$$
$$a_3 = \ \ \ .33179\ \ \ 90258$$
$$a_4 = -.24073\ \ \ 38084$$
$$a_5 = \ \ \ .16765\ \ \ 40711$$
$$a_6 = -.09532\ \ \ 93897$$
$$a_7 = \ \ \ .03608\ \ \ 84937$$
$$a_8 = -.00645\ \ \ 35442$$

Exponential Function Series Expansion

$$e^z = \exp z = 1 + \frac{z}{1!} + \frac{z^2}{2!} + \frac{z^3}{3!} + \ldots \quad (z = x + iy)$$

Fundamental Properties

$$\mathrm{Ln}\left(\exp z\right) = z + 2k\pi i \quad (k\ \text{any integer})$$
$$\ln\left(\exp z\right) = z \quad (-\pi < \oint z \le \pi)$$
$$\exp(\ln z) = \exp(\mathrm{Ln}z) = z$$
$$\frac{d}{dz}\exp z = \exp z$$

Definition of General Powers

If $N = a^z$, then $z = \log_a N$

$$a^z = \exp\left(z \ln a\right)$$

If $a = |a| \exp(i \arg a) \quad (-\pi < \arg a \le \pi)$

$$\left|a^z\right| = |a| \, xe^{-\mathrm{varg}\, a}$$

$\arg(a^z) = y \ln |a| + x \arg a$

  Ln $a^z = z \ln a$ for one of the values of Ln a^z

  $\ln a^x = x \ln a$ (a real and positive)

$$\left|e^z\right| = e^x$$

$$\arg\left(e^z\right) = y$$

$$a^{z_1} a^{z_2} = a^{z_1 + z_2}$$

$$a^z b^z = (ab)^z \quad (-\pi < \arg a + \arg b \le \pi)$$

Logarithmic and Exponential Functions.

Periodic Property

$$e^{z + 2\pi k i} = e^z \left(k \text{ any integer}\right)$$

$$e^x < \frac{1}{1 - x} \, (x < 1)$$

$$\frac{1}{1-x} < \left(1 - e^{-x}\right) < x \quad (x > -1)$$

$$x < \left(e^x - 1\right) < \frac{1}{1-x} \quad (x < 1)$$

$$1 + x > e^{\frac{x}{1+x}} \quad (x > -1)$$

$$e^x > 1 + \frac{x^n}{n!} \quad (n > 0, \quad x > 0)$$

$$e^x > \left(1 + \frac{x}{y}\right)^y > \frac{xy}{e^{x+y}} \quad (x > 0, \quad y > 0)$$

$$e^{-x} < 1 - \frac{x}{2} \quad (0 < x \le 1.5936)$$

$$\frac{1}{4}|z| < \left|e^z - 1\right| < \frac{7}{4}|z| \quad (0 < |z| < 1)$$

$$\left|e^z - 1\right| \le e|z| - 1 \le |z| e^{|z|} \quad (\text{all } z)$$

$$e^{2a \arctan \frac{1}{2}} = 1 + \frac{2a}{z-a+} \frac{a^2 + 1}{3z+} \frac{a^2 + 4}{5z+} \frac{a^2 + 9}{7z+} \cdots$$

Polynomial Approximations

$$0 \le x \le \ln 2 = .693\ldots$$
$$e^{-x} = 1 + a_1 x + a_2 x^2 + \varepsilon(x)$$
$$\left|\varepsilon(x)\right| \le 3 \times 10^{-3}$$

$a_1 = -.9664$
$a_2 = .3536$

$$0 \le x \le \ln 2$$
$$e^{-x} = 1 + a_1 x + a_2 x^2 + a_3 x^3 + a_4 x^4 + \varepsilon(x)$$
$$\left|\varepsilon(x)\right| \le 3 \times 10^{-5}$$

$a_1 = -.99986 \quad 84$
$a_1 = .49829 \quad 26$
$a_3 = -.15953 \quad 32$
$a_4 = .02936 \quad 41$

$$0 \le x \le \ln 2$$
$$e^{-x} = 1 + a_1 x + a_2 x^2 + a_3 x^4 + a_5 x^5 + a_6 x^6 + a_7 x^7 + \varepsilon(x)$$
$$\left|\varepsilon(x)\right| \le 2 \times 10^{-10}$$

$a_1 = -.99999 \quad 99995$
$a_2 = .49999 \quad 99206$
$a_3 = -.16666 \quad 53019$
$a_4 = .04165 \quad 73475$
$a_5 = -.00830 \quad 13598$
$a_6 = .00132 \quad 98820$
$a_7 = -.00014 \quad 13161$

$$0 \le x \le 1$$
$$10^x = \left(1 + a_1 x + a_2 x^2 + a_3 x^3 + a_4 x^4\right)^2 + \varepsilon(x)$$
$$\left|\varepsilon(x)\right| \le 7 \times 10^{-4}$$

$a_1 = 1.14991$ 96
$a_2 = .67743$ 23
$a_3 = .20800$ 30
$a_4 = .12680$ 89

$$0 \leq x \leq$$

$$10^x = \left(1 + a_1 x + a_2 x^2 + a_3 x^3 + a_4 x^4 + a_5 x^5 + a_6 x^6 + a_7 x^7\right)^2 + \varepsilon(x)$$

$$\left|\varepsilon(x)\right| \leq 5 \times 10^{-8}$$

$a_1 = 1.15129$ 277603
$a_2 = .66273$ 088429
$a_3 = .25439$ 357484
$a_4 = .07295$ 173666
$a_5 = .01742$ 111988
$a_6 = .00255$ 491796
$a_7 = .00093$ 264267

$$\text{Surface area of cylinder} = 2\pi rh + 2\pi r^2$$

$$\text{Volume of cylinder} = \pi r^2 h$$

Surface area of a cone $= \pi r^2 + \pi rs$

$$\text{Volume of a cone} = \frac{\pi r^2 h}{3}$$

$$\text{Volume of a pyramid} = \frac{Bh}{3}$$

$$\left(B = \text{area of base}\right)$$

Slopes

$$\text{Equation of a straight line:} \quad y - y_1 = m\left(x - x_1\right)$$

$$\text{Where } m = \text{slope} = \frac{\text{rise}}{\text{run}}$$

$$= \frac{\Delta y}{\Delta x} = \frac{y_2 - y_1}{x_2 - x_1}$$

or

$$y = mx + b$$

where m = slope, b = y-intercept

TRIGONOMETRIC RATIOS

$$\tan\theta = \frac{\sin\theta}{\cos\theta}$$

$$\sin^2\theta + \cos^2\theta = 1$$

$$1 + \tan^2\theta = \sec^2\theta$$

$$1 + \cot^2\theta = \csc^2\theta$$

$$\cos^2\theta - \sin^2\theta = \cos 2\theta$$

$$\sin 45^o = \frac{1}{\sqrt{2}}$$

$$\cos 45^o = \frac{1}{\sqrt{2}}$$

$$\tan 45^o = 1$$

$$\sin(A+B) = \sin A \cos B + \cos A \sin B$$

$$\sin(A-B) = \sin A \cos B - \cos A \sin B$$

$$\cos(A+B) = \cos A \cos B - \sin A \sin B$$

$$\cos(A-B) = \cos A \cos B + \sin A \sin B$$

$$\tan(A+B) = \frac{\tan A + \tan B}{1 - \tan A \tan B}$$

$$\tan(A-B) = \frac{\tan A - \tan B}{1 + \tan A \tan B}$$

$$\sin\theta = \frac{y}{r}\left(\text{opposite / hypotenuse}\right) = 1/\csc\theta$$

$$\cos\theta = \frac{x}{r}\left(\text{adjacent / hypotenuse}\right) = 1/\sec\theta$$

$$\tan\theta = \frac{y}{x}\left(\text{opposite / adjacent}\right) = 1/\cot\theta$$

$$\sin 30^o = \frac{1}{2} \quad \sin 60^o = \frac{\sqrt{3}}{2}$$

$$\cos 30° = \frac{\sqrt{3}}{2} \quad \cos 60° = \frac{1}{2}$$

$$\tan 30° = \frac{1}{\sqrt{3}} \quad \tan 60° = \sqrt{3}$$

Sine Law

$$\frac{a}{\sin A} = \frac{b}{\sin B} = \frac{c}{\sin C}$$

Cosine Law

$$a^2 = b^2 + c^2 - 2bc \cos A$$

$$b^2 = a^2 + c^2 - 2ac \cos B$$

$$c^2 = a^2 + b^2 - 2ab \cos C$$

$$\theta = 1\,\text{radian}$$

$$2\pi \,\text{radians} = 360°$$

Expansions

$$a(b + c) = ab + ac$$

$$(a + b)^2 = a^2 + 2ab + b^2$$

$$(a - b)^2 = a^2 - 2ab + b^2$$

$$(a + b)(c + d) = ac + ad + bc + bd$$

$$(a + b)^3 = a^3 + 3a^2b + 3ab^2 + b^3$$

$$(a - b)^3 = a^3 - 3a^2b + 3ab^2 - b^3$$

Factoring

$$a^2 - b^2 = (a + b)(a - b)$$

$$a^2 + 2ab + b^2 = (a + b)^2$$

$$a^3 + b^3 = (a + b)(a^2 - ab + b^2)$$

$$a^3b - ab = ab(a + 1)(a - 1)$$

$$a^2 - 2ab + b^2 = (a-b)^2$$

$$a^3 - b^3 = (a-b)(a^2 + ab + b^2)$$

Roots of Quadratic

The solution for a quadratic equation $ax^2 + bx + c = 0$

$$x = \frac{-b \pm \sqrt{b^2 - 4ac}}{2a}$$

Law of exponents

$$a^r . a^s = a^{r+s}$$

$$\frac{a^p a^q}{a^r} = a^{p+q-r}$$

$$\frac{a^r}{a^s} = a^{r-s}$$

$$\left(a^r\right)^s = a^{rs}$$

$$\left(ab\right)^r = a^r b^r$$

$$\left(\frac{a}{b}\right)^r = \frac{a^r}{b^r} \left(b \neq 0\right)$$

$$a^0 = 1 \left(a \neq 0\right)$$

$$a^{-r} = \frac{1}{a^r} \left(a \neq 0\right)$$

$$a^{\frac{r}{s}} = \sqrt[s]{a^r} \quad a^{\frac{1}{2}} = \sqrt{a} \quad a^{\frac{1}{3}} = \sqrt[3]{a}$$

Logarithms

Example:

$$\text{Log}(xy) = \text{Log } x + \text{Log } y \quad \text{Log}\left(\frac{x}{y}\right) = \text{Log } x - \text{Log } y$$

$$\text{Log } x^r = r\,\text{Log } x$$

$$\text{Log } x = n \leftrightarrow x = 10^n \left(\text{Common log}\right) \quad \pi \simeq 3.14159265$$

$$\log_a x = n \leftrightarrow x = a^n \left(\text{Log to the base } a\right) \quad e \simeq 2.71828183$$

$$\text{Ln } x = n \leftrightarrow x = e^n \quad \left(\text{Natural Log}\right)$$

SIX SIMPLE MACHINES FOR MATERIALS HANDLING

Material handling design and implementation constitute one of the basic functions in the practice of industrial engineering. Calculations related to the six simple machines are useful for assessing mechanical advantage for material handing purposes. The mechanical advantage is the ratio of the force of resistance to the force of effort:

$$MA = \frac{F_R}{F_E}$$

where

> MA = mechanical advantage
> F_R = force of resistance (N)
> F_E = force of effort (N)

MACHINE 1: THE LEVER

A lever consists of a rigid bar that is free to turn on a pivot, which is called a fulcrum
The law of simple machines as applied to levers is

$$F_R \cdot L_R = F_E \cdot L_E$$

MACHINE 2: WHEEL AND AXIE

A wheel and axle consist of a large wheel attached to an axle so that both turn together:

$$F_R \cdot r_R \cdot = F_E \cdot r_E$$

where

> F_R = force of resistance (N)
> F_E = force of effort (N)
> r_R = radius of resistance wheel (m)
> r_E = radius of effort wheel (m)

The mechanical advantage is

$$MA_{\text{wheel and axle}} = \frac{r_E}{r_R}$$

MACHINE 3: THE PULLEY

If a pulley is fastened to a fixed object, it is called a fixed pulley. If the pulley is fastened to the resistance to be moved, it is called a moveable pulley. When one continuous cord is used, the ratio reduces according to the number of strands holding the resistance in the pulley system.

The effort force equals the tension in each supporting stand. The mechanical advantage of the pulley is given by formula:

$$MA_{\text{pulley}} = \frac{F_R}{F_E} = \frac{nT}{T} = n$$

where

T = tension in each supporting strand
N = number of strands holding the resistance
F_R = force of resistance (N)
F_E = force of effort (N)

MACHINE 4: THE INCLINED PLANE

An inclined plane is a surface set at an angle from the horizontal and used to raise objects that are too heavy to lift vertically:

The mechanical advantage of an inclined plane is

$$MA_{\text{inclined plane}} = \frac{F_R}{F_E} = \frac{1}{h}$$

where

F_R = force of resistance (N)
R_E = force of effort (N)
1 = length of plane (m)
h = height of plane (m)

MACHINE 5: THE WEDGE

The wedge is a modification of the inclined plane. The mechanical advantage of a wedge can be found by dividing the length of either slope by the thickness of the longer end.

As with the inclined plane, the mechanical advantage gained by using a wedge requires a corresponding increase in distance.

The mechanical advantage is:

$$MA = \frac{s}{T}$$

where

MA = mechanical advantage
s = length of either slope (m)
T = thickness of the longer end (m)

MACHINE 6: THE SCREW

A screw is an inclined plane wrapped around a circle. From the law of machines,

$$F_R \cdot h = F_E \cdot U_E$$

However, for advancing a screw with a screwdriver, the mechanical advantage is:

$$MA_{\text{screw}} = \frac{F_R}{F_E} = \frac{U_E}{h}$$

where

F_R = force of resistance (N)
F_E = effort force (N)
h = pitch of screw
U_E = circumference of the handle of the screw

MECHANICS: KINEMATICS

SCALARS AND VECTORS

The mathematical quantities that are used to describe the motion of objects can be divided into two categories: Scalars and vectors.

a) Scalars
   Scalars are quantities that can be fully described by a magnitude alone.
b) Vectors
   Vectors are quantities that can be fully described by both a magnitude and direction.

DISTANCE AND DISPLACEMENT

a) Distance
   Distance is a scalar quantity that refers to how far an object has gone during its motions.
b) Displacement
   Displacement is the change in position of the object. It is a vector that includes the magnitude as a distance, such as five miles, and a direction, such as north.

ACCELERATION

Acceleration is the change in velocity per unit of time. Acceleration is a vector quality.

Speed and Velocity

a) Speed
   The distance traveled per unit of time is called the speed, for example 35 miles per hour. Speed is a scalar quantity.
b) Velocity
   The quantity that combines both the speed of an object and its direction of motion is called velocity. Velocity is a vector quantity.

Frequency

Frequency is the number of complete vibrations per unit time in simple harmonic or sinusoidal motion.

Period

Period is the time required for one full cycle. It is the reciprocal of the frequency.

Angular Displacement

Angular displacement is the rotational angle through which any point on a rotating body moves.

Angular Velocity

Angular velocity is the ratio of angular displacement to time.

Angular Acceleration

Angular acceleration is the ratio of angular velocity with respect to time.

Rotational Speed

Rotational speed is the number of revolutions (a revolution is one complete rotation of a body) per unit of time.

Uniform Linear Motion

A path is a straight time. The total distance traveled corresponds with the rectangular area in the diagram v – t.

a) Distance:

$$S = Vt$$

b) Speed:

$$V = \frac{S}{t}$$

where

s = distance (m)
v = speed (m/s)
t = time (s)

Uniform Accelerated Linear Motion

1. If $V_0 > 0$; $a > 0$, then
   a. Distance:

$$s = v_0 t + \frac{at^2}{2}$$

   b. Speed:

$$v = v_0 + at$$

where
s = distance (m)
v = speed (m/s)
t = time (s)
v_0 = initial speed (m/s)
a = acceleration (m/s^2)

2. If $v_0 = 0$; $a > 0$, then
   a. Distance:

$$s = \frac{at^2}{2}$$

The shaded areas in diagram v–t represent the distance s traveled during the time period t.

b. Speed:

$$v = a \cdot t$$

where
s = distance (m)
v = speed (m/s)
v_0 = initial speed (m/s)
a = acceleration (m/s^2)

ROTATIONAL MOTION

Rotational motion occurs when the body itself is spinning. The path is a circle about the axis.

a. Distance:

$$s = I\varphi$$

b. Velocity:

$$v = I\omega$$

c. Tangential acceleration:

$$a_t = r \cdot \alpha$$

d. Centripetal acceleration:

$$a_n = \omega^2 r = \frac{v^2}{r}$$

where
$\hat{\varphi}$ = angle determined by s and r (rad)
ω = angular velocity (s^{-1})
α = angular acceleration $(1/s^2)$
a_t = tangential acceleration $(1/s^2)$
a_n = centripetal acceleration $(1/s^2)$

Distance s, velocity v, and tangential acceleration a_t are proportional to radius r.

Uniform Rotation and a Fixed Axis

$$\omega_0 = \text{constant}; \alpha = 0,$$

a) Angle of rotations:

$$\varphi = \omega \cdot t$$

b) Angular velocity:

$$\omega = \frac{\varphi}{t}$$

where
φ = angle of rotation (rad)
ω = angular velocity (s^{-1})
α = angular acceleration $(1/s^2)$
ω_0 = initial angular speed (s^{-1})

The shade area in the diagram $\omega - t$ represents the angle of rotation $\varphi = 2\pi n$ covered during time period t.

Uniform Accelerated Rotation about a Fixed Axis

1. If $\omega_0 > 0; \alpha > 0$, then
   a. Angle of rotation:

$$\varphi = \frac{1}{2}(\omega_0 + \omega) = \omega_0 t + \frac{1}{2}\alpha t^2$$

   b. Angular velocity:

$$\omega = \omega_0 + \alpha t = \sqrt{\omega_0^2 + 2\alpha\varphi}$$

$$\omega_0 = \omega - \alpha t = \sqrt{\omega^2 - 2\alpha\varphi}$$

   c. Angular acceleration:

$$\alpha = \frac{\omega - \omega_0}{t} = \frac{\omega^2 - \omega_0^2}{2\varphi}$$

d. Time:

$$t = \frac{\omega - \omega_0}{\alpha} = \frac{2\varphi}{\omega_0 - \omega}$$

2. If $\omega_0 = 0$; $a = $ constant, then
a. Angle of rotations:

$$\varphi = \frac{\omega \cdot t}{2} = \frac{a \cdot t}{2} = \frac{\omega^2}{2a}$$

b. Angular velocity:

$$\omega = \sqrt{2a\varphi} = \frac{2\varphi}{t} = a \cdot t; \omega_0 = 0$$

c. Angular acceleration:

$$a = \frac{\omega}{t} = \frac{2\varphi}{t^2} = \frac{\omega^2}{2\varphi}$$

d. Time:

$$t = \sqrt{\frac{2\varphi}{a}} = \frac{\omega}{a} = \frac{2\varphi}{\omega}$$

SIMPLE HARMONIC MOTION

Simple harmonic motion occurs when an object moves repeatedly over the same path in equal time intervals.

The maximum deflection from the position of rest is called "amplitude."

A mass on a spring is an example of an object in simple harmonic motion. The motion is sinusoidal in time and demonstrates a single frequency.

a) Displacement:

$$s = A \sin(\omega \cdot t + \varphi_0)$$

b) Velocity:

$$v = A\omega \, \cos(\omega \cdot t + \varphi_0)$$

c) Angular acceleration:

$$a = \| A\alpha\omega^2 \sin(\omega \cdot t + \varphi_0)$$

where

s = displacement
A = amplitude
φ_0 = angular position at time $t = 0$
φ = angular position at time t
T = period

Pendulum

A pendulum consists of an object suspended so that it swings freely back and forth about a pivot.

a) Period:

$$T = 2\pi\sqrt{\frac{1}{g}}$$

where

T = period (s)
l = length of pendulum (m)
g = 9.81 (m/s²) or 32.2 (ft/s²)

Free Fall

A free-falling object is an object that is falling due to the sole influence of gravity.

a) Initial speed:

$$v_0 = 0$$

b) Distance:

$$h = -\frac{gt^2}{2} = -\frac{vt}{2} = -\frac{v^2}{2g}$$

c) Speed:

$$v = +gt = -\frac{2h}{t} = \sqrt{-2gh}$$

d) Time:

$$t = +\frac{v}{g} = -\frac{2h}{v} = \sqrt{-\frac{2h}{g}}$$

Vertical Project

a) Initial speed:

$$v_0 > 0, (\text{upwards}); \quad v_0 < 0, (\text{downwards})$$

b) Distance:

$$h = v_0 t - \frac{gt^2}{2} = (v_0 + v)\frac{t}{2}; \quad h_{max} = \frac{v_0^2}{2g}$$

c) Time:

$$t = \frac{v_0 - v}{g} = \frac{2h}{v_0 + v}; \quad t_{hmax} = \frac{v_0}{g}$$

where

v = velocity (m/s)
h = distance (m)
g = acceleration due to gravity (m/s²)

Angled Projections
Upwards $(\alpha > 0)$; downwards $(\alpha < 0)$

1) Distance:

$$s = v_0 \cdot t \cos \alpha$$

2) Altitude:

$$h = v_0 t \sin \alpha - \frac{g \cdot t^2}{2} = s \tan \alpha - \frac{g \cdot s^2}{2v_0^2 \cos \alpha}$$

$$h_{max} = \frac{v_0^2 \sin^2 \alpha}{2g}$$

3) Velocity:

$$v = \sqrt{v_0^2 - 2gh} = \sqrt{v_0^2 + g^2 t^2 - 2gv_0 t \sin \alpha} \tag{11.1}$$

4) Time:

$$t_{hmax} = \frac{v_0 \sin \alpha}{g}; \quad t_{s1} = \frac{2v_0 \sin \alpha}{g}$$

Horizontal Projection: ($\alpha = 0$)

a) Distance:

$$s = v_0 t = v_0 \sqrt{\frac{2h}{g}}$$

b) Altitude:

$$h = -\frac{gt^2}{2}$$

c) Trajectory velocity:

$$v = \sqrt{v_0^2 + g^2 t^2}$$

where

v_0 = initial velocity (m/s)
v = trajectory velocity (m/s)
s = distance (m)
h = height (m)

Sliding Motion on an Inclined Plane

1) If excluding friction ($\mu = 0$), then
   a. Velocity:

$$v = at = \frac{2s}{t} = \sqrt{2as}$$

   b. Distance:

$$s = \frac{at^2}{2} = \frac{vt}{2} = \frac{v^2}{2a}$$

c. Acceleration:

$$a = g \sin \alpha$$

2) If including friction ($\mu > 0$) then
   a. Velocity:

$$v = at = \frac{2s}{t} = \sqrt{2as}$$

b. Distance:

$$s = \frac{at^2}{2} = \frac{vt}{2} = \frac{v^2}{2a}$$

c. Accelerations:

$$s = \frac{at^2}{2} = \frac{vt}{2} = \frac{v^2}{2a}$$

where

μ = coefficient of sliding friction
g = acceleration due to gravity
$g = 9.81$ (m/s^2)
v_0 = initial velocity (m/s)
v = trajectory velocity (m/s)
s = distance (m)
a = acceleration (m/s^2)
α = inclined angle

Rolling Motion on an Inclined Plane

1) If excluding friction ($f = 0$), then
   a. Velocity:

$$v = at = \frac{2s}{t} = \sqrt{2as}$$

b. Acceleration:

$$a = \frac{gr^2}{I^2 + k^2} \sin \alpha$$

c. Distance:

$$s = \frac{at^2}{2} = \frac{vt}{2} = \frac{v^2}{2a}$$

d. Tilting angle:

$$\tan \alpha = \mu_0 \frac{r^2 + k^2}{k^2}$$

2) If including friction ($f > 0$), then
   a. Distance:

$$s = \frac{at^2}{2} = \frac{vt}{2} = \frac{v^2}{2a}$$

b. Velocity:

$$v = at = \frac{2s}{t} = \sqrt{2as}$$

c. Accelerations:

$$a = gr^2 \frac{\sin \alpha - (f/r)\cos \alpha}{I^2 + k^2}$$

d. Tilting angle:

$$\tan \alpha_{min} = \frac{f}{r}; \quad \tan \alpha_{max} = \mu_0 \frac{r^2 + k^2 - fr}{k^2}$$

The value of k can be the calculated by the formulas below:

Ball	Solid Cylinder	Pipe with low wall thickness
$k^2 = \dfrac{2r^2}{5}$	$k^2 = \dfrac{r^2}{2}$	$k^2 = \dfrac{r_i^2 + r_0^2}{2} \approx r^2$

where

s = distance (m)
v = velocity (m/s)
a = acceleration (m/s^2)

α = tilting angle (0)

f = lever arm of rolling resistance (m)
k = radius of gyration (m)
μ_0 = coefficient of static friction
g = acceleration due to gravity (m/s²)

MECHANICS: DYNAMICS

Newton's First Law of Motion

Newton's First Law or the Law of Inertia:

An object that is in motion continues in motion with the same velocity at constant speed and in a straight line, and an object at rest continues at rest unless an unbalanced (outside) force acts upon it.

Newton's Second Law

The second law of motion, called the Law of Accelerations. The total force acting on an object equals the mass of the object times its acceleration.

In equation form, this law is

$$F = ma$$

where

F = total force (N)
m = mass (kg)
a = acceleration (m/s^2)

Newton's Third Law

The Third Law of Motion, called the Law of Action and Reaction, can be stated as follows:

For every force applied by object A to object B (action), there is a force exerted by object B on object A (the reaction) which has the same magnitude but is opposite in direction.

In equation form this law is

$$F_B = -F_A$$

where

F_B = force of action (N)
F_A = force of reaction (N)

Momentum of Force

The momentum can be defined as mass in motion. Momentum is a vector quantity; in other words, the direction is important:

$$p = mv$$

Impulse of Force

The impulse of a force is equal to the change in momentum that the force causes in an object:

$$I = Ft$$

where

p = momentum (N s)
m = mass of object (kg)
v = velocity of object (m/s)
I = impulse of force (N s)
F = force (N)
t = time (s)

Law of Conservation of Momentum

One of the most powerful laws in physics is the law of momentum conservation, which can be stated as follows:

In the absence of external forces, the total momentum of the system is constant.

If two objects of mass m_1 and mass m_2, having velocity v_1, and v_2 collide and then separate with velocity v'_1 and v'_2, the equation for the conservation of momentum is

$$m_1v_1 + m_2v_2 = m_1v_1 + m_2v_2$$

Friction

Friction is a force that always acts parallel to the surface in contact and opposite to the direction of motion. Starting friction is greater than moving friction. Friction increases as the force between the surfaces increases.

The characteristics of friction can be described by the following equation:

$$F_f = \mu F_n$$

where

F_f = frictional force (N)
F_n = normal force (N)
μ = coefficient of friction ($\mu = \tan \alpha$)

General Law of Gravity

Gravity is a force that attracts bodies of matter toward each other. Gravity is the attraction between any two objects that have mass.

The general formula for gravity is

$$F = \Gamma \frac{m_A m_B}{r^2}$$

where

m_A, m_B = mass of objects A and B (kg)
F = magnitude of attractive force between objects A and B (N)
r = distance between object A and B (m)
Γ = gravitational constant ($N\ m^2/kg^2$)
$\Gamma = 6.67 \times 10^{-11}\ N\ m^2/kg^2$

Gravitational Force

The force of gravity is given by the equation

$$F_G = g\frac{R_e^2 m}{\left(R_e + h\right)^2}$$

On the earth surface, $h = 0$; so

$$F_G = mg$$

where

F_G = force of gravity (N)
R_e = radius of the Earth ($R_e = 6.37 \times 10^6$ m)
m = mass (kg)
g = acceleration due to gravity (m/s²)
$g = 9.81$ (m/s^2) or $g = 32.2$ (ft/s²)

The acceleration of a falling body is independent of the mass of the object. The weight F_w on an object is actually the force of gravity on that object:

$$F_w = mg$$

Centrifugal Force

Centrifugal force is the apparent force drawing a rotating body away from the center of rotation, and it is caused by the inertia of the body. Centrifugal force can be calculated by the formula:

$$F_c = \frac{mv^2}{r} = m\omega^2 r$$

Centripetal Force

Centripetal force is defined as the force acting on a body in curvilinear motion that is directed toward the center of curvature or axis of rotation. Centripetal force is equal in magnitude to centrifugal force but in the opposite direction.

$$F_{cp} = -F_c = \frac{mv^2}{r}$$

where

F_c = centrifugal force (N)
F_{cp} = centripetal force (N)
m = mass of the body (kg)
v = velocity of the body (m/s)
r = radius of curvature of the path of the body (m)
ω = angular velocity (s^{-1})

Torque

Torque is the ability of a force to cause a body to rotate about a particular axis. Torque can have either a clockwise or a counterclockwise direction. To distinguish between the two possible directions of rotation, we adopt the convention that a counterclockwise torque is positive and that a clockwise torque is negative. One way to quantify a torque is

$$T = F \cdot l$$

where

T = torque (N m or lb ft)
F = applied force (N or lb)
l = length of torque arm (m or ft)

Work

Work is the product of a force in the direction of the motion and the displacement.

a) Work done by a constant force:

$$W = F_s \cdot s = F \cdot s \cdot \cos\alpha$$

where

W = work (Nm = J)
Fs = component of force along the direction of movement (N)
s = distance the system is displaced (m)

b) Work done by a variable force

If the force is not constant along the path of the object, we need to calculate the force over very tiny intervals and then add them up. This is exactly what the integration over differential small intervals of a line can accomplish:

$$W = \int_{si}^{sf} F_s(s) \cdot ds = \int_{si}^{sf} F(s) \cos\alpha \cdot ds$$

where

Fs(s) = component of the force function along the direction of movement (N)
F(s) = function of the magnitude of the force vector along the displacement curve (N)
s_i = initial location of the body (m)
s_f = final location of the body (m)
α = angle between the displacement and the force

Energy

Energy is defined as the ability to do work. The quantitative relationship between work and mechanical energy is expressed by the equation:

$$TME_i + W_{ext} = TME_f$$

where

TME_i = initial amount of total mechanical energy (J)
W_{ext} = work done by external forces (J)
TME_f = final amount of total mechanical energy (J)

There are two kinds of mechanical energy: kinetic and potential.

a) Kinetic energy
   Kinetic energy is the energy of motion. The following equation is used to represent the kinetic energy of an object:

$$E_k = \frac{1}{2} m v^2$$

where

m = mass of moving object (kg)
ν = velocity of moving object (m/s)

b) Potential energy
   Potential energy is the stored energy of a body and is due to its internal characteristics or its position. Gravitational potential energy is defined by the formula

$$E_{pg} = m \cdot g \cdot h$$

where

E_{pg} = gravitational potential energy (J)
m = mass of object (kg)
h = height above reference level (m)
g = acceleration due to gravity (m/s^2)

CONSERVATION OF ENERGY

In any isolated system, energy can be transformed from one kind of another, but the total amount of energy is constant (conserved):

$$E = E_k + E_p + E_e + \ldots = \text{constant}$$

Conservation of mechanical energy is given by

$$E_k + E_p = \text{constant}$$

Power

Power is the rate at which work is done, or the rate at which energy is transformed from one form to another. Mathematically, it is computed using the following equation:

$$P = \frac{W}{t}$$

where

P = power (W)
W = work (J)
t = time (s)

The standard metric unit of power is the watt (W). As is implied by the equation for power, a unit of power is equivalent to a unit of work divided by a unit of time. Thus, a watt is equivalent to Joule/second (J/s). Since the expression for work is

$$W = F \cdot s,$$

the expression for power can be rewritten as

$$P = F \cdot v$$

where

s = displacement (m)
v = speed (m/s)

Common Statistical Distributions

Distribution of Random Variable x	Functional Form	Parameters	Mean	Variance	Range
Binomial	$P_x(k) = \dfrac{n!}{k!(n-k)!} p^k (1-p)^{n-k}$	n, p	np	$np(1-p)$	$0, 1, 2, \ldots, n$
Poisson	$P_x(k) = \dfrac{\lambda^k e^{-\lambda}}{k!}$	λ	λ	λ	$0, 1, 2, \ldots$
Geometric	$P_x(k) = p(1-p)^{k-1}$	p	$1/p$	$\dfrac{1-p}{p^2}$	$1, 2, \ldots$
Exponential	$f_x(y) = \dfrac{1}{\theta} e^{-y/\theta}$	θ	θ	θ^2	$(0, \infty)$
Gamma	$f_x(y) = \dfrac{1}{\Gamma(\alpha)\beta^\alpha} y^{(\alpha-1)} e^{-y/\beta}$	α, β	$\alpha\beta$	$\alpha\beta^2$	$(0, \infty)$
Beta	$f_x(y) = \dfrac{\Gamma(\alpha+\beta)}{\Gamma(\alpha)\Gamma(\beta)} y^{(\alpha-1)}(1-y)^{(\beta-x)}$	α, β	$\dfrac{\alpha}{\alpha+\beta}$	$\dfrac{\alpha\beta}{(\alpha+\beta)^2(\alpha+\beta+1)}$	$(0, 1)$
Normal	$f_x(y) = \dfrac{1}{\sqrt{2\pi}\,\sigma} e^{-(y-\mu)^2/2\sigma^2}$	μ, σ	μ	σ^2	$(-\infty, \infty)$
Student t	$f_x(y) = \dfrac{1}{\sqrt{\pi v}} \dfrac{\Gamma\!\left(\dfrac{v+1}{2}\right)}{\Gamma(v/2)} (1+y^2/v)$	v	0 for $v>1$	$\dfrac{v}{v-2}$ for $v>2$	$(-\infty, \infty)$
Chi Square	$f_x(y) = \dfrac{1}{2^{v/2}\Gamma(v/2)} y^{(v-2)/2} e^{-y/2}$	v	v	$2v$	$(0, \infty)$
F	$f_x(y) = \dfrac{\Gamma\!\left(\dfrac{v_1+v_2}{2}\right) v_1^{v_1/2} v_2^{v_2/2}}{\Gamma\!\left(\dfrac{v_1}{2}\right)\Gamma\!\left(\dfrac{v_2}{2}\right)} \dfrac{(y)}{(v_2-v_1)}$	v_1, v_2	$\dfrac{v_2}{v_2-2}$ for $v_2>2$	$\dfrac{v_2^2(2v_2+2v_1-4)}{v_1(v_2-2)(v_2-4)}$ for $v_2>4$	$(0, \infty)$

Discrete Distributions

Probability Mass Function, $p(x)$
Mean, μ
Variance, σ^2
Coefficient of Skewness, β_1
Coefficient of Kurtosis, β_2
Moment-generating Function, $M(t)$
Characteristic Function, $\phi(t)$
Probability-generating Function, $P(t)$

Bernoulli Distribution

$$p(x) = p^x q^{x-1} \qquad x = 0,1 \qquad 0 \le p \le 1 \qquad q = 1-p$$

$$\mu = p \qquad \sigma^2 = pq \qquad \beta_1 = \frac{1-2p}{\sqrt{pq}} \qquad \beta_2 = 3 + \frac{1-6pq}{pq}$$

$$M(t) = q + pe^t \qquad \phi(t) = q + pe^{it} \qquad P(t) = q + pt$$

Beta Binomial Distribution

$$p(x) = \frac{1}{n+1} \frac{B(a+x, b+n-x)}{B(x+1, n-x+1)B(a,b)} \; x = 0,1,2,\ldots,n \; a > 0 \; b > 0$$

$$\mu = \frac{na}{a+b} \qquad \sigma^2 = \frac{nab(a+b+n)}{(a+b)^2(a+b+1)} \qquad B(a,b) \text{ is the Beta function.}$$

Beta Pascal Distribution

$$p(x) = \frac{\Gamma(x)\Gamma(v)\Gamma(\rho+v)\Gamma(v+x-(\rho+r))}{\Gamma(r)\Gamma(x-r+1)\Gamma(\rho)\Gamma(v-\rho)\Gamma(v+x)} \quad x = r, r+1,\ldots \quad v > p > 0$$

$$\mu = r\frac{v-1}{\rho-1}, \rho > 1 \qquad \sigma^2 = r(r+\rho-1)\frac{(v-1)(v-\rho)}{(\rho-1)^2(\rho-2)}, \rho > 2$$

Binomial Distribution

$$p(x) = \binom{n}{x} p^x q^{n-x} \quad x = 0,1,2,\ldots,n \quad 0 \le p \le 1 \quad q = 1-p$$

$$\mu = np \quad \sigma^2 = npq \qquad \beta_1 = \frac{1-2p}{\sqrt{npq}} \qquad \beta_2 = 3 + \frac{1-6pq}{npq}$$

$$M(t) = \left(q + pe^t\right)^n \quad \phi(t) = \left(q + pe^{it}\right)^n \quad P(t) = \left(q + pt\right)^n$$

Discrete Weibull Distribution

$$p(x) = (1-p)^{x^\beta} - (1-p)^{(x+1)^\beta} \quad x = 0,1,\ldots \quad 0 \le p \le 1 \quad \beta > 0$$

Geometric Distribution

$$p(x) = pq^{1-x} \qquad x = 0,1,2,\ldots \qquad 0 \le p \le 1 \qquad q = 1-p$$

$$\mu = \frac{1}{p} \qquad \sigma^2 = \frac{q}{p^2} \qquad \beta_1 = \frac{2-p}{\sqrt{p}} \qquad \beta_2 = \frac{p^2 + 6q}{q}$$

$$M(t) = \frac{p}{1 - qe^t} \qquad \phi(t) = \frac{p}{1 - qe^{it}} \qquad P(t) = \frac{p}{1 - qt}$$

Hypergeometric Distribution

$$p(x) = \frac{\binom{M}{x}\binom{N-M}{n-x}}{\binom{N}{n}} \qquad x = 0,1,2,\ldots,n \qquad x \le M \qquad n-x \le N-M$$

$$n, M, N \in N \qquad 1 \le n \le N \qquad 1 \le M \le N \qquad N = 1,2,\ldots$$

$$\mu = n\frac{M}{N} \qquad \sigma^2 = \left(\frac{N-n}{N-1}\right)n\frac{M}{N}\left(1 - \frac{M}{N}\right) \qquad \beta_1 = \frac{(N-2M)(N-2n)\sqrt{N-1}}{(N-2)\sqrt{nM(N-M)(N-n)}}$$

$$\beta_2 = \frac{N^2(N-1)}{(N-2)(N-3)nM(N-M)(N-n)}$$

$$\left\{ N(N+1) - 6n(N-n) + 3\frac{M}{N^2}(N-M)\left[N^2(n-2) - Nn^2 + 6n(N-n)\right] \right\}$$

$$M(t) = \frac{(N-M)!(N-n)!}{N!} F\left(,,e^t\right)\phi(t)$$

$$= \frac{(N-M)!(N-n)!}{N!} F\left(,,e^{it}\right)P(t)$$

$$= \left(\frac{N-M}{N}\right)^n F\left(,,t\right)$$

$F(\alpha, \beta, \gamma, x)$ is the hypergeometric function. $\alpha = -n; \quad \beta = -M; \quad \gamma = N - M - n + 1$

Negative Binomial Distribution

$$p(x) = \binom{x+r-1}{r-1} p^r q^x \quad x = 0,1,2,\ldots \quad r = 1,2,\ldots \quad 0 \le p \le 1 \quad q = 1-p$$

$$\mu = \frac{rq}{p} \qquad \sigma^2 = \frac{rq}{p^2} \qquad \beta_1 = \frac{2-p}{\sqrt{rq}} \qquad \beta_2 = 3 + \frac{p^2 + 6q}{rq}$$

$$M(t) = \left(\frac{p}{1-qe^t} \right)^r \qquad \phi(t) = \left(\frac{p}{1-qe^{it}} \right)^r \qquad P(t) = \left(\frac{p}{1-qt} \right)^r$$

Poisson Distribution

$$p(x) = \frac{e^{-\mu} \mu^x}{x!} \qquad x = 0,1,2,\ldots \qquad \mu > 0$$

$$\mu = \mu \qquad \sigma^2 = \mu \qquad \beta_1 = \frac{1}{\sqrt{\mu}} \qquad \beta_2 = 3 + \frac{1}{\mu}$$

$$M(t) = \exp\left[\mu\left(e^t - 1\right)\right] \qquad \sigma(t) = \exp\left[\mu\left(e^{it} - 1\right)\right] \qquad P(t) = \exp\left[\mu\left(t - 1\right)\right]$$

Rectangular (Discrete Uniform) Distribution

$$p(x) = 1/n \qquad x = 1,\ 2,\ \ldots\ ,n \qquad n \in N$$

$$\mu = \frac{n+1}{2} \qquad \sigma^2 = \frac{n^2-1}{12} \qquad \beta_1 = 0 \qquad \beta_2 = \frac{3}{5}\left(3 - \frac{4}{n^2-1}\right)$$

$$M(t) = \frac{e^t\left(1-e^{nt}\right)}{n\left(1-e^t\right)} \qquad \phi(t) = \frac{e^{it}\left(1-e^{nit}\right)}{n\left(1-e^{it}\right)} \qquad P(t) = \frac{t\left(1-t^n\right)}{n\left(1-t\right)}$$

Continuous Distributions

Probability Density Function, $f(x)$
Mean, μ
Variance, σ^2
Coefficient of Skewness, β_1
Coefficient of Kurtosis, β_2
Moment-generating Function, $M(t)$
Characteristic Function, $\phi(t)$

Arcsine Distribution

$$f(x) = \frac{1}{\pi\sqrt{x(1-x)}} \qquad 0 < x < 1$$

$$\mu = \frac{1}{2} \qquad \sigma^2 = \frac{1}{8} \qquad \beta_1 = 0 \qquad \beta_2 \frac{3}{2}$$

Beta Distribution

$$f(x) = \frac{\Gamma(\alpha+\beta)}{\Gamma(\alpha)\Gamma(\beta)} x^{\alpha-1} (1-x)^{\beta-1} \qquad 0 < x < 1 \qquad \alpha,\ \beta > 0$$

$$\mu = \frac{\alpha}{\alpha+\beta} \qquad \sigma^2 = \frac{\alpha\beta}{(\alpha+\beta)^2(\alpha+\beta+1)} \qquad \beta_1 = \frac{2(\beta-\alpha)\sqrt{\alpha+\beta+1}}{\sqrt{\alpha\beta}(\alpha+\beta+2)}$$

$$\beta_2 = \frac{3(\alpha+\beta+1)\left[2(\alpha+\beta)^2+\alpha\beta(\alpha+\beta-6)\right]}{\alpha\beta(\alpha+\beta+2)(\alpha+\beta+3)}$$

Cauchy Distribution

$$f(x) = \frac{1}{b\pi\left[1+\left(\dfrac{x-a}{b}\right)^2\right]} \qquad -\infty < x < \infty \qquad -\infty < a < \infty \qquad b > 0$$

$\mu, \sigma^2, \beta_1, \beta_2, M(t)$ do not exist. $\phi(t) = \exp[ait - b|t|]$

Chi Distribution

$$f(x) = \frac{x^{n-1} e^{-x^2/2}}{2^{(n/2)-1}\Gamma(n/2)} \qquad x \geq 0 \qquad n \in N$$

$$\mu = \frac{\Gamma\left(\dfrac{n+1}{2}\right)}{\Gamma\left(\dfrac{n}{2}\right)} \qquad \sigma^2 = \frac{\Gamma\left(\dfrac{n+2}{2}\right)}{\Gamma\left(\dfrac{n}{2}\right)} - \left[\frac{\Gamma\left(\dfrac{n+1}{2}\right)}{\Gamma\left(\dfrac{n}{2}\right)}\right]^2$$

Chi-Square Distribution

$$f(x) = \frac{e^{-x/2} x^{(v/2)-1}}{2^{v/2}\Gamma(v/2)} \qquad x \geq 0 \qquad v \in N$$

$$\mu = v \qquad \sigma^2 = 2v \qquad \beta_1 = 2\sqrt{2/v} \qquad \beta_2 = 3+\frac{12}{v} \qquad M(t) = (1-2t)^{-v/2}, \qquad t < \frac{1}{2}$$

$$\phi(t) = (1-2it)^{-v/2}$$

Erlang Distribution

$$f(x) = \frac{1}{\beta^n (n-1)!} x^{n-1} e^{-x/\beta} \qquad x \geq 0 \qquad \beta > 0 \qquad n \in N$$

$$\mu = n\beta \qquad \sigma^2 = n\beta^2 \qquad \beta_1 = \frac{2}{\sqrt{n}} \qquad \beta_2 = 3 + \frac{6}{n}$$

$$M(t) = (1 - \beta t)^{-n} \qquad \phi(t) = (1 - \beta i t)^{-n}$$

Exponential Distribution

$$f(x) = \lambda e^{-\lambda x} \qquad x \geq 0 \qquad \lambda > 0$$

$$\mu = \frac{1}{\lambda} \qquad \sigma^2 = \frac{1}{\lambda^2} \qquad \beta_1 = 2 \qquad \beta_2 = 9 \qquad M(t) = \frac{\lambda}{\lambda - t}$$

$$\phi(t) = \frac{\lambda}{\lambda - it}$$

Extreme-Value Distribution

$$f(x) = \exp\left[-e^{-(x-\alpha)/\beta}\right] \qquad -\infty < x < \infty \qquad -\infty < \alpha < \infty \qquad \beta > 0$$

$$\mu = \alpha + \gamma\beta, \quad \gamma \doteq .5772\ldots \text{ is Euler's constant } \sigma^2 = \frac{\pi^2 \beta^2}{6}$$

$$\beta_1 = 1.29857 \qquad \beta_2 = 5.4$$

$$M(t) = e^{\alpha t}\Gamma(1 - \beta t), \quad t < \frac{1}{\beta} \qquad \phi(t) = e^{\alpha i t}\Gamma(1 - \beta i t)$$

F Distribution

$$f(x)\frac{\Gamma\left[(v_1 + v_2)/2\right]v_1^{v_1/2} v_2^{v_2/2}}{\Gamma(v_1/2)\Gamma(v_2/2)} x^{(v_1/2)-1}(v_2 + v_1 x)^{-(v_1+v_2)/2}$$

$$x > 0 \qquad v_1, \ v_2 \in N$$

$$\mu = \frac{v_2}{v_2 - 2}, \ v_2 \geq 3 \qquad \sigma^2 = \frac{2v_2^2(v_1 + v_2 - 2)}{v_1 (v_2 - 2)^2 (v_2 - 4)}, \quad v_2 \geq 5$$

$$\beta_1 = \frac{(2v_1 + v_2 - 2)\sqrt{8(v_2 - 4)}}{\sqrt{v_1}(v_2 - 6)\sqrt{v_1 + v_2 - 2}}, \quad v_2 \geq 7$$

$$\beta_2 = 3 + \frac{12\left[(v_2-2)^2(v_2-4)+v_1\ (v_1+v_2-2)(5v_2-22)\right]}{v_1\ (v_2-6)(v_2-8)(v_1+v_2-2)}, \quad v_2 \geq 9$$

$M(t)$ does not exist.

$$\phi\!\left(\frac{v_1}{v_2}t\right) = \frac{G(v_1,v_2,t)}{B(v_1/2,v_2/2)}$$

$B(a, b)$ is the Beta function. G is defined by

$$(m+n-2)G(m,n,t) = (m-2)G(m-2,n,t) + 2itG(m,n-2,\ t), \quad m,n > 2$$

$$mG(m,n,t) = (n-2)G(m+2,n-2,t) - 2itG(m+2,n-4,t), \quad n > 4$$

$$nG(2,n,t) = 2 + 2itG(2,n-2,t), \quad n > 2$$

Gamma Distribution

$$f(x) = \frac{1}{\beta^{\alpha}\Gamma(\alpha)}x^{\alpha-1}e^{-x/\beta} \qquad x \geq 0 \qquad \alpha,\beta > 0$$

$$\mu = \alpha\beta \qquad \sigma^2 = \alpha\beta^2 \qquad \beta_1 = \frac{2}{\sqrt{\alpha}} \qquad \beta_2 = 3\!\left(1+\frac{2}{\alpha}\right)$$

$$M(t) = (1-\beta t)^{-\alpha} \qquad \phi(t) = (1-\beta it)^{-\alpha}$$

Half-Normal Distribution

$$f(x) = \frac{2\theta}{\pi}\exp\!\left[-\left(\theta^2 x^2/\pi\right)\right] \qquad x \geq 0 \qquad \theta > 0$$

$$\mu = \frac{1}{\theta} \qquad \sigma^2 = \left(\frac{\pi-2}{2}\right)\frac{1}{\theta^2} \qquad \beta_1 = \frac{4-\pi}{\theta^3} \qquad \beta_2 = \frac{3\pi^2-4\pi-12}{4\theta^4}$$

LaPlace (Double Exponential) Distribution

$$f(x) = \frac{1}{2\beta}\exp\!\left[-\frac{|x-\alpha|}{\beta}\right] \qquad -\infty < x < \infty \qquad -\infty < \alpha < \infty \qquad \beta > 0$$

$$\mu = \alpha \qquad \sigma^2 = 2\beta^2 \qquad \beta_1 = 0 \qquad \beta_2 = 6$$

$$M(t) = \frac{e^{\alpha t}}{1 - \beta^2 t^2} \qquad \phi(t) = \frac{e^{\alpha it}}{1 + \beta^2 t^2}$$

Logistic Distribution

$$f(x) = \frac{\exp\left[(x-\alpha)/\beta\right]}{\beta \left(1 + \exp\left[(x-\alpha)/\beta\right]\right)^2}$$

$$-\infty < x < \infty \qquad -\infty < \alpha < \infty \qquad -\infty < \beta < \infty$$

$$\mu = \alpha \qquad \sigma^2 = \frac{\beta^2 \pi^2}{3} \qquad \beta_1 = 0 \qquad \beta_2 = 4.2$$

$$M(t) = e^{\alpha t} \pi \beta t \ \csc(\pi \beta t) \qquad \phi(t) = e^{\alpha it} \pi \beta it \ \csc(\pi \beta it)$$

Lognormal Distribution

$$f(x) = \frac{1}{\sqrt{2\pi}\,\sigma x} \exp\left[-\frac{1}{2\sigma^2}(1nx - \mu)^2\right]$$

$$x > 0 \qquad -\infty < \mu < \infty \qquad \sigma < 0$$

$$\mu = e^{\mu + \sigma^2/2} \qquad \sigma^2 = e^{2\mu + \sigma^2}\left(e^{\sigma^2} - 1\right)$$

$$\beta_1 = \left(e^{\sigma^2} + 2\right)\left(e^{\sigma^2} - 1\right)^{1/2} \qquad \beta_2 = \left(e^{\sigma^2}\right)^4 + 2\left(e^{\sigma^2}\right)^3 + 3\left(e^{\sigma^2}\right)^2 - 3$$

Noncentral Chi-Square Distribution

$$f(x) = \frac{\exp\left[-\frac{1}{2}(x+\lambda)\right]}{2^{v/2}} \sum_{j=0}^{\infty} \frac{x^{(v/2)+j-1}\lambda^j}{\Gamma\left(\frac{v}{2}+j\right)2^{2j}j!}$$

$$x > 0 \qquad \lambda > 0 \qquad v \in N$$

$$\mu = v + \lambda \qquad \sigma^2 = 2(v + 2\lambda) \qquad \beta_1 = \frac{\sqrt{8}(v + 3\lambda)}{(v + 2\lambda)^{3/2}} \qquad \beta_2 = 3 + \frac{12(v + 4\lambda)}{(v + 2\lambda)^2}$$

$$M(t) = (1 - 2t)^{-v/2} \exp\left[\frac{\lambda t}{1 - 2t}\right] \qquad \phi(t) = (1 - 2it)^{-v/2} \exp\left[\frac{\lambda it}{1 - 2it}\right]$$

Noncentral *F* Distribution

$$f(x) = \sum_{i=0}^{\infty} \frac{\Gamma\left(\frac{2i+v_1+v_2}{2}\right)\left(\frac{v_1}{v_2}\right)^{(2i+v_1)/2} x^{(2i+v_1-2)/2} e^{-\lambda/2}\left(\frac{\lambda}{2}\right)}{\Gamma\left(\frac{v_2}{2}\right)\Gamma\left(\frac{2i+v_1}{2}\right) v_1!\left(1+\frac{v_1}{v_2}x\right)^{(2i+v_1+v_2)/2}}$$

$$x > 0 \qquad v_1, v_2 \in N \qquad \lambda > 0$$

$$\mu = \frac{(v_1+\lambda)v_2}{(v_2-2)v_1}, \qquad v_2 > 2$$

$$\sigma^2 = \frac{(v_1+\lambda)^2 + 2(v_1+\lambda)v_2^2}{(v_2-2)(v_2-4)v_1^2} - \frac{(v_1+\lambda)^2 v_2^2}{(v_2-2)^2 v_1^2}, \qquad v_2 > 4$$

Noncentral *t* Distribution

$$f(x) = \frac{v^{v/2}}{\Gamma\left(\frac{v}{2}\right)} \frac{e^{-\delta^2/2}}{\sqrt{\pi}\left(v+x^2\right)^{(v+2)/2}} \sum_{i=0}^{\infty} \Gamma\left(\frac{v+i+1}{2}\right)\left(\frac{\delta^i}{i!}\right)\left(\frac{2x^2}{v+x^2}\right)^{i/2}$$

$$-\infty < x < \infty \qquad -\infty < \delta < \infty \qquad v \in N$$

$$\mu_r' = c_r \frac{\Gamma\left(\frac{v-r}{2}\right)v^{r/2}}{2^{r/2}\Gamma\left(\frac{v}{2}\right)}, \qquad v > r, \qquad c_{2r-1} = \sum_{i=1}^{r} \frac{(2r-1)!\delta^{2r-1}}{(2i-1)!(r-i)!2^{r-i}},$$

$$c_{2r} = \sum_{i=0}^{r} \frac{(2r)!\delta^{2i}}{(2i)!(r-i)!2^{r-i}}, \qquad r = 1,2,3,\ldots$$

Normal Distribution

$$f(x) = \frac{1}{\sigma\sqrt{2\pi}} \exp\left[-\frac{(x-\mu)^2}{2\sigma^2}\right]$$

$$-\infty < x < \infty \qquad -\infty < \mu < \infty \qquad \sigma < 0$$

$$\mu = \mu \quad \sigma^2 = \sigma^2 \quad \beta_1 = 0 \quad \beta_2 = 3 \quad M(t) = \exp\left[\mu t + \frac{t^2\sigma^2}{2}\right]$$

$$\phi(t) = \exp\left[\mu it - \frac{t^2 \sigma^2}{2}\right]$$

Pareto Distribution

$$f(x) = \theta a^\theta / x^{\theta+1} \qquad x \geq a \qquad \theta > 0 \qquad a > 0$$

$$\mu = \frac{\theta a}{\theta - 1}, \quad \theta > 1 \qquad \sigma^2 = \frac{\theta a^2}{(\theta-1)^2 (\theta-2)}, \qquad \theta > 2$$

$M(t)$ does not exist.

Rayleigh Distribution

$$f(x) = \frac{x}{\sigma^2} \exp\left[-\frac{x^2}{2\sigma^2}\right] \qquad x \geq 0 \qquad \sigma = 0$$

$$\mu = \sigma\sqrt{\pi/2} \qquad \sigma^2 = 2\sigma^2\left(1 - \frac{\pi}{4}\right) \qquad \beta_1 = \frac{\sqrt{\pi}\,(\pi-3)}{4\left(1-\frac{\pi}{4}\right)^{3/2}}$$

$$\beta_2 = \frac{2 - \frac{3}{16}\pi^2}{\left(1 - \frac{\pi}{4}\right)^2}$$

t Distribution

$$f(x) = \frac{1}{\sqrt{\pi v}} \frac{\Gamma\left(\frac{v+1}{2}\right)}{\Gamma\frac{v}{2}}\left(1 + \frac{x^2}{v}\right)^{-(v+1)/2} \qquad -\infty < x < \infty \qquad v \in N$$

$$\mu = 0, \quad v \geq 2 \qquad \sigma^2 = \frac{v}{v-2}, \quad v \geq 3 \qquad \beta_1 = 0, \quad v \geq 4$$

$$\beta_2 = 3 + \frac{6}{v-4}, \quad v \geq 5$$

$M(t)$ does not exist.

$$\phi(t) = \frac{\sqrt{\pi}\,\Gamma\left(\frac{v}{2}\right)}{\Gamma\left(\frac{v+1}{2}\right)} \int_{-\infty}^{\infty} \frac{e^{itz\sqrt{v}}}{\left(1+z^2\right)^{(v+1)/2}}\,dz$$

Triangular Distribution

$$f(x) = \begin{cases} 0 & x \le a \\ 4(x-a)/(b-a)^2 & a < x \le (a+b)/2 \\ 4(b-x)/(b-a)^2 & (a+b)/2 < x < b \\ 0 & x \ge b \end{cases}$$

$$-\infty < a < b < \infty$$

$$\mu = \frac{a+b}{2} \qquad \sigma^2 = \frac{(b-a)^2}{24} \qquad \beta_1 = 0 \qquad \beta_2 = \frac{12}{5}$$

$$M(t) = \frac{4\left(e^{at/2} - e^{bt/2}\right)^2}{t^2 (b-a)^2} \qquad \phi(t) = \frac{4\left(e^{ait/2} - e^{bit/2}\right)^2}{t^2 (b-a)^2}$$

Uniform Distribution

$$f(x) = \frac{1}{b-a} \qquad a \le x \le b \qquad -\infty < a < b < \infty$$

$$\mu = \frac{a+b}{2} \qquad \sigma^2 = \frac{(b-a)^2}{12} \qquad \beta_1 = 0 \qquad \beta_2 = \frac{9}{5}$$

$$M(t) = \frac{e^{bt} - e^{at}}{(b-a)t} \qquad \phi(t) = \frac{e^{bit} - e^{ait}}{(b-a)it}$$

Weibull Distribution

$$f(x) = \frac{\alpha}{\beta^\alpha} x^{\alpha-1} e^{-(x/\beta)^\alpha} \qquad x \ge 0 \qquad \alpha, \beta > 0$$

$$\mu = \beta \Gamma\left(1 + \frac{1}{\alpha}\right) \qquad \sigma^2 = \beta^2 \left[\Gamma\left(1 + \frac{2}{\alpha}\right) - \Gamma^2\left(1 + \frac{1}{\alpha}\right)\right]$$

$$\beta_1 = \frac{\Gamma\left(1 + \frac{3}{\alpha}\right) - 3\Gamma\left(1 + \frac{1}{\alpha}\right)\Gamma\left(1 + \frac{2}{\alpha}\right) + 2\Gamma^3\left(1 + \frac{1}{\alpha}\right)}{\left[\Gamma\left(1 + \frac{2}{\alpha}\right) - \Gamma^2\left(1 + \frac{1}{\alpha}\right)\right]^{3/2}}$$

$$\beta_2 = \frac{\Gamma\left(1+\dfrac{4}{\alpha}\right) - 4\Gamma\left(1+\dfrac{1}{\alpha}\right)\Gamma\left(1+\dfrac{3}{\alpha}\right) + 6\Gamma^2\left(1+\dfrac{1}{\alpha}\right)\Gamma\left(1+\dfrac{2}{\alpha}\right) - 3\Gamma^4\left(1+\dfrac{1}{\alpha}\right)}{\left[\Gamma\left(1+\dfrac{2}{\alpha}\right) - \Gamma^2\left(1+\dfrac{1}{\alpha}\right)\right]^2}$$

DISTRIBUTION PARAMETERS

Average

$$\bar{x} = \frac{1}{n}\sum_{i=1}^{n} x_i$$

Variance

$$s^2 = \frac{1}{n-1}\sum_{i=1}^{n}\left(x_i - \bar{x}\right)^2$$

Standard Deviation

$$s = \sqrt{s^2}$$

Standard Error

$$\frac{s}{\sqrt{n}}$$

SKEWNESS

(missing if $s = 0$ or $n < 3$)

$$\frac{n\sum_{i=1}^{n}\left(x_i - \bar{x}\right)^3}{(n-1)(n-2)\,s^3}$$

Standardized Skewness

$$\frac{\text{skewness}}{\sqrt{\dfrac{6}{n}}}$$

KURTOSIS

$$(\text{missing if } s = 0 \text{ or } n < 4)$$

$$\frac{n(n+1)\sum_{i=1}^{n}(x_i - \bar{x})^4}{(n-1)(n-2)(n-3)s^4} - \frac{3(n-1)^2}{(n-2)(n-3)}$$

Standardized Kurtosis

$$\frac{\text{Kurtosis}}{\sqrt{\dfrac{24}{n}}}$$

Weighted Average

$$\frac{\sum_{i=1}^{n} x_i w_i}{\sum_{i=1}^{n} w_i}$$

Estimation and Testing

100(1−α)% Confidence Interval for Mean:

$$\bar{x} \pm t_{n-1;\, \alpha/2} \frac{s}{\sqrt{n}}$$

100(1−α)% Confidence Interval for Variance:

$$\left[\frac{(n-1)s^2}{\chi^2_{n-1;\, \alpha/2}} \, , \, \frac{(n-1)s^2}{\chi^2_{n-1;\, 1-\alpha/2}} \right]$$

$100(1-\alpha)\%$ Confidence Interval for Difference in Means:

Equal Variance

$$\left(\bar{x}_1 - \bar{x}_2\right) \pm t_{n_1+n_2-2;\ \alpha/2}\ s_p \sqrt{\frac{1}{n_1} + \frac{1}{n_2}}$$

$$\text{where} \quad s_p = \sqrt{\frac{\left(n_1-1\right) s_1^{\,2} + \left(n_2-1\right) s_2^{\,2}}{n_1+n_2-2}}$$

Unequal Variance

$$\left[\left(\bar{x}_1 - \bar{x}_2\right) \pm t_{m;\ \alpha/2} \sqrt{\frac{s_1^{\,2}}{n_1} + \frac{s_2^{\,2}}{n_2}}\,\right]$$

where

$$\frac{1}{m} = \frac{c^2}{n_1-1} + \frac{\left(1-c\right)^2}{n_2-1}$$

and

$$c = \frac{\dfrac{s_1^{\,2}}{n_1}}{\dfrac{s_1^{\,2}}{n_1} + \dfrac{s_2^{\,2}}{n_2}}$$

$100(1-\alpha)\%$ Confidence Interval for Ratio of Variances:

$$\left(\frac{s_1^{\,2}}{s_2^{\,2}}\right)\left(\frac{1}{F_{n_1-1,n_2-1;\ \alpha/2}}\right), \left(\frac{s_1^{\,2}}{s_2^{\,2}}\right)\left(\frac{1}{F_{n_1-1,n_2-1;\ \alpha/2}}\right)$$

Normal Probability Plot
The data is sorted from the smallest to the largest value to compute order statistics.
A scatter plot is then generated where
horizontal position $= x_{(i)}$

$$\text{vertical position} = \Phi\left(\frac{i-3/8}{n+1/4}\right)$$

The labels for the vertical axis are based upon the probability scale using:

$$100 \left(\frac{i - 3/8}{n + 1/4} \right)$$

Comparison of Poisson Rates

n_j = # of events in sample j
t_j = length of sample j

Rate estimates:

$$r_j = \frac{n_j}{t_j}$$

Rate ratio:

$$\frac{r_1}{r_2}$$

Test statistic:

$$z = \max \left(0, \ \frac{\left| n_1 - \dfrac{(n_1 + n_2)}{2} \right| - \dfrac{1}{2}}{\sqrt{\dfrac{(n_1 + n_2)}{4}}} \right)$$

where z follows the standard normal distribution.

DISTRIBUTION FUNCTIONS—PARAMETER ESTIMATION

Bernoulli

$$\hat{p} = \bar{x}$$

Binomial

$$\hat{p} = \frac{\bar{x}}{n}$$

where n is the number of trials

Discrete Uniform

$$\hat{a} = \min x_i$$

$$\hat{b} = \min x_i$$

Geometric

$$\hat{p} = \frac{1}{1+\bar{x}}$$

Negative Binomial

$$\hat{p} = \frac{k}{\bar{x}}$$

where k = the number of successes

Poisson

$$\hat{\beta} = \bar{x}$$

Beta

$$\hat{\alpha} = \bar{x}\left[\frac{\bar{x}(1-\bar{x})}{s^2} - 1\right]$$

$$\hat{\beta} = (1-\bar{x})\left(\frac{\bar{x}(1-\bar{x})}{s^2} - 1\right)$$

Chi-Square

$$\text{d.f. } \bar{v} = \bar{x}$$

Erlang

$\hat{\alpha}$ = round ($\hat{\alpha}$ from Gamma)

$$\hat{\beta} = \frac{\hat{\alpha}}{\bar{x}}$$

Exponential

$$\hat{\beta} = \frac{1}{\bar{x}}$$

Note: system displays $1/\hat{\beta}$

F

$$\text{num d.f.}: \hat{v} = \frac{2\hat{w}^3 - 4\hat{w}^2}{\left(s^2\left(\hat{w}-2\right)^2\left(\hat{w}-4\right)\right) - 2\hat{w}^2}$$

$$\text{den.d.f.}: \hat{w} = \frac{\max\left(1,2\bar{x}\right)}{-1+\bar{x}}$$

Gamma

$$R = \log\left(\frac{\text{arithmetic mean}}{\text{geometric mean}}\right) \qquad (11.2)$$

If $0 < R \leq 0.5772$,

$$\hat{\alpha} = R^{-1}\left(0.5000876 + 0.1648852\ R - 0.0544274\ R\right)^2$$

or if $R > 0.5772$,

$$\hat{\alpha} = R^{-1}\ (17.79728 + 11.968477\ R + R^2)^{-1}$$
$$(8.898919 + 9.059950\ R + 0.9775373\ R^2)$$

$$\hat{\beta} = \hat{\alpha}\ /\ \bar{x}$$

This is an approximation of the Method of Maximum Likelihood solution.

Log Normal

$$\hat{\mu} = \frac{1}{n}\sum_{i=1}^{n}\log x_i$$

$$\hat{\alpha} = \sqrt{\frac{1}{n-1}\sum_{i=1}^{n}\left(\log x_i - \hat{\mu}\right)^2}$$

System displays:

means: $\exp\left(\hat{\mu} + \hat{\alpha}^2 / 2\right)$

Standard deviation: $\sqrt{\exp\left(2\hat{\mu} + \hat{\alpha}^2\right)\left[\exp\left(\hat{\alpha}^2\right) - 1\right]}$

Normal

$$\hat{\mu} = \overline{x}$$

$$\hat{\sigma} = s$$

Student's *t*

If $s^2 \leq 1$ or if $\hat{v} \leq 2$, then the system indicates that the data is inappropriate.

$$s^2 = \frac{\sum_{i=1}^{n} x_i^2}{n}$$

$$\hat{v} = \frac{2s^2}{-1 + s^2}$$

Triangular

$$\hat{a} = \min x_i$$

$$\hat{c} = \max x_i$$

$$\hat{b} = 3\overline{x} - \hat{a} - \overline{x}$$

Uniform

$$\hat{a} = \min x_i$$

$$\hat{b} = \max x_i$$

Weibull

Solves the simultaneous equations:

$$\hat{\alpha} = \frac{n}{\left[\left(\dfrac{1}{\hat{\beta}}\right) \sum_{i=1}^{n} x_i^{\hat{a}} \log x_i - \sum_{i=1}^{n} \log x_i\right]}$$

$$\hat{\beta} = \left(\frac{\sum\limits_{i=1}^{n} x_i^{\hat{a}}}{n} \right)^{\frac{1}{\hat{a}}}$$

Chi-square Test for Distribution Fitting

Divide the range of data into non-overlapping classes. The classes are aggregated at each end to ensure that classes have an expected frequency of at least 5.

O_i = observed frequency in class i
E_i = expected frequency in class i from fitted distribution
k = number of classes after aggregation

Test statistic

$$\chi^2 = \sum_{i=1}^{k} \frac{\left(O_i - E_i \right)^2}{E_i}$$

follow a chi-square distribution with the degrees of freedom equal to (k-1-# of estimated parameters)

Kolmogorov − Smirnov Test

$$D_n^{+} = \max \left\{ \frac{i}{n} - \hat{F}\left(x_i \right) \right\}$$

$$1 \le i \le n$$

$$D_n^{-} = \max \left\{ \hat{F}\left(x_i \right) - \frac{i-1}{n} \right\}$$

$$1 \le i \le n$$

$$D_n = \max \left\{ D_n^{+}, D_n^{-} \right\}$$

where $\hat{F}\left(x_i \right)$ = estimated cumulative distribution at x_i

ANOVA

Notations

k = number of treatments
n_t = number of observations for treatment t
$\bar{n} = n / k$ = average treatment size

$$n = \sum_{t=1}^{k} n_t$$

x_{it} = ith observation in treatment t

$$\bar{x}_t = \text{treatment mean} = \frac{\sum_{i=1}^{n_t} x_{it}}{n_t}$$

$$s_t^2 = \text{treatment variance} = \frac{\sum_{i=1}^{n_t} \left(x_{it} - \bar{x}_t\right)^2}{n_t - 1}$$

$$\text{MSE} = \text{mean square error} = \frac{\sum_{t=1}^{k} \left(n_t - 1\right) s_t^2}{\left(\sum_{t=1}^{k} n_t\right) - k}$$

$$df = \text{degrees of freedom for the error term} = \left(\sum_{t=1}^{k} n_t\right) - k$$

Standard Error (internal)

$$\sqrt{\frac{s_t^2}{n_t}}$$

Standard Error (pooled)

$$\sqrt{\frac{MSE}{n_t}}$$

Interval Estimates

$$\bar{x}_t \pm M \sqrt{\frac{MSE}{n_t}}$$

where
confidence interval

$$M = t_{n-k;\alpha/2}$$

LSD interval

$$M = \frac{1}{\sqrt{2}} t_{n-k;\alpha/2}$$

Tukey Interval

$$M = \frac{1}{2} q_{n-k,k;\alpha}$$

where $q_{n-k,k;\alpha}$ = the value of the studentized range distribution with $n-k$ degrees of freedom and k samples such that the cumulative probability equals $1 - \alpha$

Scheffe Interval

$$M = \frac{\sqrt{k-1}}{\sqrt{2}} \sqrt{F_{k-1,n-k;\alpha}}$$

Cochran C-Test

Follow F distribution with $\bar{n} - 1$ and $(\bar{n} - 1)(k - 1)$ degrees of freedom.
   Test statistic:

$$F = \frac{(k-1)C}{1-C}$$

$$\text{where} \quad C = \frac{\max s_t^2}{\sum_{t=1}^{k} s_t^2}$$

Bartlett Test

Test Statistic:

$$B = 10^{\frac{M}{(n-k)}}$$

$$M = (n-k)\log_{10} MSE - \sum_{t=1}^{k}(n_t - 1)\log_{10} s_t^2$$

The significance test is based on

$$\frac{M(\ln 10)}{1 + \frac{1}{3(k-1)}\left[\sum_{t=1}^{k}\frac{1}{(n_1 - 1)} - \frac{1}{N-k}\right]} \quad \Big] \chi^2_{k-1}$$

which follows a chi-square distribution with $k - 1$ degrees of freedom.

Hartley's Test

$$H = \frac{\max\left(s_t^2\right)}{\min\left(s_t^2\right)}$$

Kruskal-Wallis Test

average rank of treatment:

$$\bar{R}_t = \frac{\sum\limits_{i=1}^{n_t} R_{it}}{n_t}$$

If there are no ties:

$$\text{test statistic} \quad w = \left(\frac{12}{n} \sum\limits_{i=1}^{k} n_t \bar{R}_t^{\,2}\right) - 3(n+1)$$

Adjustment for ties:

Let u_j = number of observations tied at any rank for $j = 1, 2, 3, \ldots, m$ where m = number of unique values in the sample.

$$W = \frac{w}{1 - \dfrac{\sum\limits_{j=1}^{m} u_j^3 - \sum\limits_{j=1}^{m} u_j}{n\left(n^2 - 1\right)}}$$

Significance level: W follows a chi-square distribution with $k - 1$ degrees of freedom.

Freidman test

X_{it} = observation in the ith row, tth column

$$i = 1, 2, \ldots, n \quad t = 1, 2, \ldots, k$$

R_{it} = rank of X_{it} within its row

n = common treatment size (all treatment sizes must be the same for this test)

$$R_t = \sum_{i=1}^{n} R_{it}$$

average *rank* $\bar{R}_t = \dfrac{\sum_{i=1}^{n_t} R_{it}}{n_t}$

where data are ranked within each row separately.

test statistic $Q = \dfrac{12S(k-1)}{nk(k^2-1)-\left(\sum u^3 - \sum u\right)}$

where $S = \left(\sum_{t=1}^{k} R_i^2\right) - \dfrac{n^2 k(k+1)^2}{4}$

Q follows a chi-square distribution with k degrees of freedom.

REGRESSION

Notation

Y = vector of n observation for the dependent variable $\sim$
X = n by p matrix of observations for p independent variables, including constant term, if any
$\sim$ indicates a variable is a vector or matrix

$$\bar{Y} = \dfrac{\sum_{i=1}^{n} Y_i}{n}$$

Regression Statistics

(1) Estimated coefficients
Note: estimated by a modified Gram-Schmidt orthogonal decomposition with tolerance = $1.0E - 08$.

$$b = \left(X'X\right)^{-1} X Y$$

(2) Standard errors

$$S\left(\underset{\sim}{b}\right) = \sqrt{\text{diagonal elements of}\left(X'X\right)^{-1}\text{MSE}}$$

$$\text{where} \quad SSE = \underset{\sim}{Y'}\underset{\sim}{Y} - \underset{\sim}{b'}\underset{\sim}{X'}\underset{\sim}{Y}$$

$$MSE = \frac{SSE}{n-p}$$

(3) t-values

$$\underset{\sim}{t} = \frac{\underset{\sim}{b}}{S\left(\underset{\sim}{b}\right)}$$

(4) Significance level
   t-values follow the Student's t distribution with $n - p$ degrees of freedom.
(5) R-squared

$$R^2 = \frac{SSTO - SSE}{SSTO}$$

where

$$SSTO = \begin{cases} \underset{\sim}{Y'} - n\bar{Y}^2 & \text{if constant} \\ \underset{\sim}{Y}\underset{\sim}{Y} & \text{if no constant} \end{cases}$$

Note: When the no constant option is selected, the total sum of square is uncorrected for the mean. Thus, the R^2 value is of little use, since the sum of the residuals is not zero.

(6) Adjusted R-squared

$$1 - \left(\frac{n-1}{n-p}\right)\left(1 - R^2\right)$$

(7) Standard error of estimate

$$SE = \sqrt{MSE}$$

(8) Predicted values

$$\hat{Y} = X b$$

(9) Residuals

$$e = Y - \hat{Y}$$

(10) Durbin–Watson statistic

$$D = \frac{\sum_{i=1}^{n-1}\left(e_{i+1} - e_i\right)^2}{\sum_{i=1}^{n} e_i^2}$$

(11) Mean absolute error

$$\frac{\left(\sum_{i=1}^{n}|e_i|\right)}{n}$$

Predictions

$X_h = m$ by p matrix of independent variables for m predictions

(1) predicted value

$$\hat{Y}_h = X_h b$$

(2) Standard error of predictions

$$S\left(\hat{Y}_{h(new)}\right) = \sqrt{\text{diagonal elements of } \text{MSE}\left(1 + X_h\left(X'X\right)^{-1} X_h'\right)}$$

(3) Standard error of mean response

$$S\left(\hat{Y}_h\right) = \sqrt{\text{diagonal elements of } \text{MSE}\left(X_h\left(X'X\right)^{-1} \hat{X}_h\right)}$$

(4) Prediction matrix results

Column 1 = index numbers of forecasts

$2 = \hat{\underset{\sim}{Y}}_h$

$3 = S\left(\hat{\underset{\sim}{Y}}_{h(new)}\right)$

$4 = \left(\hat{\underset{\sim}{Y}}_h - t_{n-p,\alpha/2}S\left(\hat{\underset{\sim}{Y}}_{h(new)}\right)\right)$

$5 = \left(\hat{\underset{\sim}{Y}}_h + t_{n-p,\alpha/2}S\left(\hat{\underset{\sim}{Y}}_{h(new)}\right)\right)$

$6 = \hat{\underset{\sim}{Y}}_h - t_{n-p,\alpha/2}S\left(\hat{\underset{\sim}{Y}}_h\right)$

$7 = \hat{\underset{\sim}{Y}}_h + t_{n-p,\alpha/2}S\left(\hat{\underset{\sim}{Y}}_h\right)$

Nonlinear Regression

$F\left(X,\hat{\beta}\right)$ are values of nonlinear function using parameter estimates $\hat{\beta}$

(1) Estimated coefficients
   Obtained by minimizing the residual sum of squares using a search procedure suggested by Marquardt. This is a compromise between Gauss-Newton
   and steepest descent methods. The user specifies:
   a) initial estimates $\underset{\sim}{\beta}_0$

   b) initial value of Marquardt parameter λ, which is modified at each
      iteration.
         As $\lambda \to 0$, procedure approaches Gauss-Newton $\lambda \to \infty$, procedure
      approaches steepest descent
   c) scaling factor used to multiply Marquardt parameter after each Iteration
   d) maximum value of Marquardt parameter
         Partial derivatives of F with respect to each parameter are estimated
      numerically.
(2) Standard errors
   estimated from residual sum of squares and partial derivatives
(3) Ratio

$$\text{ratio} = \frac{\text{coefficient}}{\text{standard error}}$$

(4) R-squared

$$R^2 = \frac{\text{SSTO-SSE}}{\text{SSTO}} \quad \text{where}$$

$$\text{SSTO} = \underset{\sim}{Y}'\underset{\sim}{Y} - n\bar{Y}^2$$

$$\text{SSE} = \text{residual sum of squares}$$

Ridge Regression

Additional notation:

Z = matrix of independent variables standardized so that $Z'Z$ equals the correlation matrix

θ = value of the ridge parameter

Parameter estimates

$$b(\theta) = \left(Z'Z + \theta I_p \right)^{-1} Z'Y$$

where I_p is a $p \times p$ identity matrix

QUALITY CONTROL

For all quality control formulas:

$$j = 1, 2, \ldots, k$$

k = number of subgroups
n_j = number of observations in subgroup j
x_{ij} = ith observation n subgroup j

All formulas below for quality control assume 3-sigma limits. If other limits are specified, the formulas are adjusted proportionally based on sigma for the selected limits. Also, average sample size is used unless otherwise specified.

Subgroup Statistics

Subgroup means

$$\bar{x}_j = \frac{\sum_{i=1}^{n_j} x_{ij}}{n_j}$$

Subgroup standard deviations

$$s_j = \sqrt{\frac{\sum_{i=1}^{n_j} \left(x_{ij} - \bar{x}_j \right)^2}{\left(n_j - 1 \right)}}$$

Subgroup range

$$R_j = \max\left\{x_{ij} \mid 1 \le i \le n_j\right\} - \min\left\{x_{ij} \mid 1 \le i \le n_j\right\}$$

X Bar Charts

Compute

$$\overline{\overline{x}} = \frac{\displaystyle\sum_{j=1}^{k} n_i \overline{x}_j}{\displaystyle\sum_{j=1}^{k} n_i}$$

$$\overline{R} = \frac{\left(\displaystyle\sum_{j=1}^{k} n_j R_j\right)}{\displaystyle\sum_{j=1}^{k} n_i}$$

$$s_p = \sqrt{\frac{\displaystyle\sum_{j=1}^{k} \left(n_j - 1\right) s_j^2}{\displaystyle\sum_{j=1}^{k} \left(n_j - 1\right)}}$$

$$\overline{n} = \frac{1}{k} \sum_{j=1}^{k} n_i$$

For a chart based on range:

$$UCL = \overline{\overline{x}} + A_2 \overline{R}$$
$$LCL = \overline{\overline{x}} - A_2 \overline{R}$$

For a chart based on sigma:

$$UCL = \overline{\overline{x}} + \frac{3 s_p}{\sqrt{n}}$$

$$LCL = \overline{\overline{x}} - \frac{3 s_p}{\sqrt{n}}$$

For a chart based on known sigma:

$$UCL = \bar{\bar{x}} + 3\frac{\sigma}{\sqrt{n}}$$

$$LCL = \bar{\bar{x}} - 3\frac{\sigma}{\sqrt{n}}$$

If other than 3-sigma limits are used, such as 2-sigma limits, all bounds are adjusted proportionately. If average sample size is not used, then uneven bounds are displays based on

$$1/\sqrt{n_j}$$

rather than

$$1/\sqrt{n}$$

If the data is normalized, each observation is transformed according to:

$$z_{ij} = \frac{x_{ij} - \bar{\bar{x}}}{\hat{\alpha}}$$

where $\hat{\alpha}$ = estimated standard deviation.

Capability Ratios

Note: The following indices are useful only when the control limits are placed at the specification limits. To override the normal calculations, specify a subgroup size of one and select the "known standard deviation" option. Then enter the standard deviation as half of the distance between the USL and LSL. Change the position of the center line to be the midpoint of the USL and LSL and specify the upper and lower control line at one sigma.

$$C_P = \frac{USL - LSL}{6\hat{\alpha}}$$

$$C_R = \frac{1}{C_P}$$

$$C_{PK} = \min\left(\frac{USL - \bar{\bar{x}}}{3\hat{\alpha}}, \frac{\bar{\bar{x}} - LSL}{3\hat{\alpha}}\right)$$

R Charts

$$CL = \bar{R}$$

$$UCL = D_4\bar{R}$$

$$LCL = \max\left(0, D_3\bar{R}\right)$$

S Charts

$$CL = s_P$$

$$UCL = s_P \sqrt{\frac{\chi^2_{\bar{n}-1;\alpha}}{n-1}}$$

$$LCL = s_P \sqrt{\frac{\chi^2_{\bar{n}-1;\alpha}}{n-1}}$$

C Charts

$$\bar{c} = \frac{\Sigma u_j}{\Sigma n_j} \qquad UCL = \bar{c} + 3\sqrt{\bar{c}}$$
$$LCL = \bar{c} - 3\sqrt{\bar{c}}$$

where u_j = number of defects in the jth sample

U Charts

$$\bar{u} = \frac{\text{number of defects in all samples}}{\text{number of units in all samples}} = \frac{\Sigma u_j}{\Sigma n_j}$$

$$UCL = \bar{u} + \frac{3\sqrt{\bar{u}}}{\sqrt{n}}$$

$$LCL = \bar{u} - \frac{3\sqrt{\bar{u}}}{\sqrt{n}}$$

P Charts

$$p = \frac{\text{number of defective units}}{\text{number of units inspected}}$$

$$\bar{p} = \frac{\text{number of defectives in all samples}}{\text{number of units in all samples}} = \frac{\Sigma p_j n_j}{\Sigma n_j}$$

$$UCL = \bar{p} + \frac{3\sqrt{\bar{p}(1-\bar{p})}}{\sqrt{n}}$$

$$LCL = \bar{p} - \frac{3\sqrt{\bar{p}(1-\bar{p})}}{\sqrt{n}}$$

NP Charts

$$\bar{p} = \frac{\Sigma d_j}{\Sigma n_j},$$

where d_j is the number of defectives in the jth sample.

$$UCL = \bar{n}\,\bar{p} + 3\sqrt{\bar{n}\,\bar{p}\,(1-\bar{p})}$$
$$LCL = \bar{n}\,\bar{p} - 3\sqrt{\bar{n}\,\bar{p}\,(1-\bar{p})}$$

CuSum Chart for the Mean

Control mean = μ
    Standard deviation = α
    Difference to detect = Δ
    Plot cumulative sums C_t versus t, where

$$C_t = \sum_{i=1}^{t}\left(\bar{x}_i - \mu\right) \quad \text{for } t = 1, 2, \ldots, n$$

The V-mask is located at distance

$$d = \frac{2}{\Delta}\left[\frac{\alpha^2/\bar{n}}{\Delta}\ln\frac{1-\beta}{\alpha/2}\right]$$

in front of the last data point.

$$\text{Angle of mast} = 2\tan^{-1}\frac{\Delta}{2}$$
$$\text{Slope of the lines} = \pm\frac{\Delta}{2}$$

Multivariate Control Charts

    X = matrix of n rows and k columns containing n observations for each of k variable
    S = sample covariance matrix
    X = observation vector at time t
    $\bar{X}^t$ = vector of column average

Then,

$$T_t^2 = \left(X_t - \bar{X}\right)S^{-1}\left(X_t - \bar{X}\right)$$
$$UCL = \left[\frac{k(n-1)}{n-k}\right]F_{k, n-k; \alpha}$$

TIME SERIES ANALYSIS

Notation

$$x_t \text{ or } y_t = \text{observation at time } t, t = 1,2,\ldots,n$$
$$n = \text{number of observations}$$

Autocorrelation at lag k

$$r_k = \frac{c_k}{c_0}$$

where

$$c_k = \frac{1}{n}\sum_{t=1}^{n-k}\left(y_t - \bar{y}\right)\left(y_{t+k} - \bar{y}\right)$$

and

$$\bar{y} = \frac{\left(\sum_{t=1}^{n} y_t\right)}{n}$$

$$\text{standard error} = \sqrt{\frac{1}{n}\left\{1 + 2\sum_{v=1}^{k-1} r_v^2\right\}}$$

Partial autocorrelation at lag k

$\hat{\theta}_{kk}$ is obtained by solving the Yule-Walker equations:

$$r_j = \hat{\theta}_{k1}r_{j-1} + \hat{\theta}_{k2}r_{j-2} + \cdots + \hat{\theta}_{k(k-1)}r_{j-k+1} + \hat{\theta}_{kk}r_{j-k}$$
$$j = 1,2,\ldots,k$$

$$\text{standard error} = \sqrt{\frac{1}{n}}$$

Cross Correlation at lag k

$x = $ input time series
$y = $ output time series

$$r_{xy}(k) = \frac{c_{xy}(k)}{s_x s_y} \quad k - 0, \pm 1, \pm 2, \ldots$$

$$\text{where } c_{xy}\left(k\right) = \begin{cases} \dfrac{1}{n}\displaystyle\sum_{t=1}^{n-k}\left(x_t - \bar{x}\right)\left(y_{t+k} - \bar{y}\right) & k = 0,1,2,\ldots \\[2em] \dfrac{1}{n}\displaystyle\sum_{t=1}^{n+k}\left(x_t - \bar{x}\right)\left(y_{t-k} - \bar{y}\right) & k = 0,-1,-2,\ldots \end{cases}$$

and

$$S_x = \sqrt{c_{xx}\left(0\right)}$$

$$S_y = \sqrt{c_{yy}\left(0\right)}$$

Box-Cox

$$yt = \frac{\left(y + \lambda_2\right)^{\lambda_1} - 1}{\lambda_1 g^{\left(\lambda_1 - 1\right)}} \qquad \text{if } \lambda_1 > 0$$

$$yt = g\ln\left(y + \lambda_2\right) \qquad \text{if } \lambda_1 = 0$$

where g = sample geometric mean $(y + \lambda_2)$

Periodogram (computed using Fast Fourier Transform)
If n is odd:

$$I\left(f_1\right) = \frac{n}{2}\left(a_i^2 + b_i^2\right) \qquad i = 1,2,\ldots,\left[\frac{n-1}{2}\right]$$

$$\text{where } a_i = \frac{2}{n}\sum_{t=1}^{n} t_t \cos 2\pi f_i t$$

$$b_i = \frac{2}{n}\sum_{t=1}^{n} y_t \sin 2\pi f_i t$$

$$f_i = \frac{i}{n}$$

If n is even, an additional term is added:

$$I\left(0.5\right) = n\left(\frac{1}{n}\sum_{t=1}^{n}\left(-1\right)^t Y_t\right)^2$$

CATEGORICAL ANALYSIS

Notation

r = number of rows in table
c = number of columns in table
f_{ij} = frequency in position (row i, column j)
x_i = distinct values of row variable arranged in ascending order; $i = 1,..., r$
y_j = distinct values of column variable arranged in ascending order, $j = 1,..., c$

Totals

$$R_j = \sum_{j=1}^{c} f_{ij} \qquad C_j = \sum_{i=1}^{r} f_{ij}$$

$$N = \sum_{i=1}^{r}\sum_{j=1}^{c} f_{ij}$$

Note: any row or column which totals zero is eliminated from the table before calculations are performed.

Chi-Square

$$\chi^2 = \sum_{i=1}^{r}\sum_{j=1}^{c} \frac{\left(f_{ij} - E_{ij}\right)^2}{E_{ij}}$$

$$\text{where} \quad E_{ij} = \frac{R_i C_j}{N} \sim \chi^2_{(r-1)(c-1)}$$

A warning is issued if any $E_{ij} < 2$ or if 20% or more of all $E_{ij} < 5$ For 2x2 tables, a second statistic is printed using Yate's continuity correction.

Fisher's Exact Test

Run for a 2 × 2 table, when N is less than or equal to 100. For calculation details, see standard references such as The Analysis of Contingency tables by B. S. Everitt.

Lambda

$$\lambda = \frac{\left(\sum_{j=1}^{c} f_{max,j} - R_{max}\right)}{N - R_{max}}$$

with rows dependent

$$\lambda = \frac{\left(\displaystyle\sum_{i=1}^{r} f_{i,\max} - C_{\max}\right)}{N - C_{\max}}$$

with columns dependent

$$\lambda = \frac{\left(\displaystyle\sum_{i=1}^{r} f_{i,\max} + \sum_{j=1}^{c} f_{\max,j} - C_{\max} - R_{\max}\right)}{\left(2N - R_{\max} - C_{\max}\right)}$$

when symmetric

were $f_{i\max}$ = largest value in row i
$f_{\max j}$ = largest value in column j
$R_{\max}$ = largest row total
$C_{\max}$ = largest column total

Uncertainty Coefficient

$$U_R = \frac{U(R) + U(C) - U(RC)}{U(R)}$$ with rows dependent

$$U_C = \frac{U(R) + U(C) - U(RC)}{U(C)}$$ with columns dependent

$$U = 2\left(\frac{U(R) + U(C) - U(RC)}{U(R) + U(C)}\right)$$ when symmetric

where

$$U(R) = -\sum_{i=1}^{r} \frac{R_i}{N} \log \frac{R_i}{N}$$

$$U(C) = -\sum_{j=1}^{c} \frac{C_j}{N} \log \frac{C_j}{N}$$

$$U(RC) = -\sum_{i=1}^{r}\sum_{j=1}^{c} \frac{f_{ij}}{N} \log \frac{f_{ij}}{N}$$ for $f_{ij} > 0$

Somer's D

$$D_R = \frac{2\left(P_C - P_D\right)}{\left(N^2 - \sum_{j=1}^{c} C_j^2\right)} \quad \text{with rows dependent}$$

$$D_C = \frac{2\left(P_C - P_D\right)}{\left(N^2 - \sum_{i=1}^{r} R_i^2\right)} \quad \text{with columns dependent}$$

$$D = \frac{4\left(P_C - P_D\right)}{\left(N^2 - \sum_{i=1}^{r} R_i^2\right) + \left(N^2 - \sum_{j=1}^{c} C_j^2\right)} \quad \text{when symmetric}$$

where the number of concordant pairs is

$$P_C = \sum_{i=1}^{r}\sum_{j=1}^{c} f_{ij} \sum_{h<i}\sum_{k<j} f_{hk}$$

and the number of discordant pairs is

$$P_D = \sum_{i=1}^{r}\sum_{j=1}^{c} f_{ij} \sum_{h<i}\sum_{k>j} f_{hk}$$

Eta

$$E_R = \sqrt{1 - \frac{SS_{RN}}{SS_R}} \quad \text{with rows dependent}$$

where the total corrected sum of squares for the rows is

$$SS_R = \sum_{i=1}^{r}\sum_{j-1}^{c} x_i^2 f_{ij} - \frac{\left(\sum_{i=1}^{r}\sum_{j-1}^{c} x_i f_{ij}\right)^2}{N}$$

and the sum of squares of rows within categories of columns is

$$SS_{RN} = \sum_{j=1}^{c} \left[\sum_{i=1}^{r} x_i^2 f_{ij} - \frac{\left(\sum_{i=1}^{r} x_i^2 f_{ij} \right)^2}{C_j} \right]$$

$$E_C = \sqrt{1 - \frac{SS_{CN}}{SS_C}} \quad \text{with columns dependent}$$

where the total corrected sum of squares for the columns is

$$SS_C = \sum_{i=1}^{r} \sum_{j=1}^{c} y_i^2 f_{ij} - \frac{\left(\sum_{i=1}^{r} \sum_{j=1}^{c} y_i f_{ij} \right)^2}{N}$$

and the sum of squares of columns within categories of rows is

$$SS_{CN} = \sum_{i=1}^{r} \left[\sum_{j=1}^{c} y_i^2 f_{ij} - \frac{\left(\sum_{j=1}^{c} y_i^2 f_{ij} \right)^2}{R_i} \right] j$$

Contingency Coefficient

$$C = \sqrt{\frac{\chi^2}{\left(\chi^2 + N \right)}}$$

Cramer's V

$$V = \sqrt{\frac{\chi^2}{N}} \quad \text{for 2 x 2 table}$$

$$V = \sqrt{\frac{\chi^2}{N(m-1)}} \quad \text{for all others where}$$

$$m = \min(r, c)$$

Conditional Gamma

$$G = \frac{P_C - P_D}{P_C + P_D}$$

Pearson's r

$$R = \frac{\sum_{j=1}^{c}\sum_{i=1}^{r} x_i y_j f_{ij} - \dfrac{\left(\sum_{j=1}^{c}\sum_{i=1}^{r} x_i f_{ij}\right)\left(\sum_{j=1}^{c}\sum_{i=1}^{r} y_i f_{ij}\right)}{N}}{\sqrt{SS_R SS_C}}$$

If $R = 1$, no significance is printed. Otherwise, the one-sided significance is based on

$$t = R\sqrt{\frac{N-2}{1-R^2}}$$

Kendall's Tau b

$$\tau = \frac{2(P_C - P_D)}{\sqrt{\left(N^2 - \sum_{i=1}^{r} R_i^2\right)\left(N^2 - \sum_{j=1}^{c} C_i^2\right)}}$$

Tau C

$$\tau_C = \frac{2m(P_C - P_D)}{(m-1)N^2}$$

PROBABILITY TERMINOLOGY

Experiment: An experiment is an activity or occurrence with an observable result.

Outcome: The result of the experiment.

Sample point: An outcome of an experiment.

Event: An event is a set of outcomes (a subset of the sample space) to which a probability is assigned.

Basic Probability Principles

Consider a random sampling process in which all the outcomes solely depend on chance, i.e., each outcome is equally likely to happen. If S is a uniform sample space and the collection of desired outcomes is E, the probability of the desired outcomes is

$$P(E) = \frac{n(E)}{n(S)}$$

where

$n(E)$ = number of favorable outcomes in E
$n(S)$ = number of possible outcomes in S

Since E is a subset of S,

$$0 \leq n(E) \leq n(S),$$

the probability of the desired outcome is

$$0 \leq P(E) \leq 1$$

Random Variable

A random variable is a rule that assigns a number to each outcome of a chance experiment.
   Example:

1. A coin is tossed six times. The random variable X is the number of tails that are noted. X can only take the values 1,2,..., 6, so X is a discrete random variable.
2. A light bulb is burned until it burns out. The random variable Y is its lifetime in hours. Y can take any positive real value, so Y is a continuous random variable.

Mean Value $\hat{x}$ or Expected Value μ

The mean value or expected value of a random variable indicates its average or central value. It is a useful summary value of the variable's distribution.

1. If random variable X is a discrete mean value,

$$\hat{x} = x_1 p_1 + x_2 p_2 + \ldots + x_n p_n = \sum_{i=1}^{n} x_1 p_1$$

where

p_i = probability densities

2. If X is a continuous random variable with probability density function $f(x)$, then the expected value of X is

$$\mu = E(X) = \int_{-\infty}^{+\infty} xf(x)dx$$

where

$f(x)$ = probability densities

DISCRETE DISTRIBUTION FORMULAS

Probability Mass Function, $p(x)$
Mean, μ
Variance, σ^2
Coefficient of Skewness, β_1
Coefficient of Kurtosis, β_2
Moment-generating Function, $M(t)$
Characteristic Function, $\phi(t)$
Probability-generating Function, $P(t)$

Bernoulli Distribution

$$p(x) = p^x q^{x-1} \qquad x = 0,1 \qquad 0 \le p \le 1 \qquad q = 1-p$$

$$\mu = p \qquad \sigma^2 = pq \qquad \beta_1 = \frac{1-2p}{\sqrt{pq}} \qquad \beta_2 = 3 + \frac{1-6pq}{pq}$$

$$M(t) = q + pe^t \qquad \phi(t) = q + pe^{it} \qquad P(t) = q + pt$$

Beta Binomial Distribution

$$p(x) = \frac{1}{n+1} \frac{B(a+x,b+n-x)}{B(x+1,n-x+1)B(a,b)} \qquad x = 0,1,2,\ldots,n \qquad a > 0 \qquad b > 0$$

$$\mu = \frac{na}{a+b} \qquad \sigma^2 = \frac{nab(a+b+n)}{(a+b)^2(a+b+1)} \qquad B(a,b) \text{ is the Beta function.}$$

Beta Pascal Distribution

$$p(x) = \frac{\Gamma(x)\Gamma(v)\Gamma(\rho+v)\Gamma(v+x-(\rho+r))}{\Gamma(r)\Gamma(x-r+1)\Gamma(\rho)\Gamma(v-\rho)\Gamma(v+x)} \qquad x = r, r+1, \ldots \quad v > p > 0$$

$$\mu = r\frac{v-1}{\rho-1}, \quad \rho > 1 \qquad \sigma^2 = r(r+\rho-1)\frac{(v-1)(v-\rho)}{(\rho-1)^2(\rho-2)}, \rho > 2$$

Binomial Distribution

$$p(x) = \binom{n}{x}p^x q^{n-x} \qquad x = 0,1,2,\ldots,n \qquad 0 \le p \le 1 \quad q = 1-p$$

$$\mu = np \qquad \sigma^2 = npq \qquad \beta_1 = \frac{1-2p}{\sqrt{npq}} \qquad \beta_2 = 3 + \frac{1-6pq}{npq}$$

$$M(t) = \left(q + pe^t\right)^n \qquad \phi(t) = \left(q + pe^{it}\right)^n \qquad P(t) = \left(q + pt\right)^n$$

Discrete Weibull Distribution

$$p(x) = (1-p)^{x^\beta} - (1-p)^{(x+1)^\beta} \qquad x = 0,1,\ldots \qquad 0 \le p \le 1 \quad \beta > 0$$

Geometric Distribution

$$p(x) = pq^{1-x} \qquad x = 0,1,2,\ldots \qquad 0 \le p \le 1 \quad q = 1-p$$

$$\mu = \frac{1}{p} \qquad \sigma^2 = \frac{q}{p^2} \qquad \beta_1 = \frac{2-p}{\sqrt{q}} \qquad \beta_2 = \frac{p^2+6q}{q}$$

$$M(t) = \frac{p}{1-qe^t} \qquad \phi(t) = \frac{p}{1-qe^{it}} \qquad P(t) = \frac{p}{1-qt}$$

Hypergeometric Distribution

$$p(x) = \frac{\binom{M}{x}\binom{N-M}{n-x}}{\binom{N}{n}} \qquad x = 0,1,2,\ldots,n \qquad x \le M \qquad n-x \le N-M$$

$$n, M, N, \in N \qquad 1 \le n \le N \qquad 1 \le M \le N \qquad N = 1, 2, \ldots$$

$$\mu = n\frac{M}{N} \qquad \sigma^2 = \left(\frac{N-n}{N-1}\right)n\frac{M}{N}\left(1-\frac{M}{N}\right) \qquad \beta_1 = \frac{(N-2M)(N-2n)\sqrt{N-1}}{(N-2)\sqrt{nM(N-M)(N-n)}}$$

$$\beta_2 = \frac{N^2(N-1)}{(N-2)(N-3)nM(N-M)(N-n)}$$

$$\left\{ N(N+1) - 6n(N-n) + 3\frac{M}{N^2}(N-M)\left[N^2(n-2) - Nn^2 + 6n(N-n)\right] \right\}$$

$$M(t) = \frac{(N-M)!(N-n)!}{N!} F\left(.,e^t\right)$$

$$\phi(t) = \frac{(N-M)!(N-n)!}{N!} F\left(.,e^{it}\right)$$

$$P(t) = \left(\frac{N-M}{N}\right)^n F(.,t)$$

$F(\alpha, \beta, \gamma, x)$ is the hypergeometric function. $\alpha = -n$; $\beta = -M$; $\gamma = N - M - n + 1$

Negative Binomial Distribution

$$p(x) = \binom{x+r-1}{r-1} p^r q^x \qquad x = 0, 1, 2, \ldots \qquad r = 1, 2, \ldots \qquad 0 \le p \le 1 \qquad q = 1 - p$$

$$\mu = \frac{rq}{p} \qquad \sigma^2 = \frac{rq}{p^2} \qquad \beta_1 = \frac{2-p}{\sqrt{rq}} \qquad \beta_2 = 3 + \frac{p^2 + 6q}{rq}$$

$$M(t) = \left(\frac{p}{1-qe^t}\right)^r \qquad \phi(t) = \left(\frac{p}{1-qe^{it}}\right)^r \qquad P(t) = \left(\frac{p}{1-qt}\right)^r$$

Poisson Distribution

$$p(x) = \frac{e^{-\mu}\mu^x}{x!} \qquad x = 0, 1, 2, \ldots \qquad \mu > 0$$

$$\mu = \mu \qquad \sigma^2 = \mu \qquad \beta_1 = \frac{1}{\sqrt{\mu}} \qquad \beta_2 = 3 + \frac{1}{\mu}$$

$$M(t) = \exp\left[\mu\left(e^t - 1\right)\right] \qquad \sigma(t) = \exp\left[\mu\left(e^{it} - 1\right)\right] \qquad P(t) = \exp\left[\mu\left(t - 1\right)\right]$$

Rectangular (Discrete Uniform) Distribution

$$p(x) = 1/n \qquad x = 1, 2, \ldots, n \qquad n \in N$$

$$\mu = \frac{n+1}{2} \qquad \sigma^2 = \frac{n^2-1}{12} \qquad \beta_1 = 0 \qquad \beta_2 = \frac{3}{5}\left(3 - \frac{4}{n^2-1}\right)$$

$$M(t) = \frac{e^t\left(1-e^{nt}\right)}{n\left(1-e^t\right)} \qquad \phi(t) = \frac{e^{it}\left(1-e^{nit}\right)}{n\left(1-e^{it}\right)} \qquad P(t) = \frac{t\left(1-t^n\right)}{n\left(1-t\right)}$$

CONTINUOUS DISTRIBUTION FORMULAS

Probability Density Function, $f(x)$
Mean, μ
Variance, σ^2
Coefficient of Skewness, β_1
Coefficient of Kurtosis, β_2
Moment-generating Function, $M(t)$
Characteristic Function, $\phi(t)$

Arcsin Distribution

$$f(x) = \frac{1}{\pi\sqrt{x(1-x)}} \qquad 0 < x < 1$$

$$\mu = \frac{1}{2} \qquad \sigma^2 = \frac{1}{8} \qquad \beta_1 = 0 \qquad \beta_2 \frac{3}{2}$$

Beta Distribution

$$f(x) = \frac{\Gamma(\alpha+\beta)}{\Gamma(\alpha)\Gamma(\beta)} x^{\alpha-1}(1-x)^{\beta-1} \qquad 0 < x < 1 \qquad \alpha, \beta > 0$$

$$\mu = \frac{\alpha}{\alpha+\beta} \qquad \sigma^2 = \frac{\alpha\beta}{(\alpha+\beta)^2(\alpha+\beta+1)} \qquad \beta_2 = \frac{2(\beta-\alpha)\sqrt{\alpha+\beta+1}}{\sqrt{\alpha\beta}(\alpha+\beta+2)}$$

$$\beta_2 = \frac{3(\alpha+\beta+1)\left[2(\alpha+\beta)^2+\alpha\beta(\alpha+\beta-6)\right]}{\alpha\beta(\alpha+\beta+2)(\alpha+\beta+3)}$$

Cauchy Distribution

$$f(x) = \frac{1}{b\pi\left[1+\left(\dfrac{x-a}{b}\right)^2\right]} \qquad -\infty < x < \infty \qquad -\infty < a < \infty \qquad b > 0$$

$$\mu, \sigma^2, \beta_1, \beta_2, M(t) \text{ do not exist.} \qquad \phi(t) = \exp\left[ait - b|t|\right]$$

Chi Distribution

$$f(x) = \frac{x^{n-1}e^{-x^2/2}}{2^{(n/2)-1}\Gamma(n/2)} \qquad x \ge 0 \qquad n \in N$$

$$\mu = \frac{\Gamma\left(\dfrac{n+1}{2}\right)}{\Gamma\left(\dfrac{n}{2}\right)} \qquad \sigma^2 = \frac{\Gamma\left(\dfrac{n+2}{2}\right)}{\Gamma\left(\dfrac{n}{2}\right)} - \left[\frac{\Gamma\left(\dfrac{n+1}{2}\right)}{\Gamma\left(\dfrac{n}{2}\right)}\right]^2$$

Chi-Square Distribution

$$f(x) = \frac{e^{-x/2}x^{(v/2)-1}}{2^{v/2}\Gamma(v/2)} \qquad x \ge 0 \qquad v \in N$$

$$\mu = v \qquad \sigma^2 = 2v \qquad \beta_1 = 2\sqrt{2/v} \qquad \beta_2 = 3 + \frac{12}{v} \qquad M(t) = (1-2t)^{-v/2}, t < \frac{1}{2}$$

$$\phi(t) = (1-2it)^{-v/2}$$

Erlang Distribution

$$f(x) = \frac{1}{\beta^{n(n-1)!}} x^{n-1}e^{-x/\beta} \qquad x \ge 0 \qquad \beta > 0 \qquad n \in N$$

$$\mu = n\beta \qquad \sigma^2 = n\beta^2 \qquad \beta_1 = \frac{2}{\sqrt{n}} \qquad \beta_2 = 3 + \frac{6}{n}$$

$$M(t) = (1-\beta t)^{-n} \qquad \phi(t) = (1-\beta it)^{-n}$$

Exponential Distribution

$$f(x) = \lambda e^{-\lambda x} \qquad x \geq 0 \qquad \lambda > 0$$

$$\mu = \frac{1}{\lambda} \qquad \sigma^2 = \frac{1}{\lambda^2} \qquad \beta_1 = 2 \qquad \beta_2 = 9 \qquad M(t) = \frac{\lambda}{\lambda - t}$$

$$\phi(t) = \frac{\lambda}{\lambda - it}$$

Extreme-Value Distribution

$$f(x) = \exp\left[-e^{-(x-\alpha)/\beta}\right] \qquad -\infty < x < \infty \qquad -\infty < \alpha < \infty \qquad \beta > 0$$

$\mu = \alpha + \gamma\beta, \gamma \doteq .5772\ldots$ is Euler's constant $\sigma^2 = \dfrac{\pi^2\beta^2}{6}$
$\beta_1 = 1.29857\ \beta_2 = 5.4$

$$M(t) = e^{\alpha t}\Gamma(1 - \beta t), \qquad t < \frac{1}{\beta} \qquad \phi(t) = e^{\alpha it}\Gamma(1 - \beta it)$$

F Distribution

$$f(x) = \frac{\Gamma\left[(v_1 + v_2)/2\right] v_1^{v_1/2} v_2^{v_2/2}}{\Gamma(v_1/2)\Gamma(v_2/2)} x^{(v_1/2)-1}\left(v_2 + v_1 x\right)^{-x(v_1+v_2)/2}$$

$$x > 0 \qquad v_1, v_2 \in N$$

$$\mu = \frac{v_2}{v_2 - 2}, v_2 \geq 3 \qquad \sigma^2 = \frac{2v_2^2\left(v_1 + v_2 - 2\right)}{v_1\left(v_2 - 2\right)^2\left(v_2 - 4\right)}, \qquad v_2 \geq 5$$

$$\beta_1 = \frac{\left(2v_1 + v_2 - 2\right)\sqrt{8\left(v_2 - 4\right)}}{\sqrt{v_1}\left(v_2 - 6\right)\sqrt{v_1 + v_2 - 2}}, \qquad v_2 \geq 7$$

$$\beta_2 = 3 + \frac{12\left[\left(v_2 - 2\right)^2\left(v_2 - 4\right) + v_1\left(v_1 + v_2 - 2\right)\left(5v_2 - 22\right)\right]}{v_1\left(v_2 - 6\right)\left(v_2 - 8\right)\left(v_1 + v_2 - 2\right)}, \qquad v_2 \geq 9$$

$M(t)$ does not exist.

$$\phi\left(\frac{v_1}{v_2}t\right) = \frac{G\left(v_1, v_2, t\right)}{B\left(v_1/2, v_2/2\right)}$$

$B(a, b)$ is the Beta function. G is defined by

$$(m+n-2)G(m,n,t)=(m-2)G(m-2,n,t)+2itG(m,n-2,t), m,n>2$$
$$mG(m,n,t)=(n-2)G(m+2,n-2,t)-2itG(m+2,n-4,t), n>4$$
$$nG(2,n,t)=2+2itG(2,n-2,t), \quad n>2$$

Gamma Distribution

$$f(x)=\frac{1}{\beta^{\alpha}\Gamma(\alpha)}x^{\alpha-1}e^{-x/\beta} \qquad x\geq 0 \qquad \alpha,\beta>0$$

$$\mu=\alpha\beta \qquad \sigma^2=\alpha\beta^2 \qquad \beta_1=\frac{2}{\sqrt{\alpha}} \qquad \beta_2=3\left(1+\frac{2}{\alpha}\right)$$

$$M(t)=(1-\beta t)^{-\alpha} \qquad \phi(t)=(1-\beta it)^{-\alpha}$$

Half-Normal Distribution

$$f(x)=\frac{2\theta}{\pi}\exp\left[-\left(\theta^2 x^2/\pi\right)\right] \qquad x\geq 0 \qquad \theta>0$$

$$\mu=\frac{1}{\theta} \qquad \sigma^2=\left(\frac{\pi-2}{2}\right)\frac{1}{\theta^2} \qquad \beta_1=\frac{4-\pi}{\theta^3} \qquad \beta_2=\frac{3\pi^2-4\pi-12}{4\theta^4}$$

LaPlace (Double Exponential) Distribution

$$f(x)=\frac{1}{2\beta}\exp\left[-\frac{|x-\alpha|}{\beta}\right] \qquad -\infty<x<\infty \qquad -\infty<\alpha<\infty \qquad \beta>0$$

$$\mu=\alpha \qquad \sigma^2=2\beta^2 \qquad \beta_1=0 \qquad \beta_2=6$$

$$M(t)=\frac{e^{\alpha t}}{1-\beta^2 t^2} \qquad \phi(t)=\frac{e^{\alpha it}}{1+\beta^2 t^2}$$

Logistic Distribution

$$f(x)=\frac{\exp\left[(x-\alpha)/\beta\right]}{\beta\left(1+\exp\left[(x-\alpha)/\beta\right]\right)^2}$$

$$-\infty < x < \infty \qquad -\infty < \alpha < \infty \qquad -\infty < \beta < \infty$$

$$\mu = \alpha \qquad \sigma^2 = \frac{\beta^2 \pi^2}{3} \qquad \beta_1 = 0 \qquad \beta_2 = 4.2$$

$$M(t) = e^{\alpha t} \pi \beta t \csc(\pi \beta t) \phi(t) = e^{\alpha it} \pi \beta it \csc(\pi \beta it)$$

Lognormal Distribution

$$f(x) = \frac{1}{\sqrt{2\pi}\sigma x} \exp\left[-\frac{1}{2\sigma^2}(\ln x - \mu)^2\right]$$

$$x > 0 \quad -\infty < \mu < \infty \quad \sigma > 0$$

$$\mu = e^{\mu + \sigma^2/2}\sigma^2 = e^{2\mu + \sigma^2}\left(e^{\sigma^2} - 1\right)$$

$$\beta_1 = \left(e^{\sigma^2} + 2\right)\left(e^{\sigma^2} - 1\right)^{1/2} \quad \beta_2 = \left(e^{\sigma^2}\right)^4 + 2\left(e^{\sigma^2}\right)^3 + 3\left(e^{\sigma^2}\right)^2 - 3$$

Noncentral Chi-Square Distribution

$$f(x) = \frac{\exp\left[-\frac{1}{2}(x+\lambda)\right]}{2^{\nu/2}} \sum_{j=0}^{\infty} \frac{x^{(\nu/2)+j-1}\lambda^j}{\Gamma\left(\frac{\nu}{2}+j\right)2^{2j}j!}$$

$$x > 0 \quad \lambda > 0 \quad \nu \in N$$

$$\mu = \nu + \lambda \quad \sigma^2 = 2(\nu + 2\lambda) \quad \beta_1 = \frac{\sqrt{8}(\nu + 3\lambda)}{(\nu + 2\lambda)^{3/2}} \quad \beta_2 = 3 + \frac{12(\nu + 4\lambda)}{(\nu + 2\lambda)^2}$$

$$M(t) = (1-2t)^{-\nu/2} \exp\left[\frac{\lambda t}{1-2t}\right] \phi(t) = (1-2it)^{-\nu/2} \exp\left[\frac{\lambda it}{1-2it}\right]$$

Noncentral F Distribution

$$f(x) = \sum_{i=0}^{\infty} \frac{\Gamma\left(\frac{2i + \nu_1 + \nu_2}{2}\right)\left(\frac{\nu_1}{\nu_2}\right)^{(2i+\nu_1)/2} x^{(2i+\nu_1-2)/2}e^{-\lambda/2}\left(\frac{\lambda}{2}\right)}{\Gamma\left(\frac{\nu_2}{2}\right)\Gamma\left(\frac{2i+\nu_1}{2}\right)\nu_1!\left(1+\frac{\nu_1}{\nu_2}x\right)^{(2i+\nu_1+\nu_2)/2}}$$

$$x > 0 \quad \nu_1, \nu_2 \in N \quad \lambda > 0$$

$$\mu = \frac{(v_1 + \lambda)v_2}{(v_2 - 2)v_1}, v_2 > 2$$

$$\sigma^2 = \frac{(v_1 + \lambda)^2 + 2(v_1 + \lambda)v_2^2}{(v_2 - 2)(v_2 - 4)v_1^2} - \frac{(v_1 + \lambda)^2 v_2^2}{(v_2 - 2)^2 v_1^2}, \quad v_2 > 4$$

Noncentral t Distribution

$$f(x) = \frac{v^{v/2}}{\Gamma\left(\dfrac{v}{2}\right)} \frac{e^{-\delta^2/2}}{\sqrt{\pi}\left(v + x^2\right)^{(v+1)/2}} \sum_{i=0}^{\infty} \Gamma\left(\frac{v+i+1}{2}\right)\left(\frac{\delta^i}{i!}\right)\left(\frac{2x^2}{v+x^2}\right)^{i/2}$$

$$-\infty < x < \infty \quad -\infty < \delta < \infty \quad v \in N$$

$$\mu_r' = c_r \frac{\Gamma\left(\dfrac{v-r}{2}\right)v^{r/2}}{2^{r/2}\Gamma\left(\dfrac{v}{2}\right)}, \quad v > r, c_{2r-1} = \sum_{i=1}^{r} \frac{(2r-1)!\delta^{2r-1}}{(2i-1)!(r-i)!2^{r-i}},$$

$$c_{2r} = \sum_{i=0}^{r} \frac{(2r)!\delta^{2i}}{(2i)!(r-i)!2^{r-i}}, r = 1,2,3,\ldots$$

Normal Distribution

$$f(x) = \frac{1}{\sigma\sqrt{2\pi}} \exp\left[-\frac{(x-\mu)^2}{2\sigma^2}\right]$$

$$-\infty < x < \infty \quad -\infty < \mu < \infty \quad \sigma > 0$$

$$\mu = \mu \quad \sigma^2 = \sigma^2 \quad \beta_1 = 0 \quad \beta_2 = 3 \quad M(t) = \exp\left[\mu t + \frac{t^2\sigma^2}{2}\right]$$

$$\phi(t) = \exp\left[\mu i t - \frac{t^2\sigma^2}{2}\right]$$

Pareto Distribution

$$f(x) = \theta a^\theta / x^{\theta+1} \quad x \geq a \quad \theta > 0 \quad a > 0$$

$$\mu = \frac{\theta a}{\theta - 1}, \quad \theta > 1 \quad \sigma^2 = \frac{\theta a^2}{(\theta-1)^2(\theta-2)}, \quad \theta > 2$$

$M(t)$ does not exist.

Rayleigh Distribution

$$f(x) = \frac{x}{\sigma^2} \exp\left[-\frac{x^2}{2\sigma^2}\right] x \geq 0 \quad \sigma = 0$$

$$\mu = \sigma\sqrt{\pi/2} \quad \sigma^2 = 2\sigma^2\left(1-\frac{\pi}{4}\right) \quad \beta_1 = \frac{\sqrt{\pi}}{4} \frac{(\pi-3)}{\left(1-\frac{\pi}{4}\right)^{3/2}}$$

$$\beta_2 = \frac{2 - \frac{3}{16}\pi^2}{\left(1-\frac{\pi}{4}\right)^2}$$

t-Distribution

$$f(x) = \frac{1}{\sqrt{\pi v}} \frac{\Gamma\left(\frac{v+1}{2}\right)}{\Gamma\frac{v}{2}} \left(1+\frac{x^2}{v}\right)^{-(v+1)/2} \quad -\infty < x < \infty \quad v \in N$$

$$\mu = 0, \quad v \geq 2 \quad \sigma^2 = \frac{v}{v-2}, \quad v \geq 3 \quad \beta_1 = 0, \quad v \geq 4$$

$$\beta_2 = 3 + \frac{6}{v-4}, \quad v \geq 5$$

$M(t)$ does not exist.

$$\phi(t) = \frac{\sqrt{\pi}\Gamma\left(\frac{v}{2}\right)}{\Gamma\left(\frac{v+1}{2}\right)} \int_{-\infty}^{\infty} \frac{e^{itz\sqrt{v}}}{\left(1+z^2\right)^{(v+1)/2}} dz$$

Triangular Distribution

$$f(x) = \begin{cases} 0 & x \leq a \\ 4(x-a)/(b-a)^2 & a < x \leq (a+b)/2 \\ 4(b-x)/(b-a)^2 & (a+b)/2 < x < b \\ 0 & x \geq b \end{cases}$$

$$-\infty < a < b < \infty$$

$$\mu = \frac{a+b}{2} \qquad \sigma^2 = \frac{(b-a)^2}{24} \qquad \beta_1 = 0 \qquad \beta_2 = \frac{12}{5}$$

$$M(t) = -\frac{4\left(e^{at/2} - e^{bt/2}\right)^2}{t^2(b-a)^2} \qquad\qquad \phi(t) = \frac{4\left(e^{ait/2} - e^{bit/2}\right)^2}{t^2(b-a)^2}$$

Uniform Distribution

$$f(x) = \frac{1}{b-a} \qquad a \le x \le b \qquad -\infty < a < b < \infty$$

$$\mu = \frac{a+b}{2} \qquad \sigma^2 = \frac{(b-a)^2}{12} \qquad \beta_1 = 0 \qquad \beta_2 = \frac{9}{5}$$

$$M(t) = \frac{e^{bt} - e^{at}}{(b-a)t} \qquad \phi(t) = \frac{e^{bit} - e^{ait}}{(b-a)it}$$

Weibull Distribution

$$f(x) = \frac{\alpha}{\beta^\alpha} x^{\alpha-1} e^{-(x/\beta)^\alpha} \quad x \ge 0 \qquad \alpha, \beta > 0$$

$$\mu = \beta\Gamma\left(1+\frac{1}{\alpha}\right) \qquad \sigma^2 = \beta^2\left[\Gamma\left(1+\frac{2}{\alpha}\right) - \Gamma^2\left(1+\frac{1}{\alpha}\right)\right]$$

$$\beta_1 = \frac{\Gamma\left(1+\frac{3}{\alpha}\right) - 3\Gamma\left(1+\frac{1}{\alpha}\right)\Gamma\left(1+\frac{2}{\alpha}\right) + 2\Gamma^3\left(1+\frac{1}{\alpha}\right)}{\left[\Gamma\left(1+\frac{2}{\alpha}\right) - \Gamma^2\left(1+\frac{1}{\alpha}\right)\right]^{3/2}}$$

$$\beta_2 = \frac{\Gamma\left(1+\frac{4}{\alpha}\right) - 4\Gamma\left(1+\frac{1}{\alpha}\right)\Gamma\left(1+\frac{3}{\alpha}\right) + 6\Gamma^2\left(1+\frac{1}{\alpha}\right)\Gamma\left(1+\frac{2}{\alpha}\right) - 3\Gamma^4\left(1+\frac{1}{\alpha}\right)}{\left[\Gamma\left(1+\frac{2}{\alpha}\right) - \Gamma^2\left(1+\frac{1}{\alpha}\right)\right]^2}$$

VARIATE GENERATION TECHNIQUES*

*From Leemis, L. M. (1987), Variate Generation for Accelerated Life and Proportional Hazards Models, Operations Research, Vol. 35, No. 6, Nov–Dec 1987, pp. 892-894.

Let $h(t)$ and $H(t) = \int_0^t h(\tau)\,d\tau$ be the hazard and cumulative hazard functions, respectively, for a continuous nonnegative random variable T, the lifetime of the item under study. The $q \times 1$ vector z contains covariates associated with a particular item or individual. The covariates are linked to the lifetime by the function $\Psi(z)$, which satisfies $\Psi(0 = 1)$ and $\Psi(z) \geq 0$ for all z. A popular choice is $\Psi(z) = e\beta'\,z$, where β is a $q \times 1$ vector of regression coefficients.

The cumulative hazard function for T in the *accelerated life* model (Cox and Oakes 1984) is

$$H(t) = H_0\left(t\Psi(z)\right),$$

where H_0 is a baseline cumulative hazard function. Note that when $z = 0$, $H_0 \equiv H$. In this model, the covariates accelerate ($\Psi(z) > 1$) or decelerate ($\Psi(z) < 1$), the rate at which the item moves through time. The *proportional* hazards model

$$H(t) = \Psi(z) H_0(t)$$

increases ($\Psi(z) > 1$) or decreases ($\Psi(z) < 1$) the failure rate of the item by the factor $\Psi(z)$ for all values of t.

GENERATION ALGORITHMS

The literature shows that the cumulative hazard function, $H(T)$, has a unit exponential distribution. Therefore, a random variate t corresponding to a cumulative hazard function $H(t)$ can be generated by

$$t = H^{-1}\left(-\log(u)\right),$$

Where u is uniformly distributed between 0 and 1. In the accelerated life model, since time is being expanded or contracted by a factor $\Psi(z)$, variates are generated by

$$t = \frac{H_0^{-1}\left(-\log(u)\right)}{\Psi(z)}$$

In the proportional hazards model, equating $-\log(u)$ to $H(t)$ yields the variate generation formula

$$t = H_0^{-1}\left(\frac{-\log(u)}{\Psi(z)}\right).$$

FORMULAS FOR GENERATING EVENT TIMES FROM A RENEWAL OR NONHOMOGENEOUS POISSON PROCESS

	Renewal	NHPP
Accelerated life	$t = a + \dfrac{H_0^{-1}\left(-\log(u)\right)}{\Psi(z)}$	$t = \dfrac{H_0^{-1}\left(H_0\left(a\Psi(z)\right)-\log(u)\right)}{\Psi(z)}$
Proportional hazards	$t = a + H_0^{-1}\left(\dfrac{-\log(u)}{\Psi(z)}\right)$	$t = H_0^{-1}\left(H_0(a)-\dfrac{\log(u)}{\Psi(z)}\right)$

In a NHPP, the hazard function, $h(t)$, is equivalent to the intensity function, which governs the rate at which events occur. To determine the appropriate method for generating values from an NHPP, assume that the last even in a point process has occurred at time a. The cumulative hazard function for the time of the next event conditioned on survival to time a is

$$H_{T|T\rangle a}(t) = H(t) - H(a) \quad t > a.$$

In the accelerate life model, where $H(t) = H_0(t\Psi(z))$, the time of the next event is generated by

$$t = \frac{H_0^{-1}\left(H_0\left(a\Psi(z)\right)-\log(u)\right)}{\Psi(z)}.$$

If we equate the conditional cumulative hazard function to $-\log(u)$, the time of the next event in the proportional hazards case is generated by

$$t = H_0^{-1}\left(H_0(a)-\frac{\log(u)}{\Psi(z)}\right).$$

EXAMPLE

The exponential power distribution (Smith and Bain 1975) is a flexible two-parameter distribution with cumulative hazard function

$$H(t) = e^{(t/\alpha)^\gamma} - 1 \quad \alpha > 0, \quad \gamma > 0, \quad t > 0$$

and inverse cumulate hazard function

$$H^{-1}(y) = \alpha \left[\log(y+1)\right]^{1/\gamma}.$$

Assume that the covariates are linked to survival by the function $\Psi(z) = e^{\beta' z}$ in the accelerated life model. If an NHPP is to be simulated, the baseline hazard function has the exponential power distribution with parameters α and γ, and the previous event has occurred at time a, then the next event is generated at time

$$t = \alpha e^{-\beta' z} \left[\log\left(e^{\left(a e^{\beta' z}/\alpha\right)^{\gamma}} - \log(u) \right) \right]^{1/\gamma},$$

where u is uniformly distributed between 0 and 1.

Appendix B
Cantor Set Sectioning

Mathematically, the cantor set is denoted as:

$$C = \left\{ \chi \in \Omega \,\Big|\, \chi \in \bigcup_{k=0}^{\infty} \delta_k \right\}$$

where $\Omega = [0,1]$. The interval, δk, is as explained below:

Consider the closed interval $\Omega = [0,1]$ and the open intervals generated by successive removal of the middle thirds of intervals left after previous removals. The interval deletions are shown geometrically in Figure B.1. Note that:

$$\bigcup_{k=0}^{\infty} \delta_k = [0,1]$$

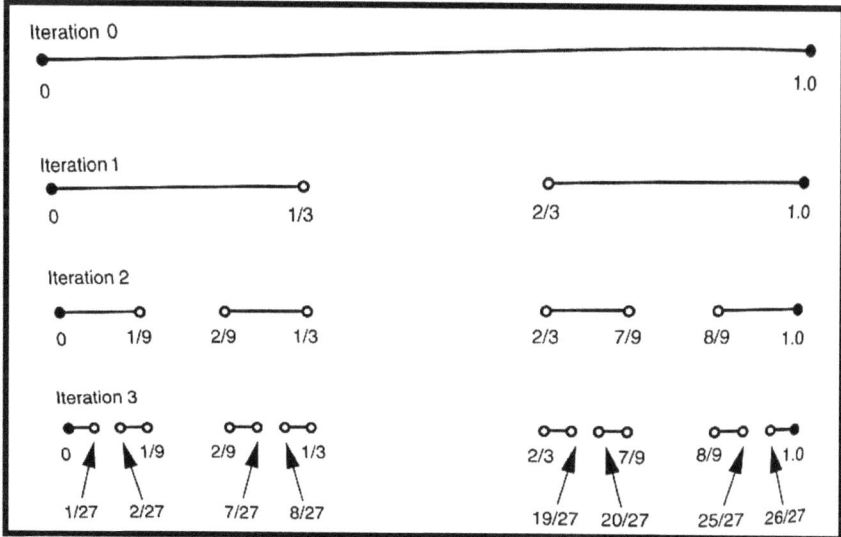

FIGURE B.1 Cantor set representation.

The interval δ_k is the union of the open intervals deleted from Ω after the kth search iteration. The deleted intervals are represented mathematically below:

$$\delta_0 = \phi\,(\text{null set})$$

$$\delta_1 = \left[\frac{1}{3}, \frac{2}{3}\right]$$

$$\delta_2 = \left[\frac{1}{9}, \frac{2}{9}\right] \cup \left[\frac{1}{3}, \frac{2}{3}\right] \cup \left[\frac{7}{8}, \frac{8}{9}\right]$$

$$\vdots$$

If Ω is considered as the universal set, then we may also express the Cantor set as the complement of the original set C. That is, alternately,

$$C = \left(\bigcup_{k=0}^{\infty} \delta_k\right)^c,$$

which, by DeMorgan's Law, implies

$$C = \bigcap_{k=0}^{\infty} (\delta_k)^c$$

If we use the following simplifying notation for the complement, $(\delta_k)^c = \lambda_k$, we would obtain the following alternate representation:

$$C = \bigcap_{k=0}^{\infty} \lambda_k,$$

where

$$\lambda_0 = [0,1]$$

$$\lambda_1 = \left[0, \frac{1}{3}\right] \cup \left[\frac{2}{3}, 1\right]$$

$$\lambda_2 = \left[0, \frac{1}{9}\right] \cup \left[\frac{2}{9}, \frac{1}{3}\right] \cup \left[\frac{2}{3}, \frac{7}{9}\right] \cup \left[\frac{8}{9}, 1\right]$$

$$\vdots$$

It should be noted that λ_k is the remaining search space available for the kth search iteration. Also note that λ_k consists of 2^k closed and nonoverlapping intervals each of real length $(1/3^k)$.

APPLICATION OF THE CANTOR SET APPROACH

Knowledge bases for expert systems consist of pieces of information on the basis of which inferences are drawn for a particular problem situation. For large domain problems, the knowledge base "lookup" or search can easily lead to a combinatorial explosion of possibilities. For example, if we have 50 pieces of evidence each of which is either true or false, then there are 250 possible combinations. From a practical point of view, we need search procedures that can considerably reduce the dimensionality of the search space.

In a manufacturing context, two physical objects are exactly alike only if they are fully interchangeable. In an actual manufacturing situation, items in a group will not necessarily have characteristics that are fully identical. Recalling the earlier example of shafts, a group of objects may consist of items that are related by their classification as "shafts." Differences within the group may pertain to the items' diameters or any other characteristic of interest. For example, we may be interested in diameters that range from 3 inches to 7 inches. Arranging shaft designs in increasing or decreasing order of shaft diameter can be used to indicate the degree of relationship or the level of property inheritance of the items in the group. Thus, in a knowledge base, inferences can be drawn to relate to certain subsets of a given set of the knowledge elements. Graphically, the shaft example may be represented as shown in Figure B.2.

Suppose we are interested in a shaft that meets a certain quality characteristic. We can conduct an exhaustive search to check if each shaft meets the desired quality characteristic. But exhaustive searches are very costly and time-consuming particularly where we have a large number of items to search. An efficient search strategy would be helpful in reducing the time and expense of finding the item that meets the specified characteristic. Suppose we know the distribution of the shaft diameters over the range of 3 inches to 7 inches. If the distribution can be reasonably expected to follow a bell-shaped curve such as the normal distribution shown in Figure B.3, then the Cantor set strategy may be employed. The search strategy would proceed as shown below:

Step 1: Identify a known property of the items to be searched (e.g., diameter sizes).
Step 2: Determine the range of values of the known property. This establishes the search space.
Step 3: Specify the desired characteristic of the item to be searched (e.g., quality characteristic).

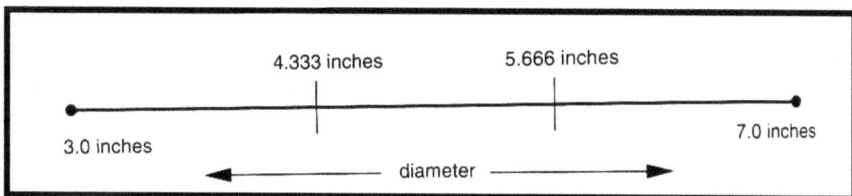

FIGURE B.2 Range of shaft diameter.

Step 4: Determine the distribution of the items based on the known property (e.g., bell- shaped).
Step 5: Sort the items in increasing or decreasing order of the known property.
Step 6: Apply the Cantor set search procedure iteratively until an item matching the specified characteristic is found.

Instead of conducting an exhaustive search over the entire search interval, we would check the middle third first. If the item that meets the requirement is not found in that interval, we would delete the interval from further consideration. The middle thirds of the remaining intervals are then searched in successive iterations.

COMMENTS ON THE SEARCH PROCEDURE

1. If the distribution of the items is bell-shaped, then searching the first middle third before any other interval is logical since that is where the majority of the items are located.
2. The largest search effort will involve the first middle third. The search process becomes less efficient as more iterations are needed to find the desired item.
3. In the second and subsequent iterations, a decision must be made concerning which middle third interval to search next. For example, Figure B.3 shows the search space left after deleting the first middle third. We have the option of first searching the middle third of Interval A and then the middle third of Interval B and vice versa. Since the intervals are equally likely to contain the desired item, one can flip a coin to determine which interval to search first. The decision becomes more difficult in the third iteration since, as shown in Figure B.4, the remaining four intervals are not equi-probable. Figure B.5 shows the search intervals for the first three iterations of the Cantor search strategy.

In the Cantor search procedure, the desired item is found only when it is located in the middle third of some interval. If the value of interest is in the interval (4.333, 5.666), then only one interval search would be needed to find it. If it is in the interval (3.444, 3.888) ∪ (6.110, 6.554), then at most three interval searches would be

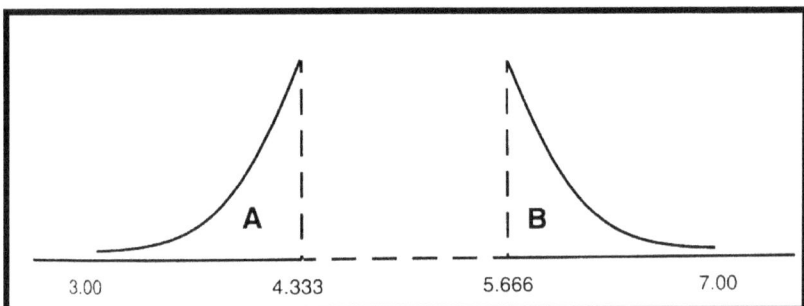

FIGURE B.3 Second search iteration.

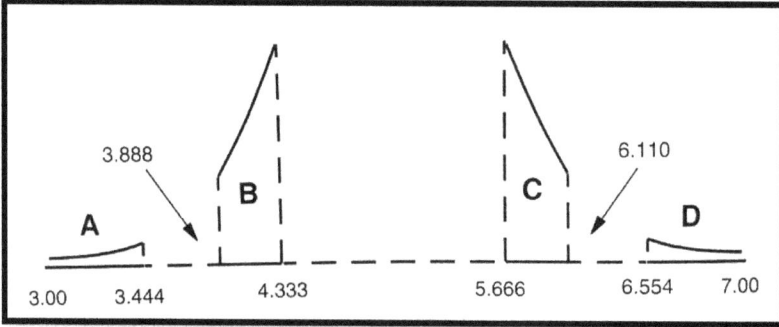

FIGURE B.4 Third search iteration.

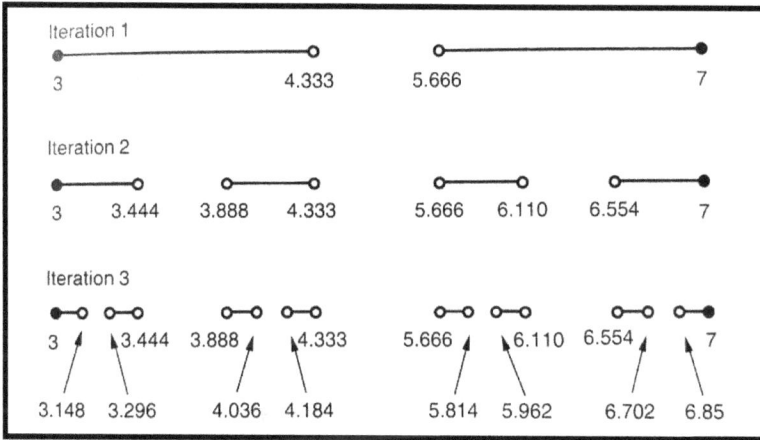

FIGURE B.5 Successive search iterations.

needed. If the value is in the interval $(3.148, 3.296) \cup (4.036, 4.184) \cup (5.814, 5.962) \cup (6.702, 6.85)$, then at most seven interval searches would be needed. In general, the maximum number of interval searches, N, needed to locate an item in a Cantor-set search strategy is one of the following:

$$N_k = 1, 3, 5, 7, 15, 31, \ldots,$$

where k is the iteration number. That is,

$$N_0 = 0$$
$$N_k = N_{k-1} + 2^{k-1}$$
$$= \sum_{j=0}^{k-1} 2^j$$
$$= \left(2^k\right) - 1$$

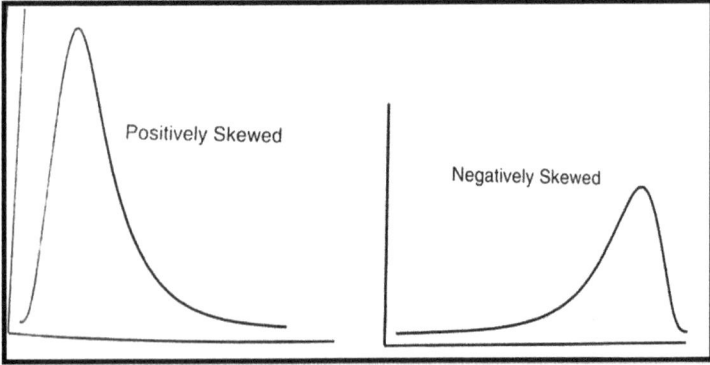

FIGURE B.6   Skewed distribution search space.

It should be noted that if the distribution of the items to be searched is skewed to the right or left (e.g., Chi-squared or Lognormal distributions), then the basic Cantor set search strategy would not be appropriate (Figure B.6).

MODIFIED SEARCH PROCEDURE

The deficiency mentioned in the third comment on the Cantor search procedure can be overcome by using the following modification of the procedure. The modification improves the efficiency of the search strategy.

Consider the search intervals to be used for the second iteration (shown earlier in Figure B.3). Instead of considering the intervals [3.00, 4.333] and [5.666, 7.00] as separate search intervals, we can merge the intervals as shown in Figure B.7. Then, the next middle third to be searched during the second iteration would be [3.89, 6.11].

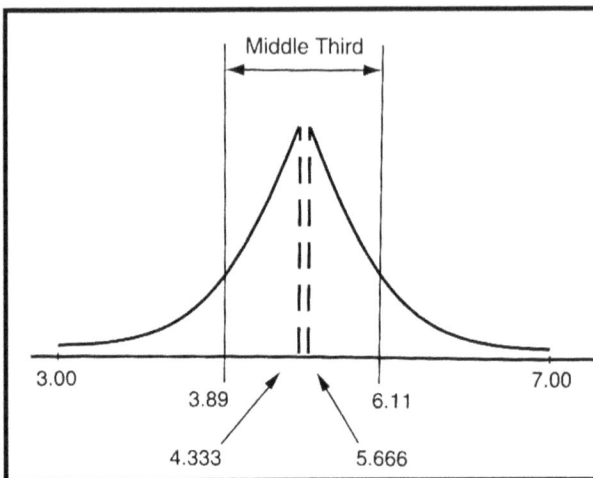

FIGURE B.7   Merged search interval.

Recall that the items in the interval [4.333, 5.666] have already been deleted in the first iteration and are not contained within the modified middle third interval of [3.89, 6.11]. This process is repeated consecutively until the desired item is found. Figure B.8 shows the search interval for the third iteration using the merged interval modification.

ALTERNATE SEARCH PREFERENCE

The conventional Cantor set search strategy gives first preference to the middle third of the ordered set of items to be searched. As mentioned previously, this is suitable if the distribution of the property of interest is bell-shaped. If, by contrast, the distribution is bimodal "end-heavy," then an alternate ordering of the items may be required. An example of a bimodal end-heavy distribution is shown in Figure B.9, which is

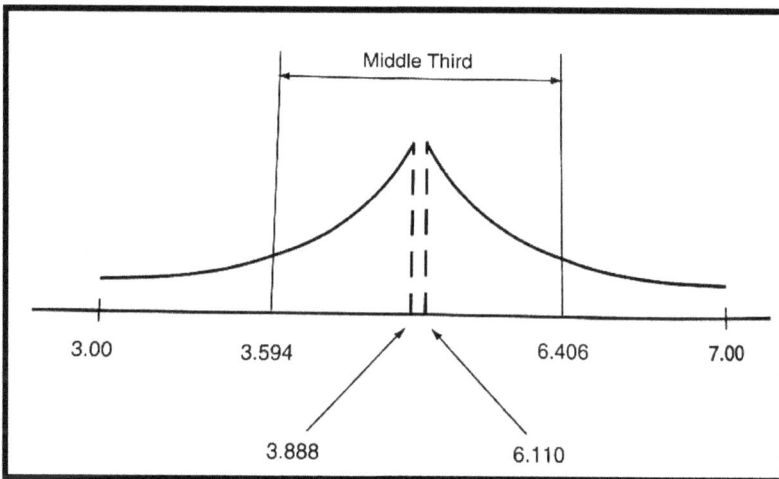

FIGURE B.8   Merged search interval.

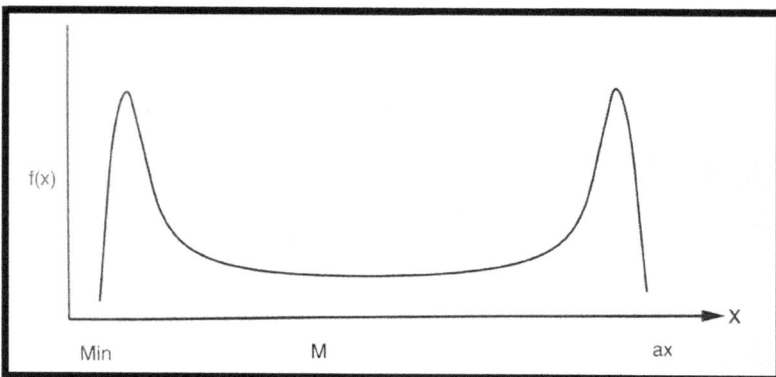

FIGURE B.9   Bathtub-shaped distribution.

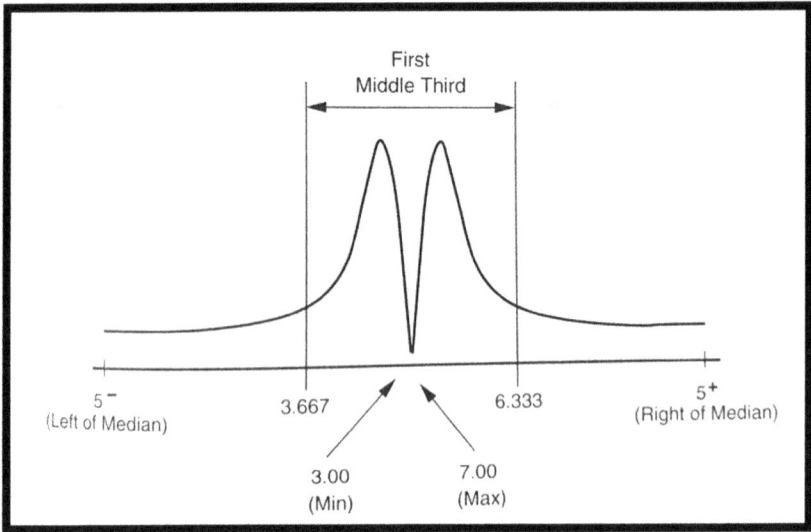

FIGURE B.10 Bisected and flipped search interval.

representative of typical hazard functions in product reliability analysis. The bath-tub shaped distribution is a special form of the beta distribution with shape parameters of $\alpha = 0.001$ and $\beta = 0.001$.

It is noted that most of the items to be searched are located in the regions close to the minimum and maximum points. An alternate arrangement of the items is achieved by bisecting the ordered set through the median and flipping over the half-sets generated. This is shown graphically in Figure B.10. This alternate arrangement gives first preference to the end points of the original set of the items to be searched.

Index

For Product Safety Concerns and Information please contact our EU
representative GPSR@taylorandfrancis.com
Taylor & Francis Verlag GmbH, Kaufingerstraße 24, 80331 München, Germany